DEPARTMENT OF DEFENSE

Glossary of Mapping, Charting, and Geodetic Terms

Fourth Edition 1981

WILDSIDE PRESS

INTRODUCTION

This new edition of the *Department of Defense Glossary of Mapping, Charting, and Geodetic Terms* was prepared under the direction of the Defense Mapping Agency. The purpose of this glossary is to provide a comprehensive and an authoritative source of current usage of mapping, charting, and geodetic terms for all levels of users to help them communicate effectively.

The terms and definitions in this publication were selected from authoritative glossaries and dictionaries, and from technical publications and papers concerned with the many disciplines associated with mapping, charting, and geodesy. Numerous changes, additions, and deletions were made after a thorough review of the Third Edition by Defense Mapping Agency components and by the Departments of the Army, the Navy, and the Air Force.

This publication is not a substitute for the *Department of Defense Dictionary of Military and Associated Terms* (*JCS Pub. 1*), which the Secretary of Defense has directed to be used throughout the Department of Defense. Terms included herein which are designated "(JCS)" were extracted from and defined as stated in *JCS Pub. 1*. In some instances, the JCS definition has been expanded to include more detailed or supplementary information. This additional matter is set off by brackets and is not to be construed as changing or conflicting with the JCS definition. JCS terms which have been accepted by NATO and by the Inter-American Defense Board are so designated in *JCS Pub. 1*. Only those NATO terms which do not appear in *JCS Pub. 1* are so indicated in this glossary.

The designation "(USPLS)" indicates U.S. Public Land Survey terms similarly defined by the Bureau of Land Management, U.S. Department of Interior.

Alphabetization of terms in this glossary follows the standard A through Z order, except that multiword terms are alphabetized according to the initial word.

Multiple definitions for a single term are numbered and, wherever applicable, are identified with the appropriate science, discipline, or function in parentheses. At the end of some definitions the user's attention is directed to related terms by the expression "See also."

When two or more terms have identical meaning, the definition has been applied only to the preferred term, followed by the expression "Also called" and a list of the synonyms. The synonyms are shown in alphabetical order in the glossary, and are referenced to the preferred term. Antonyms are listed after the expression "Opposite of."

Users are encouraged to submit constructive comments. Proposed changes or additions should be accompanied by justification or explanation. Send comments to: Director, Defense Mapping Agency Hydrographic/Topographic Center, ATTN: PPIC, 6500 Brookes Lane, Washington, D.C. 20315.

A

ABAC—A nomogram for obtaining the conversion angle to apply when plotting great-circle bearings on a Mercator projection.

A1 time—A particular atomic time scale, established by the U.S. Naval Observatory, with the origin on 1 January 1958, at zero hours UT2 and with the unit (second) equal to 9,192,631,770 cycles of cesium at zero field. See also **UT2 time.**

A-station—1. (traverse) Subsidiary stations established between principal stations of a survey traverse, for convenience of measuring, to obtain the distance between principal stations. A-stations are so-called because in a given series, these stations are designated by the name of a principal station followed by the letters A, B, C, etc., in order of distance from the principal station. **2.** (loran) The designation applied to the transmitting station of a pair, the signal of which always occurs less than half a repetition period after the next preceding signal of the other station of the pair, designated a **B-station.**

abeam—(JCS) Bearing approximately 090° or 270° relative; at right angles to the longitudinal axis of a vehicle.

aberration—1. (astronomy) The apparent angular displacement of the position of a celestial body in the direction of motion of the observer, caused by the combination of the velocity of the observer and the velocity of light. See also **annual aberration; constant of aberration; differential aberration; diurnal aberration; planetary aberration; secular aberration; stellar aberration. 2.** (optics) Failure of an optical system to bring all light rays received from a point object to a single image point or to a prescribed geometric position. See also **astigmatism; circle of confusion; coma; curvature of field; distortion; lateral chromatic aberration; lens distortion; longitudinal chromatic aberration; spherical aberration.**

aberration of fixed stars—See **secular aberration.**

Abney level—An instrument suitable for direct leveling or for measuring vertical angles or percent of slope. It consists of a simple hand level on one side of which a reversible graduated arc with pivoted bubble tube is mounted. See also **hand level.**

abnormal magnetic variation—Any anomalous deflection, whose cause is unknown, of the compass needle from the magnetic meridian.

abridged spectrophotometry—Measurement of spectral transmittance or reflectance at a limited number of wavelengths.

abscissa—The horizontal coordinate of a set of rectangular coordinates. Also used in a similar sense in connection with oblique coordinates. Also called **total departures; x-coordinate.**

absolute accuracy—The degree of perfection in a value determined through evaluation of all error sources and error propagations.

absolute altimeter—An instrument which directly determines the height of an aircraft above the terrain. See also **radar altimeter.**

absolute altitude—(JCS) The height of an aircraft directly above the surface or terrain over which it is flying. See also **radar altitude.**

absolute error—Absolute deviation, the value taken without regard to sign, from the corresponding true value.

absolute gravity—The acceleration of gravity directly determined by an experiment of time and length measurements. See also **gravity.**

absolute gravity station—A marked point, usually in a laboratory, where the value of absolute gravity has been determined. See also **absolute gravity.**

absolute orientation—The scaling and leveling to ground control (in a photogrammetric instrument) of a relatively oriented stereoscopic model or group of models. See also **relative orientation.**

absolute parallax—See **absolute stereoscopic parallax.**

absolute positioning—Determination of the position of a point with respect to the center of mass of the Earth as defined in the DoD World Geodetic System.

absolute stereoscopic parallax—Considering a pair of aerial photographs of equal principal distance, the absolute stereoscopic parallax of a point is the algebraic difference of the distances of the two images from their respective photograph nadirs, measured in a horizontal plane and parallel to the air base. Also called **absolute parallax; horizontal parallax; linear parallax; parallax; stereoscopic parallax; x-parallax.**

absolute term—A term (usually only one) in an equation, which represents a known numerical value and does not contain any unknown or variable elements.

absolute unit—Any unit in a system that is based directly upon associated fundamental units of length, mass, and time. See also **dynamic number.**

absolute value—A mathematical quantity taken without regard to its associated plus or minus sign. Used often with residuals.

absolute vector—A directed line segment whose end-points are measured in absolute units from a point designated as the origin.

absorption—Conversion of radiant energy into other forms by passage through or reflection from matter.

acceleration—1. The rate of change of velocity. **2.** The act or process of accelerating, or the state of being accelerated.

acceleration of gravity—The acceleration of a freely falling body.

accelerometer—1. A device that measures the rate of change of speed of an object. **2.** An instrument, specially designed for carrying in aircraft or missiles, which measures the rate of change in velocity, direction, and/or altitude.

accepted indicator—An airborne indicator which has been proven to be capable of accurate and reliable measurement.

accidental error—See **random error.**

accommodation—1. (JCS) The ability of the human eye to adjust itself to give sharp images of objects of different distances. In stereoscopy, the ability of the human eyes to bring two images into superimposition for stereoscopic viewing. **2.** The limits or range within which a stereoplotting instrument is capable of operating. For example, the multiplex can adjust (or accommodate) for small tilts in the projectors ranging from about 10° about the x-axis to 20° about the y-axis.

accumulated discrepancy—The algebraic sum of the separate discrepancies which occur in the various steps of making a survey or of the computation of a survey.

accumulated divergence—(leveling) The algebraic sum of the divergences for the sections of a line of levels, from the beginning of the line to any section end at which it is desired to compute the total divergence.

accumulative error—A constant error which is always plus or always minus. A number of readings under these conditions will have an accumulated error equal to the number of readings multiplied by the error in one reading. See also **constant error; systematic error.**

accuracy—1. The degree of conformity with a standard, or the degree of perfection attained in a measurement. Accuracy relates to the quality of a result, and is distinguished from precision which relates to the quality of the operation by which the result is obtained. **2.** The degree of conformity with which horizontal positions and vertical values are represented on a map, chart, or related product in relation to an established standard.

accuracy checking—The procurement of presumptive evidence of a map's compliance with specified accuracy standards. Accuracy checking generally indicates the relative (rather than the absolute) accuracy of map features.

accuracy evaluation—The comparison of the quality of an MC&G product with maintenance criteria to ascertain its adequacy with respect to its intended use.

accuracy method—The method used in determining the stated accuracy of a product. The method, which can range from a system employing highly sophisticated techniques to a highly subjective judgment, is essential for proper use of the stated accuracy.

accuracy review—The comparison of an existing MC&G product against source material or data more accurate than that from which it was produced, for the purpose of determining the accuracy of its horizontal and vertical values.

accuracy testing—The procurement of confirmed evidence, on a sampling basis, of a map's compliance with specified accuracy standards. Accuracy testing is designed to indicate both the relative and absolute accuracy of map features.

accurate contour—A contour line, the accuracy of which lies within one-half of the basic vertical interval. Also called **normal contour.**

acetate—A nonflammable plastic sheeting used as a base for photographic films or as a drafting base for overlays where critical registration is not required.

achromatic color—Color that does not elicit hue.

achromatic lens—A lens that has been partly corrected for chromatic aberration. Such a lens is usually of a multi-element design to bring green and red light rays to approximately the same point of focus.

achromatic telescope—A telescope having a mirror or an achromatic lens for its objective.

aclinal—Without dip; horizontal.

aclinic—Without magnetic dip.

aclinic line—See **magnetic equator.**

acoustic navigation—Navigation by means of sound waves whether or not they are within the audible range. Also called **sonic navigation.** See also **Doppler sonar navigation.**

actinic light—Light which is capable of causing photo-chemical changes in a sensitized emulsion.

active satellite—A satellite which transmits an electromagnetic signal. A satellite with the capability to transmit, repeat, or re-transmit electromagnetic information. See also **passive satellite.**

active tracking system—A satellite tracking system which operates by transmission of signals to and receipt of responses from the satellite.

actual error—The difference between the true value and the measured value of a physical quantity.

acutance—An objective measure of the ability of a photographic system to show a sharp edge between contiguous areas of low and high illuminance.

adaptation—The faculty of the human eye to adjust its sensitivity to varying intensities of illumination.

adding tape—A surveyor's tape which is calibrated and has an additional foot (or meter) beyond the zero end which is graduated in tenths or hundredths. See also **subtracting tape.**

additive color mixture—Superimposition or other nondestructive combination of light of different chromaticities.

additive color viewer—Projector for positive transparencies obtained through multiband photography. Each image is superimposed by use of a different colored light.

additivity of luminance—Luminance produced with a mixture of light from several sources is the sum of the luminances produced by the light from lack of the sources acting separately.

adequate—A term used to describe a product which meets all of the accuracy and currency standards established by its most stringent use, and thus, is suitable for all its intended uses.

adjoining sheets—Adjacent maps to one or all sides and corners of a particular map sheet.

adjusted angle—An adjusted value of an angle. An adjusted angle may be derived either from an observed angle or from a deduced angle.

adjusted elevation—The elevation resulting from the application of an adjustment correction to an orthometric elevation. Also, the elevation resulting from the application of both an orthometric correction and an adjustment correction to a preliminary elevation.

adjusted position—An adjusted value of the coordinate position of a point.

adjusted value—A value of a quantity derived from observed data by some orderly process which eliminates discrepancies arising from errors in those data.

adjustment—1. (general) The determination and application of corrections to observations, for the purpose of reducing errors or removing internal inconsistencies in derived results. The term may refer either to mathematical procedures or to corrections applied to instruments used in making observations. **2.** (leveling) The determination and application of corrections to orthometric differences of elevation or to orthometric elevations, to make the elevation of all bench marks consistent and independent of the circuit closures. **3.** (cartography) Placing detail or control stations in their positions relative to other detail or control stations. See also **adjustment of observations; angle method of adjustment; balancing a survey; direction method of adjustment; figure adjustment; instrument adjustment; land-line adjustment; least squares; map adjustment; short arc geodetic adjustment; station adjustment; track adjustment.**

adjustment for collimation—See **collimate,** definition 2.

adjustment of observations—The determination and application of corrections corresponding to errors affecting the observations, making the observations consistent among themselves, and coordinating and correlating the derived data.

administrative map—(JCS) A map on which is graphically recorded information pertaining to administrative matters, such as supply and evacuation installations, personnel installations, medical facilities, collecting points for stragglers and prisoners of war, train bivouacs, service and maintenance areas, main supply roads, traffic circulation, boundaries, and other details necessary to show the administrative situation in relation to the tactical situation.

aerial camera—A camera specifically designed for use from an airborne station.

aerial cartographic photography—See **mapping photography.**

aerial film—Specially designed roll-film supplied in many lengths and widths, with various emulsion types for use in aerial cameras.

aerial film speed (AFS)—A measure of speed for aerial film which replaces the formerly used aerial exposure index. It is defined as 3/2E, where E is the exposure in meter-candle-seconds at the point on the characteristic curve where the density is 0.3 above base plus fog density on black-and-white film.

aerial mosaic—See **mosaic,** definition 1.

aerial photogrammetry—The use of aerial photographs in the science of photogrammetry.

aerial photograph—(JCS) Any photograph taken from the air. Also called **air photograph.**

aerial photographic reconnaissance—The obtaining of information by aerial photography—divided into three types: (1) strategic photographic reconnaissance; (2) tactical photographic reconnaissance; and (3) survey/cartographic photography, which is aerial photography taken for survey/cartographic purposes and to survey/cartographic standards of accuracy.

aerial photography—The art, science, or process of taking aerial photographs. See also **mapping photography; reconnaissance photography.**

aerial platform—A term referring to the support of an aerial camera at the air station.

aerial reconnaissance—The collection of information by visual, electronic, photographic, or other means from the air.

aerial survey—A survey utilizing photographic, electronic, or other data obtained from an airborne station.

aerial survey team (AST)—A composite force drawn from several organizations to form a deployable unit capable of accomplishing an assigned aerial cartographic/geodetic mission.

aerial triangulation—See **phototriangulation.**

aeroleveling—As applied to model orientation during phototriangulation, barometric height measurements of the camera air stations which have been recorded during the photographic mission are used to present the *bz* values during the orientation of the successive models on the stereoplotting instrument. Only differences in flight height are required and these are provided by the statoscope. See also **orientation,** definition 7.

aerometeorograph—An instrument that records

the pressure and temperature of the air, the amount of moisture in the air, and the rate of motion of the wind.

aeronautical chart—(JCS) A specialized representation of mapped features of the Earth, or some part of it, produced to show selected terrain, cultural, and hydrographic features, and supplemental information required for air navigation, pilotage, or for planning air operations. Also called **navigation chart.**

aeronautical information overprint—(JCS) Additional information which is printed or stamped on a map or chart for the specific purpose of air navigation.

aeronautical pilotage chart—An aeronautical chart designed primarily for air navigation.

aeronautical planning chart—An aeronautical chart of small scale designed to satisfy long range air navigation and mission planning requirements.

aeropause—(JCS) Region in which functional effects of the atmosphere on man and aircraft cease to exist.

aerospace—(JCS) Of, or pertaining to, the Earth's envelope of atmosphere and the space above it; two separate entities considered as a single realm for activity in launching, guidance, and control of vehicles which will travel in both entities.

aerotriangulation—See **phototriangulation.**

affine deformation—One in which the scale along one axis or reference plane is different from the scale along the other axis or plane.

affine transformation—A transformation in which straight lines remain straight and parallel lines parallel. Angles may undergo changes and differential scale changes may be introduced.

age of diurnal inequality—The time interval between the maximum semimonthly north or south declination of the Moon and the time that the maximum effect of the declination upon the range of tide or speed of the tidal current occurs. Also called **age of diurnal tide; diurnal age.**

age of diurnal tide—See **age of diurnal inequality.**

age of parallax inequality—The time interval between the perigee of the Moon and the maximum effect of the parallax (distance of the Moon) upon the range of tide or speed of tidal current. Also called **parallax age.**

age of phase inequality—The time interval between the new or full Moon and the maximum effect of these phases upon the range of tide or speed of tidal current. Also called **age of tide; phase age.**

age of the Moon—The elapsed time, usually expressed in days, since the last new Moon.

age of tide—See **age of phase inequality.**

agonic line—(JCS) A line drawn on a map or chart joining points of zero magnetic declination for a specified epoch.

aiming line—See **line of sight,** definition 2; **line of collimation.**

Air Almanac—A joint publication of the United States Naval Observatory and Her Majesty's Nautical Almanac Office. It covers a 6-month period. It contains tabulated values of the Greenwich hour angle and declination of selected celestial bodies, plus additional celestial data used in navigation.

air base—**1.** (JCS) (photogrammetry) The line joining two air stations, or the length of that line. **2.** The distance, at the scale of the stereoscopic model, between adjacent perspective centers as reconstructed in the plotting instrument. See also **air station.**

air coordinates—See **rectangular space coordinates.**

air photograph—See **aerial photograph.**

air plot—(JCS) **1.** A continuous plot used in air navigation of a graphic representation of true headings steered and air distances flown. **2.** A continuous plot of the position of an airborne object represented graphically to show true headings steered and air distances flown. **3.** Within ships, a display which shows the positions and movements of an airborne object relative to the plotting ship.

air station—(JCS) The point in space occupied by the camera lens at the moment of exposure. Also called **camera station.** See also **air base.**

air surveillance plotting board—(JCS) A gridded, small scale, air defense map of an appropriate area. It is maintained at the air control center. On it are posted current locations, number, and altitudes of all friendly or enemy aircraft within range of radar or ground observer facilities.

Air Target Chart—(JCS) A display of pertinent air target intelligence on a specialized graphic base. It is designed primarily to support operations against designated air targets by various weapon systems. [The charts provide graphic overprint and textual data relative to radar return information and installations within the area. Air Target Charts are prepared at various scales and are produced under the Air Target Materials Program as a series of geographically integrated charts.]

air target materials—See **target materials.**

Air Target Materials Program (ATMP)—(JCS) A Department of Defense program established for the production of medium and large scale air target materials and related items in support of long range, worldwide requirements of the Unified and Specified Commands, Military Departments, and allied participants. It is under the management control of the Defense Intelligence Agency and encompasses the determination of production and coverage requirements, standardization of products, establishment of production priorities and schedules, and the production, distribution, storage, and release/exchange of the air target materials and related products.

Air Target Materials Program (ATMP) Production Management System—A system which is predicated on, and guided by, the principle that an action is initiated and controlled by the exchange of a sequence of punched cards among those who establish target material requirements, those who produce or manage the production of target materials, and the users of target materials.

Air Target Mosaic—(JCS) A large scale mosaic providing photographic coverage of an area and permitting comprehensive portrayal of pertinent target detail. These mosaics are used for intelligence study and in planning and briefing for air operations. [One of the target graphics in the Air Target Materials Program.]

Airborne Control (ABC) system—A survey system for horizontal and vertical control surveys involving electromagnetic distance measurements and horizontal and vertical angle measurements from two or more known positions to a helicopter hovering over the unknown position. The elevation of the unknown position is determined by the use of a special plumbline cable.

airborne electronic survey control—Control surveys accomplished by electronic means from an airborne vehicle or platform, such as hiran and shoran.

airborne-landing model—A specially designed assault model for use in briefing airborne troops and support personnel. These models emphasize the aspects of objects as seen from the air rather than from the ground.

Airborne Profile Recorder (APR)—See **Terrain Profile Recorder.**

airway—(JCS) A control area or portion thereot established in the form of a corridor marked with radio navigational aids.

Airy spheroid (ellipsoid)—A reference ellipsoid used in Great Britain and having the following dimensions: semimajor axis—6,377,563.396 meters; semiminor axis—6,356,256.910 meters; and the flattening or ellipticity—1/299.3249646.

Airy theory of isostasy—The theory that the continents and islands are resting hydrostatically on highly plastic or liquid material, with roots or projections penetrating the inner material of the Earth, just as icebergs extend downward into the water. The greater the elevation, the deeper the penetration. It has been called the **roots of mountain theory,** and has the support of some geologists. See also **Pratt-Hayford theory of isostasy.**

Aitoff equal-area map projection—A Lambert equal-area azimuthal projection of a hemisphere converted into a map projection of the entire sphere by a manipulation suggested by Aitoff. It is a projection bounded by an ellipse in which the line representing the Equator (major axis) is double the length of the line representing the central meridian (minor axis).

albedo—The ratio of radiant energy reflected to that received by a surface, usually expressed as a percentage; reflectivity. The term generally refers to energy within a specific frequency range, as the visible spectrum. Its most frequent application is to the light reflected by a celestial body.

Albers conical equal-area map projection—An equal-area projection of the conical type, on which the meridians are straight lines that meet in a common point beyond the limits of the map, and the parallels are concentric circles whose center is at the point of intersection of the meridians. Meridians and parallels intersect at right angles and the arcs of longitude along any given parallel are of equal length. The parallels are spaced to retain the condition of equal area. On two selected parallels, the arcs of longitude are represented in their true length. Between the selected parallels the scale along the meridians will be a trifle too large, and beyond them, too small.

albumin (albumen) process—A process of making photolithographic pressplates utilizing bichromated albumin as the photosensitive coating, and requiring a true negative to make the printing plate. See also **pressplate.**

Aldis signaling lamp—A signaling lamp used in some cases for night observations of distant stations in triangulation.

alerts—An ephemeris prepared for one or more satellites, predicting rise and set times referred to universal time coordinated, maximum angle of elevation above the observer's horizon, and azimuth from the observer. Used to identify specific satellite passes. See also **look angles.**

alidade—The part of a surveying instrument which consists of a sighting device, with index, and reading or recording accessories. See also **peepsight alidade; pendulum alidade; photoalidade; telescopic alidade.**

alignment (alinement)—1. (JCS) (cartography) Representation of a road, railway, etc., on a map or chart in relation to surrounding topographic detail. **2.** (general surveying) The placement or location of points along a straight line. **3.** (highway and route surveying) The ground plan showing the direction (center line) of the route to be followed, as distinguished from **profile,** which shows the vertical element.

alignment correction—(taping) A correction applied to the measured length of a line to allow for the tape not being held exactly in a vertical plane containing the line.

almanac—A periodical publication of astronomic coordinates useful to a navigator. It contains less information than an ephemeris and values are generally given to less precision. See also **ephemeris.**

almucantar—See **parallel of altitude.**

alphanumeric grid—See **atlas grid.**

alt-azimuth instrument—An instrument equipped with both horizontal and vertical graduated circles, for the simultaneous observation of horizontal and vertical directions or angles. Also called **astronomic theodolite; universal instrument.**

alternate depository—A file of originals, duplicate copies, computer tapes, reproduced materials, etc., of selective, current, and evaluated MC&G data indexed and stored at an appropriately secure location, physically separated from basic libraries, but available for use in the event of destruction of the primary DoD library file.

altimeter—An instrument that indicates the height above a reference surface. See also **barometric altimeter; precision altimeter; radar altimeter; surveying altimeter.**

altimetry—The art and science of measuring altitudes by barometric means and interpreting the results.

altitude—1. The vertical distance of a point, or an object considered as a point, measured from a reference surface, as mean sea level (the geoid), ellipsoid, mean terrain. **2.** Angular distance above the horizon; the arc of a vertical circle between the horizon and a point on the celestial sphere, measured upward from the horizon. See also **absolute altitude; angular altitude; apparent altitude; circum-meridian altitudes; computed altitude; density altitude; elevation; ellipsoid height; ex-meridian altitude; flight altitude; geoidal height; high altitude; meridian altitude; negative altitude; observed altitude; orbital altitude; parallel of altitude; photo altitude; positive altitude; pressure altitude; radar altitude; sextant altitude; simultaneous altitudes; solar altitude; true altitude.**

altitude circle—See **parallel of altitude.**

altitude-contour ratio—See **C-factor.**

altitude datum—(JCS) The arbitrary level from which vertical displacement is measured. The datum for height measurement is the terrain directly below the aircraft or some specified datum; for pressure altitude, the level at which the atmospheric pressure is 29.92 inches of mercury (1013.2 mbs); and for true altitude, mean sea level.

altitude difference—The difference between computed and observed altitudes, or between precomputed and sextant altitudes. Also called **altitude intercept.**

altitude hole—(JCS) The blank area at the origin of a radial display, on a radar tube presentation, the center of the periphery of which represents the point on the ground immediately below the aircraft. In side-looking airborne radar, this is known as the **altitude slot.**

altitude intercept—See **altitude difference.**

altitude slot—See **altitude hole.**

altitude tints—See **hypsometric tinting.**

amphibious-assault landing model—See **assault-landing model.**

amphidromic point—A no-tide or nodal point on a chart of cotidal lines from which the cotidal lines radiate.

amphidromic region—An area surrounding an amphidromic point in which the cotidal lines radiate from the no-tide point and progress through all hours of the tidal cycle.

amplitude—1. The maximum value of the displacement of a wave or other periodic phenomenon from a reference position. **2.** Angular distance north or south of the prime vertical; the arc of the horizon or the angle at the zenith between the prime vertical and a vertical circle, measured north or south from the prime vertical to the vertical circle. The term is customarily used only with reference to bodies whose centers are on the celestial horizon, and is prefixed E or W, as the body is rising or setting, respectively, and suffixed N or S to agree with the declination See also **compass amplitude; grid amplitude; magnetic amplitude; true amplitude.**

amplitude of vibration—(pendulum) The length of the arc passed over by a pendulum in moving from its mean position to the position of maximum displacement.

anaglyph—A stereogram in which the two views are printed or projected superimposed in complementary colors, usually red and blue. By viewing through filter spectacles of corresponding complementary colors, a stereoscopic image is formed.

analemma—1. A figure-eight shaped diagram drawn across the Torrid Zone on a terrestrial globe to show the declination of the Sun throughout the year and also the equation of time. **2.** A sundial.

analog instruments—Devices that represent numerical quantities by means of physical variables; for example, by translation; by rotation as in a mechanical gear system; and by voltage or current as in analog networks that use resistance to represent mechanical losses, capacitors and inductors to store energy and simulate the action of springs, etc. Stereoscopic plotters are examples of photogrammetric analog instruments.

analytical aerotriangulation—See **analytical pho-totriangulation.**

analytical nadir-point triangulation—Radial tri-angulation performed by computational rou-tines in which nadir points are utilized as radia[l] centers.

analytical orientation—Those computationa[l] steps required to determine tilt, direction o[f] principal line, flight height, preparation o[f] control templets at rectification scale, angula[r] elements, and linear elements in preparin[g] aerial photographs for rectification. Develope[d] data are converted to values to be set on circle and scales of rectifier or transforming printer.

Analytical Photogrammetric Positioning System (APPS)—A computer-assisted measuring system used in field units to derive very precise coordinates (geographic or grid) using Point Positioning Data Bases. APPS consists of a stereographic measuring device and an electronic calculator system with an interface unit which mates the calculator to the measuring device. See also **Point Positioning Data Base.**

analytical photogrammetry—Photogrammetry in which solutions are obtained by mathematical methods.

analytical photography—Photography, either motion picture or still, accomplished to determine (by qualitative, quantitative, or any other means) whether a particular phenomenon does or does not occur.

analytical phototriangulation—A phototriangulation procedure in which the spatial solution is obtained by computational routines. When performed with aerial photographs, the procedure is referred to as **analytical aerotriangulation.**

analytical radar prediction—Prediction based on proven formulas, power tables, graphs, and/or other scientific principles. An analytical prediction considers surface height, structural and/or terrain information and criteria for radar reflectivity together with the aspect angle and range to the target.

analytical radial triangulation—Radial triangulation performed by computational routines. See also **graphical radial triangulation.**

analytical three-point resection radial triangulation—A method of computing the coordinates of the principal points of overlapping aerial photographs by resecting on three horizontal control points appearing in the overlap area.

anastigmatic lens—A lens which has been corrected for astigmatism and, therefore, focuses vertical and horizontal lines with equal brightness and definition. Anastigmatic lenses are also free of most common aberrations.

anchorage chart—A nautical chart showing prescribed or recommended anchorages.

aneroid altimeter—See **barometric altimeter.**

aneroid barometer—A barometer which balances the atmospheric pressure against a mechanically elastic device. The usual form of an aneroid barometer consists of a thin box of corrugated metal, almost exhausted of air. When the atmospheric pressure increases, the box contracts; when the pressure lessens, the box expands. By mechanical means these movements are amplified and communicated to an index hand which registers the changes on a graduated dial.

angle—The inclination to each other of two intersecting lines, measured by the arc of a circle intercepted between the two lines forming the angle, the center of the circle being the point of intersection. See also **adjusted angle; altitude; azimuth; azimuth angle; break angle; conversion angle; counterclockwise angle; concluded angle; conversion angle; counterclockwise angle; crab angle; critical angle; crossing angle; deflection angle; dihedral angle; dip angle; direct angle; direction angle; distance angle; double zenith distance; drift angle; Eulerian angles; Greenwich hour angle; grid magnetic angle; horizontal angle; hour angle; interlocking angle; local hour angle; locking angle; look angles; measured angle; meridian angle; oblique ascension; observed angle; parallactic angle; phase angle; reciprocal vertical angle; refraction angle; repetition of angles; right ascension; screen angle; sidereal hour angle; slope angle; solid angle; spherical angle; spheroidal angle; traverse angle; vectorial angle; vertical angle; zenith distance.**

angle equation—A condition equation which expresses the relationship between the sum of the measured angles of a closed figure and the theoretical value of that sum, the unknowns being the corrections to the observed directions or angles, depending on which are used in the adjustment.

angle method of adjustment—(triangulation and traverse) A method of adjustment of observations which determines correction to observed angles. The angle method of adjustment may be used where a chain of single triangles is to be adjusted.

angle of convergence—(JCS) The angle subtended by the eyebase of an observer at the point of focus. Also called **angular parallax; parallactic angle.**

angle of coverage—See **angle of field.**

angle of current—(hydrography) In stream gaging, the angle of current is the angular difference between 90° and the angle made by the current with a measuring section.

angle of depression—(JCS) **1.** The angle in a vertical plane between the horizontal and a descending line. **2.** In air photography, the angle between the axis of an obliquely mounted air camera and the horizontal. Also called **depression angle; descending vertical angle; minus angle.** See also **angle of elevation; true depression angle.**

angle of deviation—(optics) The angle through which a ray is bent by refraction.

angle of elevation—The angle in a vertical plane between the horizontal and an ascending line, as from an observer to an object. Also called **plus angle; ascending vertical angle.** See also **angle of depression.**

angle of field—A property of a lens. The angle subtended by lines that pass through the center of the lens and locate the diameter of the maximum image area within the specified definition of the lens. Lenses are generally classified according to their angles of coverage, as follows: narrow-angle; wide-angle; normal-angle; and super-wide- angle or ultra-wide-angle. Also called **angle of coverage; angular field.**

angle of incidence–(optics) As measured from the normal, the angle at which a ray of light strikes a surface.

angle of inclination–An angle of elevation or angle of depression.

angle of reflection–(optics) As measured from the normal, the angle at which a reflected ray of light leaves a surface.

angle of refraction–The angle which the refracted ray makes with the normal to the surface separating two transparent media.

angle of tilt–See **tilt.**

angle of view– (JCS) **1.** The angle between two rays passing through the perspective center (rear nodal point) of a camera lens to two opposite corners of the format. **2.** Photogrammetrically, it is twice the angle whose tangent is one-half the length of the diagonal of the format divided by the calibrated focal length. Also called **covering power; field of view.**

angle of yaw–The angle between a line in the direction of flight and a plane through the longitudinal and vertical axes of an aircraft. It is considered positive if the nose is displaced to the right. Also called **yaw angle.**

angle point–A term applied to a marker at each point to indicate a change in the direction of a survey line.

angle to right–The horizontal angle measured clockwise from the preceding line to the following one. Also called **clockwise angle.**

angle-to-right traverse–In surveying, a technique applicable to either open or closed traverses, wherein all angles are measured in a clockwise direction after the transit has been oriented by a backsight to the preceding station.

angular altitude–A measure in degrees of a given object above the horizon, taken from a given or assumed point of observation, and expressed by the angle between the horizontal and the observer's line of sight.

angular calibration constants–In a multiple-lens camera, or multiple-camera assembly, the values of angular orientation of the lens axes of the several lens-camera units to a common reference line. For example, in a trimetrogon camera, the angular relationships of the wing-camera axes with respect to the axis of the central (vertical) camera.

angular distance–**1.** The angular difference between two directions, numerically equal to the angle between two lines extending in the given directions. **2.** The arc of the great circle joining two points, expressed in angular units. **3.** Distance between two points, expressed in angular units of a specified frequency. It is equal to the number of waves between the points multiplied by 2π if expressed in radians, or multiplied by 360° if expressed in degrees.

angular distortion–**1.** (cartography) Distortion in a map projection because of non-conformality. **2.** (optics) The failure of a lens to reproduce accurately in the image space the angle subtended by two points in the object space.

angular error of closure–See **error of closure,** definition 2.

angular field–See **angle of field.**

angular magnification–The ratio of the angle subtended at the eye by the image formed by an optical device, to the angle subtended at the eye by the object itself without the optical device. This is convenient where a distance in the object cannot be measured for expressing a linear magnification, as in using a telescope.

angular momentum–The quantity obtained by multiplying the moment of inertia of a body by its angular speed.

angular parallax–See **angle of convergence.**

angular rate–See **angular speed.**

angular speed–Change of direction per unit time. Also called **angular rate.**

angular velocity–A representation of the rate of rotation of a particle about the axis of rotation, with magnitude equal to the time rate of angular displacement of any point of the body.

angulator–An instrument for converting angles measured on an oblique plane to their corresponding projections on a horizontal plane. A rectoblique plotter and photoangulator are types of angulators. See also **equiangulator; topoangulator.**

annex point–A point used to assist in the relative orientation of vertical and oblique photographs, selected in the overlap area between the vertical and its corresponding oblique about midway between the pass points. Alternate sets of photographs only will contain annex points. See also **pass point.**

annotated photograph–A photograph on which hypsographic, geologic, cultural, hydrographic, vegetation, or place name information has been added to identify, classify, outline, clarify, or describe features that would not otherwise be apparent in examination of an unmarked photograph. Generally, the term does not apply to photographs marked only with geodetic control or pass points.

annotation–(JCS) A marking placed on imagery or drawings for explanatory purposes. Annotations are used to indicate items or areas of special importance.

annotation overprint–The outline delimiting a target or installation, or a symbol which locates its position together with an identifying reference number as depicted on a target graphic.

annotation text–A descriptive text containing the identification, function, location, physical characteristics, and other information concerning a target or installation. Descriptive texts are also prepared for special areas, radar significant power lines, and precise radar significant location points.

annual aberration—Aberration caused by the velocity of the Earth's revolution about the Sun.

annual change—See **magnetic annual change.**

annual inequality—Seasonal variation in water level or tidal current speed, more or less periodic, due chiefly to meteorological causes.

annual magnetic change—See **magnetic annual change.**

annual magnetic variation—See **magnetic annual variation.**

annual parallax—The angle subtended at a celestial body by the radius of the Earth's orbit. Also called **heliocentric parallax; stellar parallax.**

annual rate—See **magnetic annual change.**

annual rate of change—See **magnetic annual change.**

annular eclipse—An eclipse in which a thin ring of the source of light appears around the obscuring body. Annular solar eclipses occur, but never annular lunar eclipses.

anomalistic drift—The variation or drift of a frequency source. For example, the frequency changes of a crystal oscillator due to a variety of causes, such as temperature variation and component aging, none of which can be predicted in advance or completely controlled.

anomalistic month—The interval of time between two successive passages of the Moon in her orbit through perigee. The length of the anomalistic month is 27.55455 mean solar days.

anomalistic period—The interval between two successive perigee passes of a satellite in orbit about its primary. Also called **perigee-to-perigee period.**

anomalistic tide cycle—The average period of about 27½ days, measured from perigee to perigee, during which the Moon completes one revolution around the Earth.

anomalistic year—The period of one revolution of the Earth around the Sun, from perihelion to perihelion, 365 days, 6 hours, 13 minutes, 53.16 seconds in 1955, and increasing at the rate of 0.002627 second annually.

anomalous magnetic variation—See **local magnetic anomaly.**

anomaly—**1.** (general) A deviation from the norm. **2.** (geodesy) A deviation of an observed value from a theoretical value due to a corresponding irregularity in the Earth's structure at the area of observation. **3.** (astronomy) The angle between the radius vector to an orbiting body from its primary and the line of apsides of the orbit, measured in the direction of travel, from the point of closest approach to the primary. This term is also called the **true anomaly** when it is desired to distinguish it from the **eccentric anomaly,** which is the corresponding angle at the center of the orbit; or from the **mean anomaly,** which is what the true anomaly would become if the planet had a uniformly angular motion. See also **Bouguer anomaly; deflection anomaly; free-air anomaly; gravity anomaly; Hayford gravity anomalies; height anomaly; isostatic anomaly; local magnetic anomaly; magnetic disturbance; mean free-air anomaly; point anomaly; surface anomalies.**

Antarctic Circle—The geographic parallel having a south latitude equal to the complement of the declination of the winter solstice. The obliquity of the ecliptic is steadily changing so that the winter solstice is not a point of fixed declination, and the Antarctic Circle, as defined, is not a line of fixed position. When the Antarctic Circle is to be shown on a map, however, it is desirable that it be treated as a line of fixed position, and that a conventional value be adopted for its latitude. For this purpose, the value 66°33′ south latitude is used. Also called **south polar circle.**

antihalation coating—(photography) A light-absorbing coating applied to the back side of the support of a film or plate (or between the emulsion and the support) to suppress halation.

antipode—Anything exactly opposite to something else. Particularly, that point on the Earth 180° from a given place.

antisolar point—That point on the celestial sphere 180° from the Sun.

antivignetting filter—(JCS) A filter bearing a deposit which is graduated in density to correct for the uneven illumination given by certain lenses, particularly wide-angle types.

apareon—The point on a Mars-centered orbit where a satellite is at its greatest distance from Mars.

apastron—That point of the orbit of one member of a double star system at which the stars are farthest apart. Opposite of **periastron.**

aperiodic compass—Literally "a compass without a period," or a compass that, after being deflected, returns by one direct movement to its proper reading, without oscillation. Also called **deadbeat compass.**

aperture—**1.** (JCS) The opening in a lens diaphragm through which light passes. **2.** The diameter of the objective lens of a telescope or other optical instrument, usually expressed in inches, but sometimes as the angle between lines from the principal focus to opposite ends of a diameter of the objective lens. See also **relative aperture.**

aperture ratio—See **relative aperture.**

aperture stop—(optics) The physical element (such as a stop, diaphragm, or lens periphery) of an optical system which limits the size of the pencil of rays traversing the system. The adjustment of the size of the aperture stop of a given system regulates the brightness of the image without necessarily affecting the size of the area covered. Also called **stop.**

apex—See **vertex.**

aphelion—The point in the elliptical orbit of a planet which is the farthest from the Sun, when the Sun is the center of attraction. Opposite of **perihelion.**

aphylactic map projection—A map projection which does not possess any of the three special properties of equivalence, conformality, or equidistance. Also called **arbitrary projection.**

aplanatic lens—A lens which transmits light without spherical aberration.

aplune (apolune)—The point on the elliptical orbit of a satellite of the Moon which is farthest from the Moon. Also called **apocynthion.** Opposite of **perilune; pericynthion.**

apoapsis—See **apocenter.**

apocenter—In an elliptical orbit, the point in the orbit which is the greatest distance from the focus where the attracting mass is located. Also called **apoapsis; apofocus.** Opposite of **periapsis; pericenter; perifocus.**

apochromatic lens—A lens that has been corrected for chromatic aberration for three colors.

apocynthion—See **aplune (apolune).**

apofocus—See **apocenter.**

apogean tides—Tides of decreased range occurring when the Moon is near apogee.

apogee—(JCS) The point at which a missile trajectory or a satellite orbit is farthest from the center of the gravitational field of the controlling body or bodies.

apparent—A term used to designate certain measured or measurable astronomic quantities to refer them to the observed position of celestial bodies.

apparent altitude—The observed vertical angle of a celestial object corrected for instrumental errors, personal errors, and inaccuracies in the reference level (principally dip), but not for refraction, parallax, or semidiameter. Also called **rectified altitude.**

apparent horizon—(JCS) The visible line of demarcation between land/sea and sky. Also called **local horizon; topocentric horizon; visible horizon.**

apparent motion—Motion relative to a specified or implied reference point which may itself be in motion. The expression usually refers to movement of celestial bodies as observed from the Earth. Also called **relative motion.**

apparent noon—Twelve o'clock apparent time, or the instant the apparent Sun is over the upper branch of the meridian.

apparent place—(astronomy) See **apparent position.**

apparent position—An astronomic term applied to the observable position of a star, planet, or the Sun. The position on the celestial sphere at which a heavenly body (or a space vehicle) would be seen from the center of the Earth at a particular time. See also **astrometric position.** Also called **apparent place.**

apparent precession—(JCS) The apparent deflection of the gyro axis, relative to the Earth, due to the rotating effect of the Earth and not due to any applied force. Also called **apparent wander; wander.**

apparent sidereal time—The local hour angle of the true vernal equinox. Also called **true sidereal time.**

apparent solar day—The interval of time from a transit of the apparent Sun across a given meridian to its next successive transit across the same meridian.

apparent solar time—Time measured by the apparent diurnal motion of the true Sun. Also called **apparent time; true solar time.**

apparent Sun—The actual Sun as it appears in the sky. Also called **true Sun.**

apparent time—See **apparent solar time.**

apparent wander—See **apparent precession.**

appearance ratio—See **hyperstereoscopy.**

approach chart—An aeronautical chart providing essential information for making an approach to an airfield under either visual or instrument flight conditions.

approximate contour—A contour substituted for a normal contour whenever there is a question as to its reliability (reliability is defined as being accurate within one-half the contour interval).

appulse—The near approach of one celestial body to another on the celestial sphere, as in occultation or conjunction.

apse line—See **line of apsides.**

apsis—Either of the two orbital points nearest or farthest from the center of attraction, the perihelion and aphelion in the case of an orbit about the Sun, and the perigee and apogee in the case of an orbit about the Earth.

arbitrary grid—Any reference system developed for use where no grid is available or practical, or where military security for the reference is desired.

arbitrary projection—See **aphylactic map projection.**

arc correction—(pendulum) The quantity which is applied to the period of vibration of a pendulum to allow for the pendulum's departure from simple harmonic motion.

arc measurement—A survey method used to determine the size of the Earth. A long arc is measured on the Earth's surface and the angle which subtends this measured arc is determined. By assumptions and mathematical formula the size and shape of the Earth can then be determined.

arc navigation—A navigation system in which the position of an airplane or ship is maintained along an arc measured from a control station by means of electronic distance-measuring equipment, such as **shoran.** See also **loran.**

arc of parallel—A part of an astronomic or geodetic parallel of latitude.

arc of visibility—The arc of a light sector, designated by its limiting bearings as observed from seaward.

arc triangulation—A system of triangulation of limited width designed to progress in a single general direction. Arc triangulation is executed for the purpose of connecting independent and widely separated surveys, coordinating, and correlating local surveys along the arc, furnishing data for the determination of a geodetic datum, providing a network of control points for a country-wide survey, etc.

Arctic Circle—The geographical parallel having a north latitude equal to the complement of the declination of the summer solstice. The obliquity of the ecliptic is steadily changing so that the summer solstice is not a point of fixed declination, and the Arctic Circle, as defined, is not a line of fixed position. When the Arctic Circle is to be shown on a map, however, it is desirable that it be treated as a line of fixed position, and that a conventional value be adopted for its latitude. For this value 66° 33′ north latitude is used. Also called **north polar circle.**

area—(surveying) In general, an area is any aggregate of plane spaces to be considered in an investigation; especially the quantity projected on a horizontal plane enclosed by the boundary of any polygonal figure.

area analysis intelligence—Intelligence data relative to a specific geographic area.

area coverage—**1.** Complete coverage of an area by aerial photography having parallel overlapping flight lines and stereoscopic overlap between exposures in the line of flight. **2.** When applied to shoran, the term implies that recorded shoran distances are available for each exposure. **3.** Complete coverage of a geographical area by maps or other graphic material.

area pattern screen—A photographic negative or positive containing repetitively arranged small feature symbols which have been designed to present a visual portrayal of a map or chart feature; i.e., swamp, orchard, sand, etc. See also **contact screen; line pattern.**

area survey—**1.** A survey of areas large enough to require a network of control stations. **2.** An extension and densification of survey control.

area target—(JCS) A target consisting of an area rather than a single point. See also **pinpoint target.**

area triangulation—A system of triangulation designed to progress in every direction. Area triangulation is executed to provide survey control points over an area, as of the city or county; or for filling in the areas between arcs of triangulation which form a network extending over a county or state. See also **survey net; triangulation net.**

area weighted average resolution (AWAR)—A single average value for the resolution over the picture format for any given focal plane.

areal feature—A topographic feature, such as sand, swamp, vegetation, etc., which extends over an area. It is represented on the published map or chart by a solid or screened color, by a prepared pattern of symbols, or by a delimiting line.

areodesy—(JCS) That branch of mathematics which determines by observations and measurements, the exact positions of points and the figures and areas of large portions of the surface of the planet Mars, or the shape and size of the planet Mars.

aerodetic—(JCS) Of or pertaining to, or determined by aerodesy.

argument—In astronomy, an angle or arc, as in **argument of perigee.**

argument of latitude—In celestial mechanics, the angular distance measured in the orbit plane from the ascending node to the orbiting object; the sum of the argument of perigee and the true anomaly.

argument of perigee—An orbital element defined as the angle at the center of attraction from the ascending node to the perigee point measured in the direction of motion of the orbiting body.

artificial asteroid—A manmade object placed in orbit about the Sun.

artificial Earth satellite—A manmade Earth satellite, as distinguished from the Moon.

artificial horizon—(JCS) A device that indicates attitude with respect to the true horizon. A substitute for a natural horizon, determined by a liquid level bubble pendulum, or gyroscope.

artificial monument—A relatively permanent object used to identify the location of a survey station or corner. Objects include manmade structures such as abutments, stone markers, concrete markers, and railroad rails.

art-work prediction—See **experience radar prediction.**

Arundel method—A combination of graphical and analytical methods, based on radial triangulation, for point-by-point topographic mapping from aerial photographs.

ascending node—That point at which a planet, planetoid, or comet crosses the ecliptic from south to north, or a satellite crosses the equator of its primary from south to north. Opposite of **descending node.** Also called **northbound node.**

ascending vertical angle—See **angle of elevation.**

ascensional difference—The difference between right ascension and oblique ascension.

aspect change—(JCS) The different appearance of a reflecting object viewed by radar from varying directions. It is caused by the change in the effective reflecting area of the target.

aspects—The apparent positions of celestial

bodies relative to one another; particularly the apparent positions of the Moon or a planet relative to the Sun.

aspherical lens—A lens in which one or more surfaces depart from a true spherical shape.

assault-landing model—A special form of assault model designed specifically for planning amphibious landings. Also called **amphibious-assault landing model.** See also **assault models.**

assault models—Large-scale models giving a particular representation of vegetation, lesser landforms, prominent manmade features, and a detailed representation of specific or sensitive objectives such as airfields, radar installations, and the like. These models emphasize the aspects of objects as seen from surface approach.

associated Legendre function—A solution of the Legendre equation, which is a special case of the Laplace equation, in the form of a power series of a special kind; used in the spherical harmonic expansion of the gravitational potential.

assumed ground elevation—The elevation assumed to prevail in the local area covered by a particular photograph or group of photographs. Used especially to denote the elevation assumed to prevail in the vicinity of a critical point, such as a peak or other feature having abrupt local relief.

assumed latitude—The latitude at which an observer is assumed to be located for an observation or computation, as the latitude of an assumed position or the latitude used for determining the longitude by time sight.

assumed longitude—The longitude at which an observer is assumed to be located for an observation or computation, as the longitude of an assumed position or the longitude used for determining the latitude by meridian altitude.

assumed plane coordinates—A local plane-coordinate system set up at the convenience of the surveyor. The reference axes are usually assumed so that all coordinates are in the first quadrant. The *y*-axis may be in the direction of astronomic north, geodetic north, magnetic north, or an assumed north.

assurance level—See **confidence interval.**

astatized gravimeter—A gravimeter, sometimes referred to as unstable, where the force of gravity is maintained in an unstable equilibrium with the restoring force. The instability is provided by the introduction of a third force which intensifies the effect of any change in gravity from the value in equilibrium.

asteroid—A minor planet; one of the many small celestial bodies revolving around the Sun, most of the orbits being between those of Mars and Jupiter. Also called **minor planet; planetoid.** See also **artificial asteroid.**

astigmatism—An aberration affecting the sharpness of images for objects off the axis in which the rays passing through different meridians of the lens come to a focus in different planes. Thus, an extra-axial point object is imaged as two mutually perpendicular short lines located at different distances from the lens.

astigmatizer—A lens which introduces astigmatism into an optical system. Such a lens is so arranged that it can be placed in or removed from the optical path at will. In a sextant, an astigmatizer may be used to elongate the image of a celestial body into a horizontal line.

astre fictif—Any of several fictitious stars assumed to move along the celestial equator at uniform rates corresponding to the speeds of the several harmonic constituents of the tide producing force. Each astre fictif crosses the meridian at the instant the constituent it represents is at a maximum.

astro compass—(JCS) An instrument used primarily to obtain true heading or true bearing by reference to celestial bodies.

astrodynamics—The practical application of celestial mechanics, astroballistics, propulsion theory, and allied fields to the problem of planning and directing the trajectories of space vehicles.

astrogeodetic datum orientation—The position of a reference ellipsoid in relation to the geoid in a specified area of a geodetic network. It may be expressed by the astrogeodetic deflection and geoidal height at the datum point or by an astrogeodetic geoid chart of the area.

astrogeodetic deflection—The angle at a point between the normal to the geoid and the normal to the ellipsoid of an astrogeodetically oriented datum. Also called **relative deflection**

astrogeodetic leveling—A method to determine variations in the separation of the geoid and the ellipsoid using astrogeodetic deflections. Also called **astronomic leveling; geoidal height profile.**

astrogeodetic undulations—The separation between an astrogeodetic geoid, defined for a particular datum, and a specified ellipsoid surface. See also **geoidal height.**

astrograph—**1.** A device for projecting a set of precomputed altitude curves onto a chart or plotting sheet, the curves moving with time such that if they are properly adjusted, they will remain in the correct position on the chart or plotting sheet. **2.** A telescope, usually of moderate focal length, which is designed specifically for the purpose of accurately recording the positions of celestial objects by photographic means.

astrograph mean time—A form of mean time used in setting an astrograph. Astrograph mean time 1200 occurs when the local hour angle of Aries is 0°.

astrographic position—See **astrometric position**

astrogravimetric leveling—A concept whereby a gravimetric map is used for the interpolation of the astrogeodetic deflections of the vertical to determine the separation of the ellipsoid and the geoid in studying the figure of the Earth.

astrogravimetric points—Astronomic positions corrected for the deflection of the vertical by gravimetric methods.

astrolabe—1. (general) Any instrument designed to measure the altitudes of celestial bodies. **2.** (surveying) An instrument designed for very accurate celestial altitude measurements. See also **equiangulator; pendulum astrolabe; planispheric astrolabe; prismatic astrolabe.**

astrometric position—The position of a heavenly body (or space vehicle) on the celestial sphere corrected for aberration but not for planetary aberration. Astrometric positions are used in photographic observation where the position of the observed body can be measured in reference to the positions of comparison stars in the field of the photograph. Also called **astrographic position.** See also **apparent position.**

astrometry—The branch of astronomy dealing with the geometric relations of the celestial bodies and their real and apparent motions. The techniques of astrometry, especially the determination of accurate position by photographic means, are used in tracking satellites and space probes.

astronomic—Of or pertaining to astronomy, the science which treats of heavenly bodies, and the arts based on that science.

astronomic arc—The apparent arc described above (diurnal arc) or below (nocturnal arc) the horizon by the Sun or another celestial body.

astronomic azimuth—The angle between the astronomic meridian plane of the observer and the plane containing the observed point and the true normal (vertical) of the observer, measured in the plane of the horizon, preferably clockwise from north.

astronomic azimuth mark—A marked point whose astronomic azimuth from a survey station is determined from direct observations on a celestial body. The mark may be a lamp or illuminated target placed especially for the purpose; it may be a well defined illuminated point on a permanent structural point.

astronomic bearing—See **true bearing.**

astronomic constants—The elements of the orbits of the bodies of the solar system, their masses relative to the Sun, their size, shape, orientation, rotation, and inner constitution, and the velocity of light. See also **system of astronomic constants.**

astronomic control—A network of control stations the positions of which have been determined by astronomic observation. Latitudes and longitudes thus determined will normally differ from the geodetic latitudes and longitudes of the same stations by amounts corresponding to components of the deflection of the vertical.

astronomic coordinates—1. Quantities defining a point on the surface of the Earth, or of the geoid, in which the local direction of gravity is used as a reference. Also called **geographic coordinates; gravimetric coordinates; terrestrial coordinates. 2.** The coordinates of an astronomic body referred to a given equinox.

astronomic date—Designation of epoch by year, month, day, and decimal fraction. For example, the astronomic date of December 21, 1978, 18^h UTC (universal time coordinated) is 1978 December 21.75 UTC. The astronomic date is also used in connection with the other time systems. The system commences every calendar year at 0^h on December 31 of the previous year. This epoch is denoted by January 0.0.

astronomic day—A mean solar day beginning at mean noon, 12 hours later than the beginning of the civil day of the same date. Astronomers now generally use the civil day.

astronomic equator—The line on the surface of the Earth whose astronomic latitude at every point is 0°. Due to the deflection of the plumbline, the astronomic equator is not a plane curve. However, the verticals at all points on it are parallel to one and the same plane, the plane of the celestial equator; that is, the zenith at every point on the astronomic equator lies in the celestial equator. When the astronomic equator is corrected for station error, it becomes the geodetic equator. Also called **terrestrial equator.** See also **geodetic equator.**

astronomic latitude—The angle between the plumbline and the plane of celestial equator. Also defined as the angle between the plane of the horizon and the axis of rotation of the Earth. Astronomic latitude applies only to positions on the Earth and is reckoned from the astronomic equator (0°) north and south through 90°. Astronomic latitude is the latitude which results directly from observations of celestial bodies, uncorrected for deflection of the vertical.

astronomic leveling—See **astrogeodetic leveling.**

astronomic longitude—The angle between the plane of the celestial meridian and the plane of an initial meridian, arbitrarily chosen. Astronomic longitude is the longitude which results directly from observations on celestial bodies, uncorrected for deflection of the vertical.

astronomic meridian—A great circle of the celestial sphere intersecting the north and south celestial poles. The local astronomic meridian is that meridian which intersects the zenith of the point.

astronomic meridian plane—A plane that contains the vertical of the observer and is parallel to the instantaneous rotation axis of the Earth.

astronomic parallel—A line on the surface of the Earth which has the same astronomic latitude at every point. Because the deflection of the vertical is not the same at all points on the Earth, an astronomic parallel is an irregular line, not lying in a single plane. See also **astronomic equator.**

astronomic position—**1.** A point on the Earth whose coordinates have been determined as a result of observations of celestial bodies. The expression is usually used in connection with positions on land determined with great accuracy for survey purposes. **2.** A point on the Earth, defined in terms of astronomic latitude and longitude.

astronomic refraction—The apparent displacement of an object that results from light rays from a source outside the atmosphere being bent in passing through the atmosphere. This results in all objects appearing to be higher above the horizon than they actually are. The magnitude of this displacement is greater when the object is near the horizon and decreases to a minimum assumed to be zero when the object is at the zenith. Also called **astronomic refraction error; celestial refraction.** See also **atmospheric refraction; refraction.**

astronomic refraction error—See **astronomic refraction.**

astronomic station—A point on the Earth whose position has been determined by observations on celestial bodies.

astronomic surveying—The celestial determination of latitude and longitude. Separations are calculated by computing distances corresponding to measured angular displacements along the reference spheroid.

astronomic theodolite—See **alt-azimuth instrument.**

astronomic tidal constituent—See **constituent.**

astronomic time—Solar time in a day (astronomic day) that begins at noon. Astronomic time may be either apparent solar time or mean solar time. Since 1925, civil time is generally used instead of astronomic time.

astronomic transit—See **transit,** definition 4.

astronomic triangle—The navigational triangle, either terrestrial or celestial, used in the solution of celestial observations. Referring to the celestial sphere it is the triangle formed by arcs of great circles connecting the celestial pole, the zenith, and a celestial body. The angles of the astronomic triangles are: at the pole, the hour angle; at the celestial body, the parallactic angle; at the zenith, the azimuth angle. The sides are: pole to zenith, the co-latitude; zenith to celestial body, the zenith distance; and celestial body to pole, the polar distance. Also called **PZS triangle.**

astronomic unit—A unit of length equal to 149,600,000 kilometers (adopted 1960) used for measuring distances within the solar system. This distance approximates the mean distance of the Earth from the Sun.

astronomic year—See **tropical year.**

asymmetry of object (target)—Lack of symmetry in the visible aspect of an object as seen from a particular point of observation. A square or rectangular pole may so face the observer that the line bisecting its tangents does not pass through its geometric center. With a square cupola or tower, the error resulting from observing tangents and taking a mean may be quite large. The error caused by asymmetry of an observed object is of the same character and requires the same treatment as the error resulting from observing an eccentric object. See also **phase.**

asymptote—A straight line or curve which some curves of infinite length approach but never reach.

atlas grid—A reference system that permits the designation of the location of a point or an area on a map, photo, or other graphic in terms of numbers and letters. Also called **alphanumeric grid.**

atmosphere—(JCS) The air surrounding the Earth. See also **ionosphere; stratosphere; tropopause; troposphere.**

atmospheric drag—A major perturbation of close artificial satellite orbits caused by the resistance of the atmosphere. The secular effects are decreasing eccentricity, major axis, and period. Also called **drag.**

atmospheric refraction—The refraction of light passing through the Earth's atmosphere. Atmospheric refraction includes both **astronomic refraction** and **terrestrial refraction.**

atomic time—Time interval based on the frequency of atomic oscillators.

atran—An acronym for "automatic terrain recognition and navigation," a navigation system which depends upon the correlation of terrain images appearing on a radar cathode-ray tube with previously prepared maps or simulated radar images of the terrain.

attenuation—(JCS) Decrease in intensity of a signal, beam, or wave as a result of absorption of energy and of scattering out of the path of a detector, but not including the reduction due to geometric spreading, i.e., the inverse square of distance effect.

attitude—**1.** (JCS) The position of a body as determined by the inclination of the axes to some frame of reference. If not otherwise specified, this frame of reference is fixed to the Earth. **2.** (JCS) Grid bearing relative to the long axis of the target. **3.** (photogrammetry) The angular orientation of a camera, or of the photograph taken with that camera, with respect to some external reference system. Usually expressed as tilt, swing, and azimuth or roll, pitch, and yaw.

augmentation—The apparent increase in the semidiameter of a celestial body as its altitude increases, due to the reduced distance from the observer. The term is used principally in refer ence to the Moon.

augmentation correction—A correction due to augmentation, particularly that sextant altitude correction due to the apparent increase in the semidiameter of a celestial body as its altitude increases.

augmenting factor—A factor used in connection with the harmonic analysis of tides or tidal currents to allow for the difference between the times of hourly tabulation and the corresponding constituent hours.

austral—Of or pertaining to south.

Australian National spheroid—A reference spheroid having the following dimensions: semimajor axis—6,378,160.0 meters; and a flattening or ellipticity of 1/298.25.

authalic (equal-area) latitude—A latitude based on a sphere having the same area as the spheroid, and such that areas between successive parallels of latitude are exactly equal to the corresponding areas on the spheroid. Authalic latitudes are used in the computation of equal-area map projections.

authalic map projection—An equal-area map projection.

auto radar plot—See **chart comparison unit.**

auto-reducing tachymeter—A class of tachymeter by which horizontal and height distances are read simultaneously. Horizontal distance is the intercept multiplied by 100 and the vertical distance is the mid-wire (curve) multiplied by a factor which appears in the optics.

autocollimation—(surveying) The procedure used to determine or transfer azimuth to an instrument or device. This procedure requires use of a specially adapted telescope, capable of bisecting the real image of its own reticule as reflected from a mirror or Porro prism. When such bisection is accomplished, the line of sight of the telescope is perpendicular to the face of the mirror or apex edge of the prism.

autocollimator—A collimator provided with a means of illuminating its crosshairs so that, when a reflecting plane is placed normal to the emergent light beam, the reflected image of the crosshairs appears to be coincident with the crosshairs themselves. This device is used in calibrating optical and mechanical instruments and transferring direction.

autofocus rectifier—A precise, vertical photo-enlarger which permits the correction of distortion in an aerial negative caused by tilt. The instrument's operations are motor driven and are interconnected by mechanical linkages to insure automatically maintained sharp focus.

Automated Tactical Target Graphic (ATTG)—A tactical target materials item which provides aerial photographic coverage of a target and a limited area surrounding it at a scale permitting optimum identification of target detail. The ATTG also includes textual intelligence on a sheet separate from the graphic portion. Each part can be revised independent of the other. ATTGs cover single targets and are produced in two forms: a lithographic sheet and a miniaturized version in an aperture card.

Automatic Digital Annotation System (ADAS)—A system used to record camera position and other information on film at time of exposure.

automatic gage—See **self-registering gage.**

automatic level—See **pendulum level.**

automatic rectifier—Any rectifier which employs mechanisms to insure automatic fulfillment of the lens law and the Scheimpflug condition. These devices, called inversors, provide a mechanical solution for the linear and angular elements of rectification. Essentially, this class of rectifier is a tilt analyzer using inversors to solve for the optical-geometric elements needed for sharp focus.

automatic rod—See **tape rod.**

automatic traverse computer—An instrument which, by mechanical linkages and gears, converts courses into latitudes and departures. It also mechanically totals the algebraic sum of the latitudes and departures.

autoscreen film—A photographic film embodying a halftone screen which automatically produces a halftone negative from continuous-tone copy.

autumnal equinox—That point of intersection of the ecliptic and the celestial equator occupied by the Sun as it changes from north to south declination, about 23 September. Also called **first point of Libra; September equinox.**

auxiliary contour—See **supplementary contour.**

auxiliary guide meridian—(USPLS) Where guide meridians have been placed at intervals exceeding the distance of 24 miles, and new governing lines are required, a new guide meridian is established, and a local name is assigned, such as "Twelfth Auxiliary Guide Meridian West," or "Grass Valley Guide Meridian." Auxiliary guide meridians are surveyed, in the same manner as guide meridians. See also **guide meridian; principal meridian.**

auxiliary meander corner—An auxiliary meander corner is established at a suitable point on the meander line of a lake lying entirely within a quarter section or on the meander line of an island falling entirely within a section and which is found to be too small to subdivide. A line is run connecting the monument to a regular corner on the section boundary.

auxiliary station—Any station connected to the main-scheme net and dependent upon it for the accuracy of its position.

average deviation—In statistics, the average or arithmetic means of the deviations, taken without regard to sign, from some fixed value, usually the arithmetic mean of the data. Also called **mean deviation.**

average terrestrial pole—The average position of the instantaneous pole of rotation of the Earth, averaged over a specified time period. See also

conventional international origin.

averaging device—A device for averaging a number of readings, as on a bubble sextant.

axis—See **camera axis; collimation axis; coordinate axes; equatorial axis; fiducial axes; horizontal axis; major axis; minor axis; optical axis; polar axis; semimajor axis; semiminor axis; spirit level axis; topple axis; transverse axis; vertical axis; x-axis; y-axis; z-axis.**

axis of homology—The intersection of the plane of the photograph with the horizontal plane of the map or the plane of reference of the ground. Corresponding lines in the photograph and map planes intersect on the axis of homology. Also called the **axis of perspective; map parallel; perspective axis.** See also **ground parallel.**

axis of lens—See **optical axis.**

axis of level—See **spirit level axis.**

axis of perspective—See **axis of homology.**

axis of the level bubble—See **spirit level axis.**

axis of tilt—A line through the perspective center perpendicular to the principal plane. The axis of tilt could be any of several lines in space (e.g., the isometric parallel or the ground line), but the present definition is the only one which permits the concept of tilting a photograph without upsetting the positional elements of exterior orientation.

azimuth—**1.** (JCS) Quantities may be expressed in positive quantities increasing in a clockwise direction or in *x, y* coordinates where south and west are negative. They may be referenced to true north or magnetic north depending on the particular weapon system used. **2.** (surveying) The horizontal direction of a line measured clockwise from a reference plane, usually the meridian. Also called **forward azimuth** to differentiate from **back azimuth. 3.** (photogrammetry) Azimuth of the principal plane. See also **astronomic azimuth; azimuth by altitude; back azimuth; computed azimuth angle; direction method of determining astronomic azimuth; geodetic azimuth; grid azimuth; inertial azimuth; Laplace azimuth; magnetic azimuth; method of repetitions (determination of astronomic azimuth); micrometer method (determination of astronomic azimuth); normal section azimuth; true azimuth.**

azimuth angle—**1.** (JCS) An angle measured clockwise in the horizontal plane between a reference direction and any other line. **2.** (astronomy) The angle 180° or less between the plane of the celestial meridian and the vertical plane containing the observed object, reckoned from the direction of the elevated pole. In astronomic work, the azimuth angle is the spherical angle at the zenith in the astronomic triangle which is composed of the pole, the zenith, and the star. In geodetic work, it is the horizontal angle between the celestial pole and the observed terrestrial object. **3.** (surveying) An angle in triangulation or in a traverse through which the computation of azimuth is carried. In a simple traverse, every angle may be an azimuth angle. Sometimes, in a traverse, to avoid carrying azimuths over very short lines, supplementary observations are made over comparatively long lines, the angles between which form azimuth angles. In triangulation, certain angles, because of their size and position in the figure, are selected for use as azimuth angles, and enter into the formation of the azimuth condition equation (azimuth equation).

azimuth bar—See **azimuth instrument.**

azimuth by altitude—An azimuth determined by solution of the navigational triangle with altitude, declination, and latitude given.

azimuth circle—A ring designed to fit snugly over a compass or compass repeater, and provided with means for observing compass bearings and azimuths.

azimuth equation—A condition equation which expresses the relationship between the fixed azimuths of two lines which are connected by triangulation or traverse.

azimuth error of closure—See **error of closure,** definition 3.

azimuth instrument—(magnetic) An instrument for measuring azimuths, particularly a device which fits over a central pivot in the glass cover of a magnetic compass. Also called **azimuth bar; bearing bar.**

azimuth line—(photogrammetry) A radial line from the principal point, isocenter, or nadir point of a photograph, representing the direction to a similar point of an adjacent photograph in the same flight line; used extensively in radial triangulation.

azimuth mark—A mark set at a significant distance from a triangulation or traverse station to mark the end of a line for which the azimuth has been determined, and to serve as a starting or reference azimuth for later use. See also **astronomic azimuth mark; geodetic azimuth mark; Laplace azimuth mark.**

azimuth resolution—(JCS) The ability of radar equipment to separate two reflectors at similar ranges but different bearings from a vehicle. Normally the minimum separation distance between the reflectors is quoted and expressed as the angle subtended by the reflectors at the vehicle.

azimuth transfer—Connecting, with a straight line, the nadir points of two vertical photographs selected from overlapping flights.

azimuth traverse—A survey traverse in which the direction of the measured course is determined by azimuth and verified by back azimuth. To initiate this type of traverse it is necessary to have a reference meridian, either true, magnetic, or assumed.

azimuthal chart—A chart on an azimuthal projection. Also called **zenithal chart.**

azimuthal equidistant chart—A chart on the azimuthal equidistant map projection.

azimuthal equidistant map projection—An azimuthal map projection on which straight lines radiating from the center or pole of projection represent great circles in their true azimuths from that center, and lengths along those lines are of exact scale. This projection is neither equal-area nor conformal.

azimuthal map projection—A map projection on which the azimuths or directions of all lines radiating from a central point or pole are the same as the azimuths or directions of the corresponding lines on the sphere. Also called **zenithal map projection.**

azimuthal orthomorphic map projection—See **stereographic map projection.**

B

B-station—**1.** (loran) The designation applied to the transmitting station of a pair, the signal of which always occurs more than half a repetition period after the next succeeding signal and less than half a repetition period before the next preceding signal from the other station of the pair, designated an **A-station.** **2.** (traverse) See **A-station,** definition 1.

BC-4 camera—A trade name for the ballistic or geodetic stellar camera consisting of a Wild Astrotar or Wild Cosmotar lens cone mounted on the modified lower part of the Wild T-4 astronomic theodolite. Originally designed for the recording of the trajectory of a rocket but since adapted for the photographic tracking of artificial Earth satellites for geodetic purposes.

Bz curve—(photogrammetry) A graphical representation of the vertical errors in a stereotriangulated strip. In a *Bz* curve, the *x*-coordinates of the vertical control points, referred to the initial nadir point as origin, are plotted as abscissas, and the differences between the known elevations of the control points and their elevations as read in the stereotriangulated strip are plotted as ordinates; a smooth curve drawn through the plotted points is the *Bz* curve. The elevation read on any pass point in the strip is adjusted by the amount of the ordinate of the *Bz* curve for an abscissa corresponding to the *x*-coordinate of the point.

Bz curve method—A method utilizing characteristics of the *Bz* curve for finding the displacement of true photo plumb points from indicated projector plumb points in multiplex strip orientation. The method also provides a means of strip leveling using only the barometric-altimeter readings of the aircraft-flying height.

Bache-Wurdeman base-line measuring apparatus—A compensating base-line measuring apparatus having a measuring element composed of a bar of iron and a bar of brass, each a little less than 6 meters in length, held together firmly at one end, with the free ends so connected by a compensating lever as to form a compensating apparatus.

back azimuth—**1.** (geodetic surveying) If the azimuth of point B from point A is given on a reference sphere or ellipsoid, the back azimuth is the azimuth of point A from point B. Because of the convergence of the meridians, the forward and backward azimuths of a line do not differ by exactly 180°, except where A and B have the same geodetic longitude or where the geodetic latitudes of both points are 0°. **2.** (plane surveying) When referred to a plane rectangular coordinate system, same as above except forward and backward azimuths differ by exactly 180°. See also **azimuth,** definition 1.

back bearing—**1.** A bearing differing by 180°, or measured in the opposite direction from a given bearing. Also called **reciprocal bearing.** **2.** (USPLS) The bearing at the opposite end of a line from the observer as measured from the true meridian at the opposite end of the line. The back bearing on all lines (other than north-south lines) are different from the bearing at the observer's station. They differ by the amount of convergency of the meridians between the two points.

back focal distance—See **back focal length.**

back focal length—The distance measured along the lens axis from the rear vertex of the lens to the plane of best average definition. Also called **back focal distance; back focus.**

back focus—See **back focal length.**

backsight—**1.** A sight on a previously established survey point or line. **2.** (traverse) A sight on a previously established survey point, which is not the closing sight of the traverse. **3.** (leveling) A reading on a rod held on a point whose elevation has been previously determined and which is not the closing sight of a level circuit; any such rod reading used to determine height of instrument prior to making a foresight. Also called **plus sight.**

backstep—The method of determining the offsets for the bottom latitude of a projection by measuring the appropriate distances down from the top latitude of a chart.

backup—An image printed on the reverse side of a map sheet already printed on one side. Also the printing of such images.

balancing a survey—Distributing corrections through any traverse to eliminate the error of closure, and to obtain an adjusted position for each traverse station. Also called **traverse adjustment.** See also **compass rule; distance prorate rule; transit rule.**

Baldwin solar chart—A chart designed for orienting a planetable by means of the Sun's shadow.

ballistic camera—A precision terrestrial camera, usually employing glass plates, used at night to photograph such objects as rockets, missiles, or satellites against a star background. Also called **tracking camera.** See also **BC-4 camera.**

band—See **latitude band.**

bar check—A method of field calibrating the sounding equipment used in hydrographic survey by suspending a bar or disc beneath the transducer at various depths.

bar scale—See **graphic scale.**

Barlow leveling rod—A speaking rod marked with triangles each 0.02 foot in height.

barometer—An instrument for measuring atmospheric pressure. See also **aneroid barometer; cistern barometer; mercury barometer; siphon barometer.**

barometric altimeter—An instrument that indicates elevation or height above sea level, or some other reference height, by measuring the weight of air above the instrument. Also called **pressure altimeter; sensitive altimeter.** See also **aneroid altimeter.**

barometric elevation—An elevation determined with a barometer or altimeter.

barometric hypsometry—The determination of elevations by means of either mercurial or aneroid barometers.

barometric leveling—A method of determining differences of elevation from differences of atmospheric pressure observed with a barometer or barometric altimeter. A type of indirect leveling.

barycenter—The center of mass of a system of masses; as the barycenter of the Earth-Moon system.

basal coplane—(photogrammetry) The condition of exposure of a pair of photographs in which the two photographs lie in a common plane parallel to the air base. If the air base is horizontal, the photographs are said to be exposed in horizontal coplane.

basal orientation—The establishment of the position of both ends of an air base with respect to a ground system of coordinates. In all, six elements are required. These are essentially the three-dimensional coordinates of each end of the base. In practice, however, it is also convenient to express these elements in one of two alternative ways: (1) The ground rectangular coordinates of one end of the base and the difference between these and the ground rectangular coordinates of the other end of the base. (2) The ground rectangular coordinates of one end of the base, the length of the base, and two elements of direction such as base direction and base tilt.

basal plane—See **epipolar plane.**

base-altitude ratio—The ratio between the air-base length and the flight altitude of a stereoscopic pair of photographs. This ratio is referred to as the K-factor. More commonly called **base-height ratio.** Also indicated functionally as **B/H.**

base apparatus—(surveying) Any apparatus designed for use in measuring with accuracy and precision the length of a base line in triangulation, or the length of a line in first- or second-order traverse. See also **Bache-Wurdeman base-line measuring apparatus; compensating base-line measuring apparatus; duplex base-line measuring apparatus; Hassler base-line measuring apparatus; iced-bar apparatus; Jaderin wires (base apparatus); optical base-line measuring apparatus; Repsold base-line measuring apparatus; Schott base-line measuring apparatus.**

base chart—See **base map.**

base color—The first color printed of a polychrome map to which succeeding colors are registered.

base construction line—The bottom line of a map projection, at right angles to the central meridian, along which other meridians are established.

base direction—The direction of the vertical plane containing the air base, which might be expressed as a bearing or an azimuth. See also **basal orientation.**

base-height ratio—See **base-altitude ratio.**

base line—**1.** (JCS) (surveying) A surveyed line established with more than usual care, to which surveys are referred for coordination and correlation. **2.** (JCS) (photogrammetry) The line between the principal points of two consecutive vertical air photographs. It is usually measured on one photograph after the principal point of the other has been transferred. **3.** (triangulation) See **triangulation base line. 4.** (USPLS) A line which is extended east and west on a parallel of latitude from an initial point, and from which are initiated other lines for the cadastral survey of the public lands within the area covered by the principal meridian that runs through the same initial point. **5.** (navigation) The line between two radio transmitting stations operating in conjunction for the determination of a line of position, as the two stations of a loran system.

base-line extension—(navigation) The continuation of the base line in both directions beyond the transmitters of a pair of radio stations operating in conjunction for determination of a line of position.

base-line levels—A level line run along a base line to determine and establish the elevation of the base-line stations.

base-line terminal stations—The monumented stations marking the end points of a base line.

base manuscript—See **compilation manuscript.**

base map—(JCS) A map or chart showing certain fundamental information, used as a base upon which additional data of specialized nature are compiled or overprinted. Also, a map containing all the information from which maps showing specialized information can be prepared.

base net—A small net of triangles and quadrilaterals, starting from a measured base line, and connecting with a line of the main-scheme of a triangulation net.

base sheet—A sheet of dimensionally stable material upon which the map projection and ground control are plotted, and upon which stereotriangulation or stereocompilation is performed.

base station—**1.** (surveying) The point from which a survey begins. **2.** (gravity) A geographic position whose absolute gravity value is known. In exploration, a reference station where quantities under investigations have known values or may be under repeated or continuous measurement in order to establish additional stations in relation to it.

base tape—A tape or band of metal or alloy, so designed and graduated and of such excellent workmanship that it is suitable for measuring lengths of lines (base lines) for controlling triangulation, and for measuring the lengths of first- and second-order traverse lines.

base tilt—The inclination of the air base with respect to the horizontal. See also **basal orientation.**

basement contours—Contours on the surface of the basement complex or basic metamorphic and volcanic rocks underlying an area.

basic control—Horizontal and vertical control of third- or higher-order accuracy, determined in the field and permanently marked or monumented, that is required to control further surveys.

basic cover—(JCS) Coverage of any installation or area of a permanent nature with which later coverage can be compared to discover any changes that have taken place. See also **comparative cover.**

bathygraphic—Descriptive of ocean depths.

bathymetric—Relating to the measurement of ocean depths.

bathymetric chart—A topographic map of the floor of the ocean.

bathymetry—The science of determining and interpreting ocean depths and topography.

battle map—(JCS) A map showing ground features in sufficient detail for tactical use by all forces, usually at a scale of 1:25,000.

beacon tracking—The tracking of a moving object by means of signals emitted from a transmitter or transponder within or attached to the object.

beam of light—A group of pencils of light, as those originating at the many points of an illuminated surface. A beam of parallel light rays is a special case in which each pencil is of such small cross section that it may be regarded as a ray.

Beaman arc—A specially graduated arc fitted to the vertical circle of a transit or alidade for the easy reduction of stadia observations. Also called **stadia circle.**

bearing—**1.** (JCS) (general) The horizontal angle at a given point measured clockwise from a specific reference datum to a second point. Also called **bearing angle. 2.** (navigation) The horizontal direction of one terrestrial point from another, expressed as the angular distance from a reference direction. It is usually measured from 000° at the reference direction clockwise through 360°. The terms **bearing** and **azimuth** are sometimes used interchangeably, but in navigation the former customarily applies to terrestrial objects and the latter to the direction of a point on the celestial sphere from a point on the Earth. **3.** (surveying) See **bearing of line.** See also **astronomic bearing; back bearing; compass bearing; computed bearing; curve of equal bearing; electronic bearing; false bearing; great circle bearing; grid bearing; Lambert bearing; magnetic bearing; polar bearing; rhumb bearing; true bearing.**

bearing angle—See **bearing,** definition 1.

bearing bar—See **azimuth instrument.**

bearing circle—A ring designed to fit snugly over a compass or compass repeater, and provided with vanes for observing compass bearings.

bearing line—A line extending in the direction of a bearing.

bearing of line—(plane surveying) The horizontal angle which a line makes with the meridian of reference adjacent to the quadrant in which the line lies. A bearing is identified by naming the end of the meridian (north or south) from which it is reckoned and the direction (east or west) of that reckoning. Thus, a line in the northeast quadrant making an angle of 50° with the meridian will have a bearing of N 50° E. In most survey work, it is preferable to use azimuths rather than bearings.

bearing tree—(USPLS) A marked tree used as a corner accessory; its distance and direction from the corner being recorded. Bearing trees are identified by prescribed marks cut into their trunks; the species and sizes of the trees are also recorded.

Bell gravity meter—A single-axis, pendulous force rebalance accelerometer mounted on a stabilized platform and interfaced to a dynamic digital filter for measuring gravity aboard a survey platform.

bench mark (BM)—A marked vertical control point which has been located on a relatively permanent material object, natural or artificial, and whose elevation above or below an adopted datum has been established. It is usually monumented to include bench mark name or number, frequently its elevation, and the name of the responsible agency. Since elevations are computed at a later time, they are seldom added to newer control bench marks. A BM (aside from a vertical angle bench mark) seldom has a surveyed latitude or longitude. See also **first-order bench mark; junction bench mark; permanent bench mark; primary bench mark; second-order bench mark; temporary bench mark; tidal bench mark; vertical-angle bench mark.**

Bessel spheroid (ellipsoid)—A reference ellipsoid having the following approximate dimensions: semimajor axis—6,377,397.2 meters; semiminor axis—6,356,078.9 meters; and the flattening or ellipticity—1/299.15.

Besselian star numbers—Constants used in the reduction of a mean position of a star to an apparent position (used to account for short-term variations in the precession, nutation, aberration, and parallax).

Besselian year—See **fictitious year.**

Bessel's method—See **triangle-of-error method.**

biangle screen—A photographic negative containing a composite of two dot screens, with the screen angles oriented 30° apart. These screens are used to print tones of color for chart features with thin lines.

Bilby steel tower—A triangulation tower consisting of two steel tripods, one within the other. The inner tripod holds the instrument platform, and the outer tripod holds the observer's platform. The tower can be easily erected and as easily disassembled and moved to a new location. See also **survey tower.**

bimargin format—(JCS) The format of a map or chart on which the cartographic detail is extended to two edges of the sheet, normally north and east, thus leaving only two margins. See also **bleed; bleeding edge.**

binocular—An optical instrument for use with both eyes simultaneously.

binocular vision—Simultaneous vision with both eyes.

bivariate normal distribution function—Mathematical function describing the behavior of two-dimensional random errors (e.g., latitude, longitude; *x, y;* easting, northing). Also called **circular normal distribution.**

blackbody—An ideal surface or body that completely absorbs all radiant energy falling upon it. Blackbodies are used as models in the design and calibration of remote sensing systems.

blaze—(USPLS) A mark made upon a tree trunk usually at about breast height. The bark and a small amount of the live wood are removed with an axe or other cutting tool, leaving a flat, smoothed surface which forever brands the tree. On rough-barked tree monuments or bearing trees the appropriate marks are scribed into a smooth, narrow, vertical blaze the lower end of which is about 6 inches above the root crown. The blaze should be just long enough to allow the markings to be made.

bleed—**1.** (lithography) A condition wherein ink pigment is dissolved by press fountain solution causing a light film of ink (scum) on the plate and impression. **2.** (cartography) Cartographic detail extending to the edge of a map or chart sheet.

bleeding edge—(JCS) That edge of a map or chart on which cartographic detail is extended to the edge of the sheet. See also **bimargin format; bleed.**

blind image—See **blueline.**

blip—(JCS) The display of a received pulse on a cathode ray tube. Also called **echo.**

blister—See **border break.**

block—(aerial photography) Two or more strips of overlapping photography. See also **flight block.**

block adjustment—The adjustment of strip coordinates or photograph coordinates for two or more strips of photographs. See also **strip adjustment.**

block-out—See **opaque,** definition 4.

bloomed lens—See **coated lens.**

blooming—**1.** The term used to describe localized overexposure caused by incoming radiant energy levels which exceed film emulsion latitude thereby causing the image to lack definition. **2.** A process for increasing the light transmission of lenses.

blowup—A photographic enlargement. Also used as a verb.

blue magnetism—The magnetism displayed by the south-seeking end of a freely suspended magnet. This is the magnetism of the Earth's north magnetic pole. See also **red magnetism.**

blueline—A nonreproducible blue image or outline usually printed photographically on paper or plastic sheeting, and used as a guide for drafting, stripping, or layout. Also called **blind image.**

blunder—A mistake generally caused by carelessness. A blunder may be large and easily detectable, or smaller and more dangerous, or very small and indistinguishable from a random error. Blunders are detected by repetition and by external checks, such as closing a traverse or substituting the solution of an equation in the original. See also **random error; systematic error.**

Board on Geographic Names (BGN)—An agency of the U.S. Government, first established by Executive Order in 1890 and currently functioning under Public Law 242-80, 25 July 1947. Nine departments and agencies enjoy Board membership. Conjointly with the Secretary of the Interior, the Board provides for "uniformity in geographic nomenclature and orthography throughout the Federal Government." It develops policies and romanization systems under which names are derived and it standardizes geographic names for use on maps and in textual materials.

boat sheet—The worksheet used in the field for plotting details of a hydrographic survey as it progresses. See also **field sheet.**

Bonne map projection—A modified equal-area map projection of the so-called conical type, having lines representing a standard parallel and a central meridian intersecting near the center of the map. The line representing the central meridian (geographic) is straight and the scale along it is exact. All geographic parallels are represented by arcs of concentric circles at their true distances apart, divided to exact scale, and all meridians, except the central one, are curved lines connecting corresponding points on the parallels.

border break—(JCS) A cartographic technique used when it is required to extend cartographic detail of a map or chart beyond the sheet lines into the margin. This technique eliminates the necessity of producing an additional sheet. Also called **blister.**

border data—See **margin data.**

border information—See **margin data.**

Boston leveling rod—A two-piece rod with fixed target on one end. The target is adjusted in elevation by moving one part of the rod on the other. Read by vernier. For heights greater than 5½ feet, the target end is up; for lesser heights, the target end is down.

Bouguer anomaly—A difference between an observed value of gravity and a theoretical value at the point of observation, which has been corrected for the effect of topography and elevation only, the topography being considered as a plate of indefinite extent.

Bouguer correction—A correction made in gravity work to take account of the altitude (elevation) of the station and the density of the mass between an infinite plane through the point of observation and the infinite plane of the reference elevation.

Bouguer plate—An imaginary layer of infinite length and of thickness equal to the height of the observation point above the reference surface (usually the geoid). In applying the Bouguer correction, the attracting layer lessens the free-air effect.

Bouguer reduction—Geophysically, the Bouguer reduction removes all masses above the reference surface (usually the geoid) and then reduces the gravity from the terrain to the reference surface.

boundary (de facto)—(JCS) An international or administrative boundary whose existence and legality is not recognized but which is a practical division between separate national and provincial administering authorities.

boundary (de jure)—(JCS) An international or administrative boundary whose existence and legality is recognized.

boundary line—A line of demarcation between contiguous political or geographical entities. The word "boundary" is sometimes omitted, as in "state line"; sometimes the word "line" is omitted, as in "international boundary," "county boundary," etc. The term **boundary line** is usually applied to boundaries between political territories, as "state boundary line," between two states. A boundary line between privately owned parcels of land is termed a property line by preference, or if a line of the United States public land surveys, is given the particular designation of that survey system, as "section line," "township line," etc.

boundary map—A map prepared specifically for the purpose of delineating a boundary line and adjacent territory.

boundary monument—A material object placed on or near a boundary line to preserve and identify the location of the boundary line on the ground.

boundary survey—A survey made to establish or to re-establish a boundary line on the ground or to obtain data for constructing a map or plat showing a boundary line. The term **boundary survey** is usually restricted to surveys of boundary lines between political territories. For the survey of a boundary line between privately owned parcels of land, the term **land survey** is preferred; except in United States public land surveys the term **cadastral survey** is used.

boundary vista—A lane cleared along a boundary line passing through a wooded area.

Bowie effect—The indirect effect on gravity due to the warping of the geoid, or the elevation of the geoid with respect to the spheroid of reference.

Bowie method of adjustment—A method for the adjustment of large networks of triangulation.

box compass—See **declinatoire.**

break angle—The deflection angle between the two vertical phases passing through the common nadir point and the principal points of the left and right oblique photographs.

break-circuit chronometer—A chronometer equipped with a device which automatically breaks an electric circuit, the breaks being recorded on a chronograph.

break tape—See **broken tape measurement.**

break-up—(JCS) In detection by radar, the separation of one solid return into a number of individual returns which correspond to the various objects or structure groupings. This separation is contingent upon a number of factors including range, beam width, gain setting, object size, and distance between objects.

breakaway method—See **breakaway strip method.**

breakaway strip method—A technique used in photomosaicking when two or more sheets are prepared. The process involves placing an extra-wide strip of masking tape along the outside edge of the neatline of one sheet before mosaicking photos. The mosaicked overedge is then cut along the neatline and transferred to the adjoining sheet. Also called **breakaway method.**

breaking tape—See **broken tape measurement.**

bridging—A photogrammetric method of establishing and adjusting control between bands of existing ground control, both horizontally and vertically. The term is usually qualified as horizontal or vertical according to its primary purpose. Also called **horizontal bridging; horizontal/vertical bridging; vertical bridging.**

brightness scale—(photography) The ratio of the brightness of highlights to the deepest shadow in the actual terrain, as measured from the camera stations, for the field of view under consideration.

British grid reference system—A system of rectangular coordinates devised or adopted by the British for use on military maps. There is no related global plan for the many grids, belts, and zones which make up the British grid system. It is being replaced by the Universal Transverse Mercator (UTM) grid system.

broadcast ephemeris—A set of parameters broadcast by satellite from which Earth-fixed satellite positions can be computed. In particular, the parameters for the Navy Navigation Satellites (NNS) are computed for each NNS by fitting 36- to 48-hour orbital arcs to Doppler data from four tracking stations and extrapolating this arc 12 to 24 hours beyond the last data used. The length of the arc fit and the extrapolation period depend on the upper atmospheric air density. The computed parameters are injected into the satellite memory and are transmitted along with time on each even minute. See also **Navy Navigation Satellite System.**

broken base—A base line for triangulation consisting of two or more lines that form a continuous traverse and have approximately the same general direction.

broken grade—(tape) The change in grade when the middle point of a tape is not on grade with its ends. If the middle support for the tape is not on the same grade as the end supports, the fact is noted with a reference "broken grade at–," naming the particular tape length which contains the broken grade.

broken tape measurement—(surveying) The short distances measured and accumulated to total a full tape length when a standard 100-foot tape cannot be held horizontally without plumbing from above shoulder level. Also called **break tape; breaking tape.**

broken-telescope transit—A precise astronomic transit in which the light entering the objective lens is reflected at right angles by a prism placed within the telescope, the reflected light ray passing to the eyepiece, which is in the horizontal axis of the telescope.

Brown gravity apparatus—An apparatus for measuring the acceleration of gravity which utilizes the Mendenhall pendulum, but has a clamping device for holding the pendulum in the receiver when being transported from station to station, and which utilizes an electrical pick-up and amplifying device for recording the oscillations (pendulum) on the chronograph sheet.

Brunton compass—An instrument combining the features of both the sighting compass and the clinometer that can be used in the hand or upon a Jacob's staff or light tripod for reading horizontal and vertical angles, for leveling, and for reading the magnetic bearing of a line. Also called **Brunton pocket transit.**

Brunton pocket transit—See **Brunton compass.**

bubble axis—See **spirit level axis.**

bubble level—See **spirit level.**

bubble sextant—A sextant in which the bubble of a spirit level serves as the horizon.

Bullard method of isostatic reduction—See **Hayford-Bullard (or Bullard) method of isostatic reduction.**

bull's-eye level—See **circular level.**

burn—(lithography) The process of exposing a pressplate.

C

C-constant—See **level constant.**

C-factor—An empirical value which expresses the vertical measuring capability of a given stereoscopic system; generally defined as the ratio of the flight height to the smallest contour interval accurately plottable. The C-factor is not a fixed constant, but varies over a considerable range, according to the elements and conditions of the photogrammetric system. In planning for aerial photography, the C-factor is used to determine the flight height required for a specified contour interval, camera, and instrument system. Also called **altitude-contour ratio.**

cadastral map—A map showing the boundaries of subdivisions of land, usually with the bearings and lengths thereof and the areas of individual tracts, for purposes of describing and recording ownership. Also called **property map.** See also **plat.**

cadastral survey—A survey relating to land boundaries and subdivisions, made to create units suitable for transfer or to define the limitations of title. The term **cadastral survey** is now used to designate the surveys of the public lands of the United States, including retracement surveys for the identification and resurveys for the restoration of property lines; the term can also be applied properly to corresponding surveys outside the public lands, although such surveys are usually termed **land surveys** through preference.

cairn—An artificial mound of rocks, stones, or masonry usually conical or pyramidal, whose purpose is to designate or to aid in identifying a point of surveying or of cadastral importance.

calculated altitude—See **computed altitude.**

calendar day—The period from midnight to midnight. The calendar day is 24 hours of mean solar time in length and coincides with the civil day unless a time change occurs during the day.

calendar month—A division of the year as determined by a calendar, approximately one-twelfth of a year in length. While arbitrary in character, the calendar month is based roughly on the synodical month. The calendar month ranges in length from 28 to 31 mean solar days.

calendar year—A conventional year based on the tropical year and adjusted by "leap years" to fit the non-integral length of the tropical year.

calibrated focal length—**1.** (JCS) An adjusted value of the equivalent focal length, so computed as to equalize the positive and negative values of distortion over the entire field used in a camera. **2.** The distance along the lens axis from the interior perspective center to the image plane, the interior center of perspective being selected so as to distribute the effect of lens distortion over the entire field.

calibration—The act or process of determining certain specific measurements in a camera or other instrument or device by comparison with a standard, for use in correcting or compensating for errors or for purposes of record. See also **camera calibration; field calibration; shop calibration.**

calibration card—A card having a list of calibration corrections or calibrated values.

calibration constants—The results obtained by calibration, which give the calibrated focal length of the lens-camera unit and the relationship of the principal point to the fiducial marks of a camera and give significant calibration corrections for lens distortions.

calibration correction—The value to be added to or subtracted from the reading of an instrument to obtain the correct reading.

calibration course—See **field comparator.**

calibration error—See **instrument error.**

calibration plate—A glass negative exposed with its emulsion side corresponding to the position of the emulsion side of the film in the camera at the time of exposure. This plate provides a record of the distance between the fiducial marks of the camera. Also called **flash plate; master glass negative.**

calibration table—A list of calibration corrections or calibrated values.

calibration templet—(photogrammetry) A templet of glass, plastic, or metal made in accordance with the calibration constants to show the relationship of the principal point of a camera to the fiducial marks; used for the rapid and accurate marking of principal points on a series of photographs. Also, for a multiple-lens camera, a templet prepared from the calibration data and used in assembling the individual photographs into one composite photograph.

call—(USPLS) A reference to, or statement of, an object, course, distance, or other matter of description in a survey or grant, requiring or calling for a corresponding object, or other matter of description, on the land.

Callippic cycle—A period of four Metonic cycles equal to 76 Julian years, or 27,759 days.

camera—A lightproof chamber or box in which the image of an exterior object is projected upon a sensitized plate or film through an opening usually equipped with a lens or lenses, shutter, and variable aperture. See also **aerial camera; BC-4 camera; ballistic camera; continuous-strip camera; convergent camera; copy camera; direct-scanning camera; fan cameras; frame camera; geodetic stellar camera; horizon camera; mapping camera; metric camera; multiple-camera assembly; multiple-lens camera; PC-1000 camera; panoramic camera; photogrammetric camera; positioning camera; precision camera; rectifier; rotating prism camera; split cameras; stellar camera; stereometric camera; terrestrial camera; trimetrogon**

camera; variable-perspective camera system; zenith camera.

camera axis—(JCS) An imaginary line through the optical center of the lens perpendicular to the negative photo plane.

camera axis direction—(JCS) Direction on the horizontal plane of the optical axis of the camera at the time of exposure. This direction is defined by its azimuth expressed in degrees in relation to true/magnetic north.

camera calibration—(JCS) The determination of the calibrated focal length, the location of the principal point with respect to the fiducial marks, and the lens distortion effective in the focal plane of the camera and referred to the particular calibrated focal length. [In a multiple-lens camera, the calibration also includes the determination of the angles between the component perspective units. The setting of the fiducial marks and the positioning of the lens are ordinarily considered as adjustments, although they are sometimes performed during the calibration process. Unless a camera is specifically referred to, distortion and other optical characteristics of a lens are determined in a focal plane located at the equivalent focal length and the process is termed **lens calibration.**]

camera lucida—A monocular instrument using a half-silvered mirror, or the optical equivalent, to permit superimposition of a vertical image of an object upon a plane. Also called **camera obscura.** See also **sketchmaster.**

camera magazine—(JCS) A removable part of a camera in which the unexposed and exposed portions of film are contained.

camera obscura—See **camera lucida.**

camera station—See **air base; air station.**

camera transit—See **phototheodolite.**

camera window—(JCS) A window in the camera compartment through which photographs are taken.

Canadian grid—See **perspective grid.**

candle—A unit of luminous intensity.

cantilever extension—Phototriangulation from a controlled area to an area of no control. Also, the connection by relative orientation and scaling of a series of photographs in a strip to obtain strip coordinates. Also called **extension.**

Cape Canaveral datum—This datum had its origin at station Central on the John F. Kennedy Space Center, Cape Canaveral, Florida, with azimuth to Central SE Base. The geodetic coordinates of these two stations were identical to those on North American datum of 1927. Datum differences for other points may be determined by subtracting North American datum of 1927 values from the Cape Canaveral datum values as established by the USC&GS transcontinental traverse of the United States. See also **North American datum of 1927.**

cardan link—A universal joint. An optical cardan link is a device for universal scanning about a point.

cardinal point effect—(JCS) The increased intensity of a line or group of returns on the radarscope occurring when the radar beam is perpendicular to the rectangular surface of a line or group of similarly aligned features in the ground pattern.

cardinal points—**1.** (JCS) The directions: north, south, east, west. **2.** (optics) Those points of a lens used as reference for determining object and image distances. They include principal planes and points, nodal points, and focal points.

Carpentier inversor—One of the inversors which corrects for the Scheimpflug condition in a rectifier if the negative, lens, or easel planes are tilted and not parallel.

carrying contour—A single contour line representing two or more contours, used to show vertical or near-vertical topographic features, such as steep slopes and cliffs.

Cartesian coordinates—(JCS) A coordinate system in which locations of points in space are expressed by reference to three mutually perpendicular planes, called coordinate planes. The three planes intersect in three straight lines called coordinate axes. [Also the values representing the location of a point in a plane in relation to two perpendicular intersecting straight lines, called axes. The point is located by measuring its distance from each axis along a parallel to the other axis.]

cartographic annotation—The delineation of additional data, new features, or deletion of destroyed or dismantled features on a mosaic to portray current details. Cartographic annotations may include elevation values for airfields, cities, and large bodies of water; new construction and destroyed or dismantled roads, railroads, bridges, dams, target installations, and cultural features of landmark significance.

cartographic compilation—See **compilation,** definition 1.

cartographic feature—The natural or cultural objects shown on a map or chart. See also **topography,** definition 1.

cartographic film—Film with a dimensionally stable base, used for map negatives and/or positives. Usually referred to by trade name.

cartographic license—The freedom to adjust, add, or omit map features within allowable limits to attain the best cartographic expression. License must not be construed as permitting the cartographer to deviate from specifications.

cartographic photography—See **mapping photography.**

cartographic scanner—A device for strip-by-strip scanning of two-dimensional copy and for digital registration of the light/dark (black/white) parts as rectangular coordinates.

cartography—The art and science of expressing

graphically, by maps and charts, the known physical features of the Earth, or of another celestial body; usually includes the works of man and his varied activities.

cartometric scaling—The accurate measurement of geographic or grid coordinates on a map or chart by means of a scale. This method may be used for plotting the positions of points, or determining the location of points.

carving—The development of the model surface by carving away the steps of the plaster step cast in the production of relief models.

cassette—(JCS) In photography, a reloadable container for either unexposed or exposed sensitized materials which may be removed from the camera or darkroom equipment under lighted conditions.

Cassini map projection—A conventional map projection constructed by computing the lengths of arcs along a selected geographic meridian and along a great circle perpendicular to that meridian, and plotting these as rectangular coordinates on a plane.

Cassini-Soldner map projection—Similar to a polyconic map projection except that it uses but one central meridian for a whole series. Best adapted for north-south belts and large-scale maps of small areas.

casting—The process of reproducing relief models in plaster or epoxy from the terrain base of the model, or after the surface of the model has been developed. Models are first cast negative, from which any number of positive castings may be made.

casual error—See **random error.**

catadioptric system—(optics) An optical system containing both refractive and reflective elements.

catenary correction—(taping) See **sag correction.**

catoptric system—(optics) An optical system in which all elements are reflective (mirrors).

cautionary note—Information calling special attention to some fact, usually a danger area, shown on a map or chart.

celestial coordinates—Any set of coordinates used to define a point on the celestial sphere.

celestial equator—The great circle on the celestial sphere whose plane is perpendicular to the axis of rotation of the Earth. Also called **equinoctial.**

celestial equator system of coordinates—A set of celestial coordinates based on the celestial equator as the primary great circle; usually declination and hour angle or sidereal hour angle. Also called **equator system; equatorial system; equinoctial system of coordinates.**

celestial fix—A position established by means of observation on one or more celestial bodies.

celestial geodesy—The branch of geodesy which utilizes observations of near celestial bodies, including Earth satellites, to determine the size and shape of the Earth.

celestial horizon—That circle of the celestial sphere formed by the intersection of the celestial sphere and a plane through the center of the Earth and perpendicular to the zenith-nadir line. Also called **rational horizon.**

celestial latitude—Angular distance north or south of the ecliptic; the arc of a circle of latitude between the ecliptic and a point on the celestial sphere, measured northward or southward from the ecliptic through 90°, and labeled N or S to indicate the direction of measurement. Also called **ecliptic latitude.**

celestial line of position—A line of position determined by means of the observation of a celestial body.

celestial longitude—Angular distance east of the vernal equinox, along the ecliptic; the arc of the ecliptic or the angle at the ecliptic pole between the circle of latitude of the vernal equinox and the circle of latitude of a point on the celestial sphere, measured eastward from the circle of latitude of the vernal equinox, through 360°. Also called **ecliptic longitude.**

celestial mechanics—The study of the theory of the motions of celestial bodies under the influence of gravitational fields.

celestial meridian—An hour circle of the celestial sphere, through the celestial poles and the zenith. The two intersections of the celestial meridian with the horizon are known as the north and south points.

celestial observation—**1.** Observation of celestial phenomena. **2.** (navigation) The measurement of the altitude or the azimuth, or both, of a celestial body. Also the data obtained by such measurement.

celestial parallel—See **parallel of declination.**

celestial pole—Either of the two points of intersection of the celestial sphere and the extended axis of the Earth.

celestial refraction—See **astronomic refraction.**

celestial sphere—(JCS) An imaginary sphere of infinite radius concentric with the Earth, on which all celestial bodies except the Earth are imagined to be projected. [For observations on bodies within the limits of the solar system, the assumed center is the center of the Earth. For bodies where the parallax is negligible, the assumed center may be the point of observation.]

celestial triangle—A spherical triangle on the celestial sphere, especially the navigational triangle.

center of gravity—The point in any body at which the force of gravity may be considered to be concentrated.

center of instrument—The point on the vertical axis of rotation of an instrument at the same elevation as the axis of collimation when that axis is in a horizontal position. In a transit or theodolite, it is close to or at the intersection of the horizontal and vertical axes of the instrument.

center of mass—The point at which all the given mass of a body or bodies may be regarded as being concentrated as far as motion is concerned.

center of oscillation—(pendulum) The position in a compound pendulum of the particle which corresponds to the heavy particle of an equivalent simple pendulum. The centers of suspension and oscillation are interchangeable. If the center of oscillation is made the center of suspension, the former center of suspension becomes the new center of oscillation. This principle is the basis of design of compound reversible pendulums.

center of projection—See **perspective center.**

center of radiation—See **radial center.**

center of suspension—(pendulum) The fixed point about which a pendulum oscillates. See also **center of oscillation.**

center point—See **radial center.**

centerline—1. (USPLS) The line connecting opposite corresponding quarter corners or opposite subdivision-of-section corners or their theoretical positions. **2.** A line extending from the true center point of overlapping aerial photos through each of the transposed center points. **3.** (engineering survey) The continuous center of a highway or railroad, with stationing indicating starting point, culverts, points of curvature, etc.

centimeter-gram-second (c.g.s.) system—A system of units based on the centimeter as the unit of length, the gram as the unit of mass, and the mean solar second as the unit of time. A part of the metric system.

central force—A force which for purposes of computation can be considered to be concentrated at one central point with its intensity at any other point being a function of the distance from the central point. Gravitation is considered as a central force in celestial mechanics.

central force field—The spatial distribution of the influence of a central force.

central force orbit—The theoretical orbit achieved by a particle of negligible mass moving in the vicinity of a point mass with no other forces acting; an unperturbed orbit.

central meridian—1. The line of longitude at the center of a projection. Generally, the basis for constructing the projection. **2.** The longitude of origin at the center of each 6-degree zone of the Universal Transverse Mercator (UTM) grid. The central meridian is arbitrarily numbered 500,000 and is called a **false easting. 3.** (state plane-coordinate system) The meridian used as the y-axis for computing projection tables for a state coordinate system. The central meridian of the system usually passes close to the center of figure of the area or zone for which the tables are computed.

central-point figure—A triangulation figure consisting of a polygon with an interior station, formed by a series of adjoining triangles with a common vertex at the interior station.

centrifugal force—The force with which a body moving under constraint along a curved path reacts to the constraint. Centrifugal force acts in a direction away from the center of curvature of the path of the moving body. As a force caused by the rotation of the Earth on its axis, centrifugal force is opposed to gravitation and combines with it to form gravity. See also **centripetal force.**

centripetal force—The force directed towards the center of curvature, which constrains a body to move in a curved path. See also **centrifugal force.**

chain—A device used by surveyors for measuring distance, or the length of this device as a unit of distance. The usual chain is 66 feet long, and consists of 100 links, each 7.92 inches long. See also **Gunter's chain; engineer's chain.**

chain gage—See **tape gage.**

chaining—See **taping.**

chaining pin—See **pin.**

challenger—See **interrogator-responsor.**

chambered spirit level—A level tube with a partition near one end which cuts off a small air reservoir so arranged that the length of the bubble can be regulated.

character—The distinctive trait, quality, property, or behavior of manmade or natural features as portrayed by a cartographer. The more character applied to detail, the more closely it will resemble these features as they appear on the surface of the Earth. See also **generalization.**

characteristic curve—(photography) A curve showing the relationship between exposure and resulting density in a photographic image, usually plotted as the density (D) against the logarithm of the exposure (log E) in candle-meter-seconds. Also called **H and D curve; sensitometric curve; D log E curve; time-gamma curve; density-exposure curve.** See also **contrast; density,** definition 1.

chart—1. A special-purpose map, generally designed for navigation or other particular purposes, in which essential map information is combined with various other data critical to the intended use. **2.** To prepare a chart, or engage in a charting operation. See also **aeronautical chart; aeronautical pilotage chart; aeronautical planning chart; Air Target Chart; anchorage chart; approach chart; azimuthal chart; azimuthal equidistant chart; Baldwin solar chart; bathymetric chart; chartlet; coastal chart; combat chart; conformal chart; conic chart; conic chart with two standard parallels; Consol chart; cotidal chart; current chart; Decca chart; enroute chart; equatorial chart; firing chart; general chart; Global Navigation Chart; gnomonic chart; great circle chart; harbor chart;**

historical chart; hydrographic chart; hypsographic map (or chart); ice chart; index chart; instrument approach chart; isobaric chart; isoclinic chart; isogonic chart; isogriv chart; isomagnetic chart; isoporic chart; Jet Navigation Chart; Lambert conformal chart; local chart; long-range navigation chart; loran chart; lunar chart; lunar earthside chart; lunar farside chart; magnetic chart; Marsden chart; mean chart; Mercator chart; meteorological chart; mileage chart; miscellaneous chart; modified Lambert conformal chart; new chart; oblique chart; oblique Mercator chart; obsolete chart; Operational Navigation Chart; orthographic chart; orthomorphic chart; perspective chart; pilot chart; pilotage chart; planning chart; plotting chart; polar chart; polyconic chart; radar chart; rectangular chart; route chart; sailing chart; search-and-rescue chart; secant conic chart; sectional chart; sextant chart; simple conic chart; star chart; stereographic chart; Tactical Pilotage Chart; tidal current chart; time zone chart; track chart; transverse chart; transverse Mercator chart; virtual PPI reflectoscope chart; visibility chart; weather map; World Aeronautical Chart.

chart comparison unit—A device permitting simultaneous viewing of navigational instrument presentation, such as a radarscope and a navigational chart, so that one appears superimposed upon the other. Also called **auto radar plot** when used with radar.

chart datum—See **hydrographic datum.**

charted depth—The vertical distance from the tidal datum to the bottom surface.

charting photography—See **mapping photography.**

chartlet—A small chart, such as those annexed to Notices to Mariners.

check point—1. (JCS) A predetermined point on the Earth's surface used as a means of controlling movement, a registration target for fire adjustment, or a reference for location. **2.** (JCS) Geographical location on land or water above which the position of an aircraft in flight may be determined by observation or by electronic means. **3.** A point, selected on obliques only, in the vicinity of each tie point and distant point for the purpose of checking the identification of these points.

check profile—A profile plotted from a field survey and used to check a profile prepared from a topographic map. The comparison of the two profiles serves as a check on the accuracy of the contours on the topographic map.

checked spot elevation—An elevation established in the field by: closed spirit leveling, trigonometric leveling by a closed circuit of barometric leveling, or any other method such that proof of accuracy is obtained.

checking positive—A composite printing on glass of the contour and drainage drawings used on the shadow projector for checking the horizontal accuracy of landforms to be developed on relief models.

chopping—(star or satellite trails) Interrupting the photographic image of a star or satellite trail by a shutter or other device to provide precise timing and orientation data for geodetic observations of aerospace vehicles against a stellar background.

chord—1. (route surveying) Chord used in highway and other surveys to indicate a straight line between two points on a curve, regardless of the distance between them. **2.** (USPLS) In surveying and geometry, a straight line joining any two points on an arc, curve, circumference, or surface.

chorographic map—Any map representing large regions, countries, or continents on a small scale. Atlas and small-scale wall maps belong in this class.

chromatic aberration—See **lateral chromatic aberration; longitudinal chromatic aberration.**

chromatic colors—Colors eliciting hue.

chromaticity—A composite of dominant wavelength and purity.

chromaticity coordinates—The proportions of standard components required for color match, used as an ordinate and abscissa to represent color in a chromaticity diagram.

chromaticity diagram—A plane diagram formed by plotting one of the chromaticity coordinates against another.

chronograph—An instrument for producing a graphical record of time as shown by a clock or other device. In use, a chronograph produces a double record: the first is made by the associated clock and forms a continuous time scale with significant marks indicating periodic beats of the timekeeper; the second is made by some external agency, human or mechanical, and records the occurrence of an event or of a series of events.

chronometer—A portable timekeeper with compensated balance, capable of showing time with extreme precision and accuracy. See also **break-circuit chronometer; hack chronometer.**

chronometer correction—See **clock correction.**

chronometer error—The amount by which the chronometer differs from the correct time.

chronometer rate—See **clock rate.**

cine-theodolite—A photographic tracking instrument which records on each film frame the target and the azimuth and elevation angles of the optical axis of the instrument.

circle of confusion—(optics) The circular image of a distant point object as formed in a focal plane by a lens. A distant point object (e.g., a star) is imaged in a focal plane of a lens as a circle of finite size, because of such conditions as (1) the focal plane's not being placed at the point of sharpest focus, (2) the effect of certain aberrations, (3) diffraction at the lens, (4) grain in a photographic emulsion, and/or (5) poor workmanship in the manufacture of the lens.

circle of declination—See **hour circle.**

circle of equal altitude—See **parallel of altitude.**

circle of equal declination—See **parallel of declination.**

circle of latitude—1. A great circle of the celestial sphere through the ecliptic poles, and hence perpendicular to the plane of the ecliptic. **2.** A meridian, along which latitude is measured.

circle of longitude—1. A circle of the celestial sphere, parallel to the ecliptic. **2.** A circle on the surface of the Earth, parallel to the plane of the Equator; a parallel, along which longitude is measured. Also called **parallel of latitude.**

circle of perpetual apparition—That circle of the celestial sphere, centered on the polar axis and having a polar distance from the elevated pole approximately equal to the latitude of the observer, within which celestial bodies do not set. See also **circle of perpetual occultation.**

circle of perpetual occultation—That circle of the celestial sphere, centered on the polar axis and having a polar distance from the depressed pole approximately equal to the latitude of the observer, within which celestial bodies do not rise. See also **circle of perpetual apparition.**

circle of position—A small circle on the globe (Earth) at every point of which, at the instant of observation, the observed celestial body (sun, star, or planet) has the same altitude and, therefore, the same zenith distance.

circle of right ascension—See **hour circle.**

circle of the sphere—A circle upon the surface of the sphere, specifically of the Earth or of the heavens, called a **great circle** when its plane passes through the center of the sphere; in all other cases, a **small circle.**

circle position—See **position,** definition 4.

circuit—(leveling or traverse) A continuous line of levels, a series of lines of levels, or a combination of lines or parts of lines of levels, such that a continuous series of measured differences of elevation extends around the circuit or loop and then back to the starting point. Also applied to a continuous line of transit traverse in a similar manner.

circuit closure—(leveling) The amount by which the algebraic sum of the measured differences of elevation around a circuit fails to equal the theoretical closure of zero. See also **error of closure,** definition 4.

circular cylindrical coordinates—See **cylindrical coordinates.**

circular error (CE)—An accuracy figure representing the stated percentage of probability that any point expressed as a function of two linear components (e.g., horizontal position) will be within the given figure. Commonly used are CEP (50 percent), CE 1σ (39.35 percent), and CE (90 percent).

circular error probable (CEP)—The 50 percent error interval based on the bivariate normal distribution function. Also called **circular probable error.**

circular level—A spirit level having the inside surface of its upper part ground to spherical shape, the outline of the bubble formed being circular, and the graduations being concentric circles. This form of spirit level is used where a high degree of precision is not required, as in plumbing a level rod or setting an instrument in approximate position. Also called **bull's-eye level; universal level.**

circular map accuracy standard (CMAS)—The 90 percent error interval based on the bivariate normal distribution function.

circular near-certainty error (3.5σ)—The 99.78 percent error interval based on the bivariate normal distribution function.

circular normal distribution—See **bivariate normal distribution function.**

circular probable error (CPE)—See **circular error probable.**

circular sigma—See **circular standard error.**

circular standard error (σ_c)—The 39.35 percent error interval based on the bivariate normal distribution function. Also called **circular sigma.**

circulation map—See **traffic-circulation map.**

circum-meridian altitudes—Ex-meridian altitudes observed for determination of latitude when a heavenly body is close to transit.

circumferentor—A type of surveyor's compass having slit sights on projecting arms.

circumlunar—Around the Moon, generally applied to trajectories.

circumpolar—Revolving about the elevated pole without setting. A celestial body is circumpolar when its polar distance is approximately equal to or less than the latitude of the observer.

cislunar—1. This side of the Moon. **2.** Of or pertaining to phenomena, projects, or activity in the space between the Earth and the Moon, or between the Earth and the Moon's orbit.

cistern barometer—A mercury barometer in which a column of mercury is enclosed in a vertical glass tube, the upper end of which is sealed and exhausted of air, and the lower end placed in a cistern or reservoir of mercury which is exposed to atmospheric pressure. The atmospheric pressure on the free surface of the mercury in the cistern determines the height to which the mercury will rise in the vertical tube. This may be measured, and the pressure reported in terms of that height, as in inches of mercury.

city graphic—See **city products.**

city plan—See **city products.**

city products—Large scale maps of populated places and environs, usually portraying street and through-route information, important buildings and other urban features, airfields, port facilities, and relief, drainage, and vegetation when important. Several different types of city products are produced by DMA, among which are city graphics, city plans, city route graphics, and military city maps. Specifications for these maps vary according to particular military requirements.

city route graphics—See **city products.**

city survey—A specialized type of land survey restricted to work completed primarily within the limits of a city.

civil day—A solar day beginning at midnight. The civil day may be based on either apparent solar time or mean solar time. See also **astronomic day.**

civil time—Solar time in a day (civil day) that begins at midnight. Civil time may be either apparent solar time or mean solar time; it may be counted in two series of 12 hours each, beginning at midnight, marked "a.m." (ante meridian), and at noon, marked "p.m." (post meridian), or in a single series of 24 hours beginning at midnight.

Clairaut's theorem—A theorem that, in its original form, relates the value of centrifugal force at the Equator to the value of gravity at the Equator. Importance to physical geodesy is that the flattening of the Earth can be obtained from gravity measurements.

clamping error—A systematic error in observations made with a repeating theodolite caused by strains set up by the clamping devices of the instrument.

Clarke spheroid (ellipsoid) of 1866—A reference ellipsoid having the following approximate dimensions: semimajor axis—6,378,206.4 meters; semiminor axis—6,356,583.8 meters; and the flattening or ellipticity—1/294.97.

Clarke spheroid (ellipsoid) of 1880—A reference ellipsoid having the following approximate dimensions: semimajor axis—6,378,249.1 meters; and the flattening or ellipticity—1/293.46.

classification copy—A specialized item of source material used as a guide by the compiler and/or draftsman in preparing a map or chart. Usually consists of detailed information pertaining to roads, railroads, city data, and the like that has been developed from field surveys. Usually furnished in the form of overlays, annotated maps, drawings, photographs, or field sheets.

classification survey—See **field inspection.**

clearing y-parallax—See **relative orientation.**

clinometer—A simple instrument used for measuring the degree of slope in percentage or in angular measure.

clock correction—The quantity which is added, algebraically, to the time shown by a clock to obtain the time of a given meridian. If the clock is slow, the correction is positive; if fast, negative. When applied to a chronometer it is called **chronometer correction.**

clock rate—The amount gained or lost by a clock in a unit of time. When applied to a chronometer, it is called **chronometer rate.**

clockwise angle—See **angle to right.**

closed traverse—A survey traverse which starts and ends upon the same station, or upon stations whose relative positions have been determined by other surveys of equal or higher order of accuracy.

closest approach—**1.** The event that occurs when two planets or other celestial bodies are nearest to each other as they orbit about the Sun or other primary. **2.** The place or time of such an event. **3.** (satellite surveying) The time and location of the satellite when it is closest to the observer/receiver antenna.

closing—The act of finishing a survey process so that the accuracy may be checked.

closing corner—(USPLS) A corner established where a survey line intersects a previously fixed boundary at a point between corners. The closing corner is located by law at the actual point of intersection without regard to its monumented location.

closing error—See **error of closure,** definition 1.

closing the horizon—Measuring the last of a series of horizontal angles at a station, required to make the series complete around the horizon. At any station, the sum of the horizontal angles between adjacent lines should equal 360°.

closing township corner—(USPLS) **1.** The point of intersection of a guide meridian or a range line with a previously fixed standard parallel, or a base line. **2.** The point of intersection of any township or range line with a previously fixed boundary at a point between previously established corners. See also **township corner.**

closure—See **error of closure,** definition 1.

closure of horizon—See **error of closure,** definition 6.

closure of traverse—See **error of closure,** definition 8.

closure of triangle—See **error of closure,** definition 7.

coaltitude—The complement of altitude, or 90° minus the altitude. The term has significance only when used in connection with altitude measured from the celestial horizon, when it is synonymous with **zenith distance.**

coast pilot—See **sailing directions.**

Coast-Survey method—See **triangle-of-error method.**

coastal chart—A nautical chart intended for inshore coastwise navigation when a vessel's course may carry her inside outlying reefs and shoals, for use in entering or leaving bays and harbors of considerable size, or for use in navigating larger inland waterways.

coastal refraction—(JCS) The change of the direction of travel of a radio ground wave as it passes from land to sea or from sea to land. Also called **land effect; shoreline effect.**

coastlining—The process of obtaining data from which the coastline can be drawn on a chart.

coated lens—A lens whose air-glass surfaces have been coated with a thin transparent film of such index of refraction as to minimize the light loss by reflection. This reflection loss for uncoated lenses amounts to approximately 4 percent per air-glass surface. Also called **bloomed lens.**

codeclination—The complement of the declination, or 90° minus the declination. When the declination and latitude are of the same name, codeclination is the same as polar distance measured from the elevated pole.

coefficient of refraction—The ratio of the refraction angle at the point of observation to the angle at the center of the Earth which is formed by the observer, the center of the Earth, and point observed.

cogeoid—See **compensated geoid.**

coincidence—1. In the measurement of angles with theodolites, the instant at which two diametrically opposed index marks on the circle are in perfect optical alignment and appear to form a continuous line across the dividing line of the circle. **2.** (surveying) A prismatic arrangement common to leveling instruments wherein one-half of opposite ends of the leveling bubble are brought into view in a single image. Coincidence is achieved when the two halves of the bubble ends match. **3.** (pendulum) An exact agreement in occurrence of a prescribed phase of the beat of a free-swinging pendulum and a prescribed phase of the beat of a clock or chronometer.

coincidence method—1. (theodolite) The procedure by which the circles of the theodolite are read. See **coincidence,** definition 1. **2.** (pendulum) The determination of the period of a free-swinging pendulum by observing the time interval between coincidences with a clock pendulum or chronometer beat.

colatitude—The complement of the latitude, or 90° minus the latitude. Colatitude forms one side, zenith to pole, of the astronomic triangle. It is the side opposite the celestial body.

collation—1. The verification of the order, number, and date of maps. **2.** The assembling of pages of publications in sequence.

collection (acquisition)—1. (JCS) The obtaining of information in any manner, to include direct observation, liaison with official agencies, or solicitation from official, unofficial, or public sources. **2.** The process of arranging for and obtaining existing data from one or more sources for a library file or a specific mapping, charting, and geodetic production program.

collection requirement—An identified gap in information or material holdings, including general requirement statements, intended for field collection action. Not intended to apply to requirements of data available from existing Department of Defense data libraries.

collimate—1. (physics and astronomy) To render parallel to a certain line or direction; to render parallel, as rays of light; to adjust the line of sight or lens axis of an optical instrument so that it is in its proper position relative to the other parts of the instrument. **2.** (photogrammetry) To adjust the fiducial marks of a camera so that they define the principal point. Also called **adjustment for collimation; collimation adjustment.** See also **collimating marks.**

collimation adjustment—See **collimate,** definition 2.

collimation axis—In an optical instrument, the line through the rear nodal point of the objective lens that is precisely parallel with the center line of the instrument.

collimation error—The angle by which the line of sight of an optical instrument differs from its collimation axis. Also called **error of collimation.**

collimation plane—The plane described by the collimation axis of a telescope of a transit when rotated around its horizontal axis.

collimating eyepiece—A prismatic eyepiece used with a collimator.

collimating marks—(JCS) Index marks, rigidly connected with the camera body, which form images on the negative. These images are used to determine the position of the optical center or principal point of the imagery. Also called **fiducial marks.**

collimator—An optical device for artificially creating a target at infinite distance, a beam of parallel rays of light; used in testing and adjusting certain optical instruments. It usually consists of a converging lens and a target, a system or arrangement of crosshairs, placed at the principal focus of the lens. See also **autocollimator; collimating eyepiece; vertical collimator.**

color composite—A composite in which the component images are shown in different colors. See also **composite.**

color gradients—See **hypsometric tinting.**

color mixture curve—A graph representing tristimulus value for unit flux of spectral energy, shown as a function of wavelength.

color mixture data—Amounts of components required in a three-color colorimeter to match various wavelengths.

color plate—A general term for the pressplate from which any given color is printed. Normally, the term is modified to reflect a special color or type of plate, such as brown plate or contour plate. See also **process plates.**

color proof—A single or composite copy of each color of a polychrome (multicolor) printing which may be produced by any method.

color-proof process—A photomechanical printing

process which makes possible the combining of negative separations by successive exposures to produce a composite color proof on a vinyl plastic sheet. The method is usually referred to by the manufacturer's trade name of the materials used.

color registration guide—A visual display on a litho copy of a chart which accurately reflects the amount and direction of misregistration between the graticule and certain significant overprint. See also **register marks.**

color separation—**1.** The process of preparing a separate drawing, engraving, or negative for each color required in the production of a lithographed map or chart. **2.** A photographic process or electronic scanning procedure using color filters to separate multicolored copy into separate images of each of the three primary colors.

color-separation drawing—One of a set of drawings which contains similar or related features, such as drainage or culture. There are as many drawings as there are colors to be shown on the lithographed copy.

color-separation guide—See **guide.**

colorimeter—An instrument designed for the direct measurement of color.

colures—The hour circles through the equinoxes and the solstices. See also **equinoctial colure; solstitial colure.**

coma—An aberration affecting the sharpness of images off the axis in which rays from a point object off the axis passing through a given circular zone of the lens come to a focus in a circle rather than a point, and the circles formed by rays through different zones are of different sizes and are located at different distances from the axis. Therefore, the image of a point object is comet-shaped.

combat chart—A special-purpose chart of a land-sea area using the characteristics of a map to represent the land area and the characteristics of a chart to represent the sea area, with such special characteristics as to make the chart most useful in military operations, particularly amphibious operations. Also called **map chart.**

combination plate—Halftone and line work on one plate. Also, two or more subjects combined on the same plate. See also **process plates.**

Command Operational Priority Requirements List (COPRL)—A list of the total Unified and Specified Command/Department requirements for ATMP validated air target materials as compiled by each of the U&S Commands and, as appropriate, the Military Departments. The COPRL includes for each required graphic, an assigned relative priority based on the "user's" need, and also an evaluation of the adequacy of intelligence/radar annotations if the required coverage is available.

common control—(JCS) Horizontal and vertical map or chart location of points in the target area and position area, tied in with the horizontal and vertical control in use by two or more units. May be established by firing, survey, or combination of both, or by assumption. See also **control point; field control; ground control.**

common establishment—See **establishment of the port.**

communications satellite—(JCS) An orbiting vehicle, which relays signals between communications stations. They are of two types: (1) active communications satellite—a satellite which receives, regenerates, and retransmits signals between stations; (2) passive communications satellite—a satellite which reflects communications signals between stations.

comparative cover—(JCS) Coverage of the same area or object taken at different times to show any changes in detail. See also **basic cover; coverage.**

comparator—**1.** An instrument or apparatus for measuring a dimension in terms of a standard. **2.** A precision optical instrument used to determine the rectangular coordinates of a point with respect to another point on any plane surface, such as a photographic plate. **3.** (surveying) An instrument for comparing standards of length; for subdividing such standards; or for determining a standard length of measuring devices (bar, tape, etc.). See also **field comparator; monocomparator; stereocomparator; Vaisala Comparator; vertical comparator.**

comparator base—See **field comparator.**

compass—An instrument for indicating a horizontal reference direction relative to the Earth. See also **aperiodic compass; astro compass; Brunton compass; circumferentor; declinatoire; declinometer; Earth inductor compass; gyrocompass; gyro-magnetic compass; lensatic compass; liquid hand compass; magnetic compass; peep sight compass; prismatic compass.**

compass amplitude—Amplitude relative to compass east or west. See also **amplitude.**

compass bearing—**1.** (navigation) Bearing relative to compass north. See also **magnetic bearing,** definition 1. **2.** (surveying) See **magnetic bearing,** definition 2.

compass index error—The instrument error in the magnetic bearing given by readings of the needle.

compass north—(JCS) The uncorrected direction indicated by the north-seeking end of a compass needle. See also **magnetic north.** [**Compass north** and **magnetic north** differ in that the former may be determined by other influences than the Earth's magnetic field.]

compass rose—(JCS) A graduated circle, usually marked in degrees indicating directions and printed or inscribed on an appropriate medium.

compass rule—A method of balancing a traverse survey. Corrections corresponding to the closing errors in latitude and departure are distributed according to the proportion: length of line to total length of traverse. The compass rule is used when it is assumed that the closing errors are as much due to errors in observed angles as to errors in measured distances.

compass survey—A traverse survey which relies on the magnetic needle for orienting the sequence as a whole or for determining the bearings of the lines individually.

compensated geoid—A surface derived from the geoid by application of computed values of the deflection of the vertical which depend upon the topographic and isostatic compensation. Also called **cogeoid.**

compensating backsights and foresights—When backsight and foresight distances are equal at a given position of a level instrument, the effects of curvature, refraction, and lack of adjustment of line of sight (if bubble is leveled when taking a rod reading) are compensated for. Backsight and foresight distances are commonly controlled by the use of pacing or stadia.

compensating base-line measuring apparatus—A base apparatus having a length element composed of two metals having different coefficients of thermal expansion, so arranged and connected that the differential expansion of its components will maintain a constant length of the element under all temperature conditions of use.

compensating error—An error that tends to offset a companion error and thus obscure or reduce the effect of each.

compensating lens—(photogrammetry) A lens introduced into an optical system to correct for radial distortion.

compensation plate—(photogrammetry) A glass plate having a surface ground to a predetermined shape, for insertion in the optical system of a diapositive printer or plotting instrument, to compensate for radial distortion introduced by the camera lens.

compilation—**1.** (JCS) Selection, assembly, and graphic presentation of all relevant information required for the preparation of a map or chart. Such information may be derived from other maps or charts or from other sources. **2.** (photogrammetry) The production of a new or recompiled map, chart, or related product from aerial photographs and geodetic control data by use of photogrammetric instruments. Also called **photogrammetric compilation; stereocompilation.** See also **recompilation.**

compilation history—Complete information regarding the development of a map or chart. It explains problems encountered and their solution, and aids in simplifying the research and analysis of source materials considered for compilation or revision of other maps or charts. The compilation history contains information on the planning factors, source materials utilized, control, compilation methods, drafting, reproduction, and edit procedures.

compilation instructions—Written directions describing cartographic sources and their use in determining information to be compiled. Compilation instructions are not to be confused with specifications.

compilation manuscript—The original drawing, or group of drawings, of a map or chart as compiled or constructed from various data on which cartographic and related detail is delineated in colors on a stable-base medium. A compilation manuscript may consist of a single drawing called a **base manuscript,** or because of congestion, several overlays may be prepared showing vegetation, relief, names, and other information. Since the latter is usually the case, the base and its appropriate overlays are collectively termed the **compilation manuscript.**

compilation scale—The scale at which a map or chart is delineated on the original manuscript. This scale may vary from that of the reproduction scale.

compiled map—A map incorporating information collected from various sources; not compiled from survey data made for the map in question.

component—**1.** One of the parts into which a vector quantity can be divided. For example, the Earth's magnetic force at any point can be divided into horizontal and vertical components. **2.** One of the parts of a complete system. See also **constituent.**

composite—Reproduction from a successive series of images. A proof made by exposing color-separation negatives one after the other on a single sheet of paper. Used in checking and editing. Also called **composite print.** See also **color composite; color proof; double burn.**

composite air photography—(JCS) Air photographs made with a camera having one principal lens and two or more surrounding and oblique lenses. The several resulting photographs are corrected or transformed in printing to permit assembly as verticals with the same scale.

composite print—See **composite.**

compound centrifugal force—See **Coriolis force.**

compound harmonic motion—The projection of two or more uniform circular motions on a diameter of the circle of such motion.

compound pendulum—Any actual pendulum. A compound pendulum may be considered as composed of an indefinitely large number of material particles, at different distances from the center of suspension, each constituting a simple pendulum. The period of vibration (oscillation) of the compound pendulum may be taken as a resultant of the periods of the simple pendulums of which it is composed.

compression—See **flattening (of the Earth).**

computed altitude—Altitude determined by computation, table, mechanical computer, or graphics.

computed azimuth angle—Azimuth angle determined by computation, table, mechanical device, or graphics for a given place and time.

computed bearing—Bearing angles determined by computation from known bearings.

computed data method—A method of rectification with an autofocus rectifier whereby tilt existing in an aerial photograph is computed and, from these computatuions, the instrument settings are established mathematically. Rectification is then accomplished without further comparison to templet or other guide base.

concave lens—See **negative lens.**

concluded angle—(triangulation) The third angle of a triangle, not measured, but computed from the two other angles.

condition equation—An equation which expresses exactly certain relationships that must exist among related quantities, which are not independent of one another, exist a priori, and are separated from relationships demanded by observation. See also **angle equation; azimuth equation; correlate equation; latitude equation; length equation; longitude equation; normal equation; observation equation; perpendicular equation; side equation; side-equation tests.**

conditions—A term used in setting up equations for computation and adjustment of triangulation, trilateration, or traverse.

cone angle banding—Technique used in analytical photogrammetry for reducing mensuration requirements on a photographic plate by segregating images into annular zones defined by specific bands subtended by, usually, 5° of arc. Areas read are then only in certain outer bands depending on the calibration of the lens cone.

confidence interval—A statement of accuracy based on a statistic whose distribution function is known; e.g., the normal distribution function or bivariate normal distribution function. Errors are stated as some percentage of the total probability of 100 percent; e.g., a 90 percent assurance level. Also called **assurance level; error interval; probability interval.**

configuration of terrain—See **topographic expression.**

conformal chart—A chart on a conformal projection.

conformal map projection—A map projection on which the shape of any small area of the surface mapped is preserved unchanged, and all angles around any point are correctly represented. Also called **orthomorphic map projection.**

conic chart—A chart on a conic projection.

conic chart with two standard parallels—A chart on the conic projection with two standard parallels. Also called **secant conic chart.**

conic map projection—A map projection produced by projecting the geographic meridians and parallels onto a cone which is tangent to (or intersects) the surface of a sphere, and then developing the cone into a plane. Conic map projections may be considered as including cylindrical map projection when the apex of the cone is at an infinite distance from the sphere, and projections on a tangent plane when that distance is zero. Conic map projections may be illustrated with a single cone which is tangent to the sphere or which cuts the sphere along two parallels; or they may be a series of tangent cones, all with apexes on an extension of the axis of the sphere, at constantly increasing (or decreasing) distances from the sphere. It is best used to show areas of large longitudinal rather than latitudinal distances. Also called **tangent conical map projection.**

conic map projection with two standard parallels—A conic map projection in which the surface of a sphere or spheroid, such as the Earth, is conceived as developed on a cone which intersects the sphere or spheroid along two standard parallels. The Lambert conformal projection is an example. Also called **secant conic map projections.**

conjugate distances—The corresponding distances of object and image from the nodal points of the lens.

conjugate image points—See **corresponding image points.**

conjugate image rays—See **corresponding image rays.**

conjugate points—The object and image points in an optical system. They are physically related according to the definition for **conjugate distances.**

conjunction—The situation of two celestial bodies having the same celestial longitude or the same sidereal hour angle. See also **inferior conjunction; opposition; superior conjunction.**

connecting traverse—A traverse which starts and ends at separate points whose relative positions have been determined by a survey of an equal- or higher-order of accuracy. Considered less subject to undetected error than a loop traverse.

connection—(geodesy) The systematic elimination of discrepancies between adjoining or overlapping triangulation networks for the purpose of establishing a common framework from which long-range measurements can be taken.

consecutive mean—A smoothed representation of a time series derived by replacing each observed value with a mean value computed over a selected interval. Consecutive means are used in smoothing to eliminate unwanted periodicities or minimize irregular variations. Also called **moving average; overlapping mean; running mean.**

Consol—(JCS) A long-range radio aid to navigation, the emissions of which, by means of their radio frequency modulation characteristics, enable bearings to be determined.

Consol chart—A chart showing Consol lines of

position.

constant error—A systematic error which is the same in both magnitude and sign throughout a given series of observations, such as an index error of an instrument. See also **accumulative error.**

constant of aberration—The maximum aberration of a star observed from the surface of the Earth, 20.496 seconds of arc.

constant of gravitation—The proportionality factor (equal to 6.67×10^{-5} m^3 kg^{-1} sec^{-2}) in the universal law of gravitation; i.e., every particle of matter attracts every other particle with a force that is directly proportional to the product of their masses and inversely proportional to the square of their distance apart. Also called **gravitational constant; law of universal gravitation.**

constant pressure chart—See **isobaric chart.**

constituent—One of the harmonic elements in a mathematical expression for the tide-producing force and in the corresponding formulas for the tide or tidal current. Each constituent represents a periodic change or variation in the relative positions of the Earth, Moon, and Sun. Also called **astronomic tidal constituent; harmonic constituent; partial tide; tidal constituent.** See also **component; diurnal constituent; semidiurnal constituent.**

constituent day—The duration of the Earth's daily rotation relative to a fictitious star which represents one of the periodic tide-producing forces; it approximates the length of the lunar or solar day and corresponds to the period of a diurnal constituent of twice the period of a semidiurnal constituent. The term is not applicable to the long-period constituents.

contact base-line measuring apparatus—A base apparatus composed of bars whose lengths are defined by the distance between their end faces or points. In use, the bars are laid end to end, one bar being kept in position while another bar is being moved ahead.

contact glass—See **focal-plane plate.**

contact plate—See **focal-plane plate.**

contact print—(JCS) A print made from a negative or a diapositive in direct contact with sensitized material.

contact printer—A device which provides a light source and a means for holding the negative and the sensitive material in contact during exposure. Also, a specialized device for exposing diapositive plates at the same scale as that of the negative.

contact printing frame—In photography and platemaking, a device for holding the negative and the sensitive material in contact during exposure. The light source may or may not be a separate element. If the frame contains a vacuum pump to exhaust all air within the frame to insure perfect contact between the negative and the sensitive material, it is known as a **contact vacuum printing frame.**

contact screen—**1.** A halftone screen made on a film base and used in direct contact with the film to obtain a halftone image from a continuous-tone original. **2.** A pattern image on a film base used in contact with an open window negative to obtain a pattern image on film or plate. See also **area pattern screen; magenta contact screen.**

contact size—In reproduction, printing to the same size as the original. Also called **one-to-one (1:1) copy.** See also **scale of reproduction.**

contact-slide base-line measuring apparatus—A modified contact base-line measuring apparatus consisting of two steel measuring bars (rods), each 4 meters in length, so mounted that contact is effected by coincidence of lines on a rod and a contact-slide. Each rod forms a metallic thermometer with two zinc tubes, one on each side of the bar; opposite ends of the bar are fastened to the ends of the tubes, the other ends of which are free to move with changes of temperature.

contact vacuum printing frame—See **contact printing frame.**

contact vernier—The usual type of vernier, having the vernier scale and the graduated circle in physical contact.

continuous processor—(JCS) Equipment which processes film or paper in continuous strips.

continuous-strip camera—(JCS) A camera in which the film moves continuously past a slit in the focal plane, producing a photograph in one unbroken length by virtue of the continuous forward motion of the aircraft.

continuous-strip photography—(JCS) Photography of a strip of terrain in which the image remains unbroken throughout its length along the line of flight.

continuous tone—An image which has not been screened and contains unbroken, gradient tones from black to white, and may be either in negative or positive form. Aerial photographs are examples of continuous-tone prints. See also **halftone; line copy.**

continuous tone gray scale—A scale of tones from white to black or from transparent to opaque, each tone of which blends imperceptibly into the next without visible texture or dot formation. Also called **continuous wedge.** See also **step wedge.**

continuous wedge—See **continuous tone gray scale.**

contour—An imaginary line on the ground, all points of which are at the same elevation above or below a specified datum surface, usually mean sea level.

contour finder—A stereomapping instrument of simple design for use with photographic prints. This instrument does not provide a method of compensating for scale changes in different parts of the model resulting from differences in relief.

contour interval—(JCS) Difference in elevation between two adjacent contour lines. [Occasionally, the interval may vary within an individual sheet.] See also **variable contour interval.**

contour line—(JCS) A line on a map or chart connecting points of equal elevation. See also **accurate contour; approximate contour; carrying contour; depression contour; depth curve; form lines, geoidal contour; index contour; intermediate contour; sea level contour; supplementary contour.**

contour map—A topographic map which portrays relief by the use of contour lines.

contour sketching—Freehand delineation of the surface relief on a map as seen in perspective view, but controlled by locations on the map corresponding to salient points on the ground.

contour value—A numerical value placed upon a contour line to denote its elevation relative to a given datum, usually mean sea level.

contrast—(photography) The actual difference in density between the highlights and the shadows on a negative or positive. Contrast is not concerned with the magnitude of density, but only with the difference in densities. Also, the rating of a photographic material corresponding to the relative density difference which it exhibits. See also **characteristic curve; density,** definition 1.

control—**1.** The coordinated and correlated dimensional data used in geodesy and cartography to determine the positions and elevations of points on the Earth's surface or on a cartographic representation of that surface. **2.** (JCS) A collective term for a system of marks or objects on the Earth or on a map or a photograph, whose positions or elevation, or both, have been or will be determined. See also **astronomic control; basic control; common control; electronic control; geodetic control; ground control; horizontal control; Laplace control; level control; photogrammetric control; recover; starting control; supplemental control; vertical control.**

control base—A surface upon which the map projection and ground control are plotted and upon which templets have been assembled or aerotriangulation has been accomplished and the control points thus determined have been marked.

control data card—A card containing positional data and descriptions of individual horizontal and/or vertical control points. Also called **geodetic data sheet.** See also **trig list.**

control flight—See **control strip.**

control marking—A note or other form of caveat shown on an MC&G product indicating a need for special handling and for controlled dissemination.

control net—See **survey net.**

control point—**1.** (photogrammetry) Any station in a horizontal and vertical control system that is identified on a photograph and used for correlating the data shown on that photograph. The term is usually modified to reflect the type or purpose, such as **ground control point, horizontal control point, photocontrol point, picture control point,** and **vertical control point.** See also **control station; secondary control point; supplemental control point.** **2.** (JCS) A point located by ground survey with which a corresponding point on a photograph is matched, as a check, in making mosaics.

control-point photography—Electronically controlled aerial photography consisting of four flight lines flown in a cloverleaf pattern from the four cardinal directions and with the flights intersecting over a target or secondary control point.

control station—An object or mark on the ground of known position or elevation, or both, in a network of ground control. Control stations constitute the framework by which map details are fixed in their correct position, azimuth, elevation, and scale with respect to the Earth's surface. Also called **ground control point.** See also **control point.**

control-station identification—See **photoidentification.**

control strip—(aerial photography) A strip of aerial photographs taken to aid in planning and accomplishing later aerial photography, or to serve as control in assembling other strips. Also called **control flight; tie flight; tie strip.** See also **cross-flight photography.**

control survey—A survey which provides positions, horizontal and vertical, of points to which supplementary surveys are adjusted. The fundamental control survey of the United States provides the geographic positions and plane coordinates of triangulation and traverse stations and the elevations of bench marks which are used as the bases for hydrographic surveys of the coastal waters, for the control of the topographic survey of the United States, and for the control of many state, city, and private surveys.

control survey classification—A series of designations to classify control surveys according to their field survey methods and accuracy. The highest prescribed order of control surveys is designated first order; the next lower prescribed classification, second order; etc.

controlled map—(JCS) A map with precise horizontal and vertical ground control as a basis. Scale, azimuth, and elevations are accurate.

controlled mosaic—(JCS) A mosaic corrected for scale, rectified, and laid to ground control to provide an accurate representation of distances and direction. See also **mosaic; semicontrolled mosaic; uncontrolled mosaic.**

controlling depth—The least depth in the approach or channel to an area, such as a port or anchorage, governing the maximum draft of craft that can enter.

conventional international origin (CIO)—The average terrestrial pole of the period 1900 to 1905. Often used as the origin to which the coordinates of the instantaneous pole of rotation of the Earth are referred. In 1967, the IUGG recommended that the CIO be used to define the direction of the geodetic north pole. Abbreviated to OIC in French publications. See also **average terrestrial pole.**

convergence constant—The angle at a given latitude between meridians 1° apart. Sometimes loosely called "convergency," a term which more properly is the equivalent of "convergence."

convergence of meridians—The angular drawing together of the geographic meridians in passing from the Equator to the poles. At the Equator, all meridians are mutually parallel; passing from the Equator, they converge until they meet at the poles, intersecting at angles that are equal to their differences of longitude. The term **convergence of meridians** is used to designate also the relative difference of direction of meridians at specific points on the meridians. Thus, for a geodetic line, the azimuth at one end differs from the azimuth at the other end by 180° plus or minus the amount of the convergence of the meridians at the end points.

convergent camera—An assembly of two aerial cameras which take simultaneous photographs and maintain a fixed angle between their optical axes. The effect is to increase the angular coverage in one direction, along the longitudinal axis of the aircraft.

convergent model datum--See **model datum.**

convergent photography—Photography taken with a convergent camera. In photogrammetry, the angle of convergence of the two lens axes is usually 40° maintaining a 1:1 base height ratio.

convergent position--A split camera installation so positioned that the plane containing the camera axis is parallel to the line of flight.

converging lens--See **positive lens.**

conversion—The changing of one system of measurement to another; e.g., converting meters to feet. Conversion is usually accomplished by the use of conversion factors, scales, and tables.

conversion angle--(JCS) The angle between a great circle (orthodromic) bearing and a rhumb line (loxodromic) bearing of a point, measured at a common origin.

conversion factor—A quantity by which the numerical value in one system of units must be multiplied to arrive at the numerical value in another system of units.

conversion scale—(JCS) A scale indicating the relationship between two different units of measurement (e.g., meters to feet).

convertible lens—A lens containing two or more elements which can be used individually or in combination.

convex lens--See **positive lens.**

cooperative mapping agreement—A formal agreement between national governments specifying responsibilities for MC&G activities such as procurement of aerial photography, execution of geodetic control surveys, and production of maps, charts, and related products. See also **map exchange agreement.**

coordinate axes—In a rectangular coordinate system, the axes of reference which intersect at right angles at the point of origin.

coordinate conversion—Changing the coordinate values from one system to those of another system; e.g., geographic coordinates to Universal Transverse Mercator grid coordinates.

coordinate protractor—A square-shaped protractor having graduations on two adjacent edges with the center at one corner. It is equipped with a movable arm turning about the center, and graduated to show linear quantities on a given scale. The protractor is covered with a grid of the same scale and units as the arm.

coordinate transformation—1. A mathematical or graphic process of obtaining a modified set of coordinates by some combination of rotation of coordinate axes at their point of origin, translocation of the point of origin, modification of scale along coordinate axes, or change of the size or geometry of the reference space. **2.** The set of parameters used to accomplish this process. See also **affine transformation; Cartesian coordinates; datum transformation; rectification; Universal Polar Stereographic grid; Universal Transverse Mercator grid.**

coordinated series—A series of geographically integrated target charts and other graphics of a uniform scale and format developed to provide continuous and complete coverage of a large area. Also called **series.**

coordinates—(JCS) Linear or angular quantities which designate the position that a point occupies in a given reference frame or system. Also used as a general term to designate the particular kind of reference frame or system, such as plane rectangular coordinates or spherical coordinates. See also **assumed plane coordinates; astronomic coordinates; Cartesian coordinates; celestial equator system of coordinates; chromaticity coordinates; curvilinear coordinates; cylindrical coordinates; Earth-fixed coordinate system; ecliptic system of coordinates; galactic system of coordinates; geocentric coordinates; geocentric geodetic coordinates; geodetic coordinates; geographic coordinates; geomagnetic coordinates; grid coordinates; grid coordinate system; ground-space coordinate system; horizon system of coordinates; hour angle system (of coordinates); inertial coordinate system; local coordi-**

nate system; model coordinates; oblique coordinates; origin of coordinates; photograph coordinates; plane polar coordinates; plane rectangular coordinates; plate coordinates; polar coordinates; rectangular coordinates; rectangular space coordinates; relative coordinate system; right ascension system; selenocentric coordinates; space coordinates; space-polar coordinates; spherical coordinates; state coordinate systems; strip coordinates; topocentric coordinates; topocentric equatorial coordinates; universal space rectangular coordinate system; Universal Transverse Mercator coordinates; vertical coordinates.

coordination—The placing of all survey data on the same coordinate system or datum. Coordination does not imply the adjustment of observations to remove discrepancies. Two field surveys over the same area may be coordinated by computation on the same datum, but there may remain between them discrepancies that can be removed only by correlation.

coordinatograph—An instrument used to plot in terms of plane coordinates. It may be an integral part of a stereoscopic plotting instrument whereby the planimetric motions (x and y) of the floating mark are plotted directly. Also called **rectangular coordinate plotter.**

coplanar—Lying in the same plane.

copy—The manuscript or text furnished for reproduction. See also **continuous tone; line copy; tone copy.**

copy (copying) camera—A precision camera used in the laboratory for copying purposes. Also called **process camera.**

corange line—A line through points of equal tidal range.

Coriolis—A fictitious force used to explain the horizontal departure from a straight line of a moving object on or near the Earth's surface caused by viewing the trajectory of the moving object while the observer is stationary with respect to the rotating Earth. This "force" causes deflections to the right in the Northern Hemisphere and to the left in the Southern Hemisphere. Coriolis deflects objects to the west if they are moving toward the Equator and to the east if they are moving away from the Equator. It affects air (wind) and water (current), and introduces an error in the bubble sextant observations made from a moving craft; the effect increasing with higher latitude and greater speed of the object. Also called **compound centrifugal force; Coriolis force; deflecting force.**

Coriolis correction—A correction applied to an assumed position, celestial line of position, celestial fix, or to a computed or observed altitude to allow for apparent acceleration due to Coriolis force.

Coriolis force—See **Coriolis.**

corner—**1.** A point on a land boundary, at which two or more boundary lines meet. **2.** (USPLS) A point on the surface of the Earth, determined by the surveying process, which defines an extremity on a boundary of the public lands. See also **auxiliary meander corner; closing corner; closing township corner; double corners; existent corner; found corner; indicated corner; lost corner; meander corner; obliterated corner; quarter-section corner; quarter-quarter section corner; section corner; sixteenth-section corner; special meander corner; standard corner; theoretical corner; township corner; witness corner.**

corner accessories—(USPLS) Nearby physical objects to which corners are referenced for their future identification or restoration. Accessories include bearing trees, mounds, pits, ledges, rocks, and other natural features, to which distances or directions (or both) from the corner or monument are known. Such accessories are actually a part of the monumentation. See also **bearing tree.**

corner marks—See **register marks.**

corner ticks—See **register marks.**

corrected establishment—The mean high water interval for all stages of the tide.

correction—A quantity, equal in absolute magnitude opposite in sign to the error, added to a calculated or observed value to obtain the true or adjusted value. See also **arc correction; augmentation correction; Bouguer correction; clock correction; Coriolis correction; curvature correction; dynamic correction; dynamic temperature correction; eccentric reduction; Eötvös correction; field correction; free-air correction; grade correction; height-of-eye correction; index correction; ionospheric correction; latitude correction; length correction; level correction; orthometric correction; Polaris correction; rod correction; sag correction; semidiameter correction; slope correction; surface corrections; tape corrections; temperature correction; tension correction; terrain correction; tidal correction; timing correction; transit micrometer contact correction; velocity correction.**

correction code—A code consisting of letters, numbers, and symbols which are used to indicate edit corrections on maps or on overlays attached thereto.

correction for datum—A conversion factor used in the prediction of tides to resolve the difference between chart datum of the reference and a secondary station.

correction for inclination of tape—See **grade correction.**

correction for inclination of the horizontal axis—A correction applied to an observed horizontal direction to eliminate any error that may have been caused by the horizontal axis of the instrument not being exactly horizontal.

correction for run of micrometer—A correction

applied to an observed reading of a graduated circle made with a micrometer microscope to compensate for run of micrometer.

correction line—See **standard parallel,** definition 1.

correction notices—A variety of notices (e.g., Notice to Mariners, Notice to Airmen, errata notices, chart update manuals, target material bulletins, etc.) utilized to transmit correction data which the user applies to an existing MC&G product.

correction overlay—A transparent material on which edit corrections are noted. The method permits an immediate location of features to be revised without the necessity of marking the drawing or map.

correlate equation—An equation derived from an observation or condition equation, using undetermined multipliers, and expressing the condition that the sum of the squares of the residuals (or corrections) resulting from the application of these multipliers to the observation or condition equations shall be a minimum. See also **condition equation; normal equation.**

correlation—1. (general) The statistical interdependence between two quantities (e.g., in geodesy, gravity anomalies are correlated with other gravity anomalies, with elevation, with elevation differences, and with geology, etc.). **2.** (surveying) The removal of discrepancies that exist among survey data so that all parts are interrelated without apparent error. The terms **coordination** and **correlation** are usually applied to the harmonizing of surveys of adjacent areas or of different surveys over the same area. Two or more such surveys are coordinated when they are computed on the same datum; they are correlated when they are adjusted together.

Correlation Tracking and Triangulation (COTAT)—A trajectory measuring system composed of several antenna base lines, each separated by large distances, used to measure direction cosines to an object. From these measurements its space position is computed by triangulation.

correspondence—(stereoscopy) The condition that exists when corresponding images on a pair of photographs lie in the same epipolar plane; the absence of *y*-parallax.

corresponding image points—The images on two or more overlapping photographs of a single object point. Sometimes incorrectly called **conjugate image points.**

corresponding image rays—Rays connecting each of a set of corresponding image points with its particular perspective center.

corresponding images—A point or line in one system of points or lines homologous to a point or line in another system. Sometimes incorrectly called **conjugate points.**

cotidal chart—A chart of cotidal lines that show approximate locations of high water at hourly intervals measured from a reference meridian, usually Greenwich.

cotidal hour—The average interval expressed in solar or lunar hours between the Moon's passage over the meridian of Greenwich and the following high water at a specified place.

cotidal line—A line on a chart passing through all points where high water occurs at the same time. The lines show the lapse of time, usually in lunar-hour intervals, between the Moon's transit over a reference meridian (usually Greenwich) and the occurrence of high water for any point lying along the line.

counter-etch—To remove, with certain diluted acids, impurities from a lithographic plate, making it receptive to an image.

counterclockwise angle—A horizontal angle measured in a counterclockwise direction; used primarily for the measurement of deflection angles.

county map—A map of the area of a county as a unit.

course—1. (land surveying) The bearing of a line; also the bearing and length of a line. **2.** (traverse) The azimuth and length of a line, considered together. **3.** (navigation) The azimuth or bearing of a line along which a ship or aircraft is to travel or does travel, without change of direction; the line drawn on a chart or map as the intended track. The direction of a course is always measured in degrees from the true meridian, and the true course is always meant unless it is otherwise qualified; e.g., as a magnetic or compass course. See also **track. 4.** (geography) A route on the Earth along which a river flows; the river itself.

covariance—A mathematical quantity σ_{xy} related to the coefficient of correlation ρ_{xy} between two variables $\sigma_{xy} = \rho_{xy} \sigma_x \sigma_y$, where σ_x^2 and σ_y^2 are the variances of x and y, respectively. Used in the variance-covariance matrix of a least-squares solution.

cover—(JCS) Photographs or other recorded images which show a particular area of ground. See also **basic cover; comparative cover.**

cover search—(JCS) In air photographic reconnaissance, the process of selection of the most suitable existing cover for a specific requirement.

cover trace—(JCS) (reconnaissance) One of a series of overlays showing all air reconnaissance sorties covering the map sheet to which the overlays refer.

coverage—(JCS) The ground area represented on imagery, photomaps, mosaics, maps, and other geographical presentation systems.

covering power—See **angle of view.**

crab—1. (aerial photography) The condition caused by failure to orient a camera with respect to the track of the aircraft. In vertical photography, crab is indicated by the edges of the photographs not being parallel to the air-base lines. **2.** (air navigation) See **yaw,** definition 1.

crab angle—(JCS) The angle between the aircraft track or flight line and the fore and aft axis of a vertical camera, which is in line with the aircraft heading.

critical angle—The minimum angle of incidence at which a ray of radiant energy impinging on the surface of a transparent medium is completely reflected, no part of it entering the medium.

critical deficiency—An item or the condition of information or data which, because of omission, misidentification, mislocation, or other such significant error, could cause serious adverse impact on navigation safety or operational mission accomplishment.

critical elevation—The highest elevation in any group of related and more-or-less contiguous relief formations on a map or chart. See also **highest elevation.**

critical range—The spread of ranges in which there is an element of uncertainty of interpretation of values.

crop—To trim or cut off parts of a photograph in order to eliminate superfluous portions and thus improve balance or composition. Usually accomplished by masking the image area during printing.

cross check lines—A series of data lines which cross the principal lines of development, preferably at right angles, which provides verification of, or reveals discrepancies in, the principal lines of the survey development.

cross-country movement study—A graphic or series of graphics and supporting text or tables portraying off-road movement conditions for specific vehicles or a group of vehicles. It is usually overprinted on a medium or large scale topographic map base.

cross-flight photography—Single photographic strips having stereoscopic overlap between exposures and having a flight direction at right angles to that of coexistent area-coverage photography. When applied to shoran, the term implies that each of the cross-flight exposures is accompanied by recorded shoran distances. See also **control strip.**

cross section—A horizontal grid system laid out on the ground for determining contours, quantities of earthwork, etc., by means of elevations of the grid points.

cross tilt—An error introduced into stereotriangulation due to the inability to recover the exact camera stations for successive pairs. This condition is generally due to variations in equipment, materials, or to imperfect relative orientation.

crosshairs—A set of wires or etched lines placed on a reticle held in the focal plane of a telescope. They are used as index marks for pointings of the telescope such as in a transit or level when pointings and readings must be made on a rod.

crossing angle—The angle at which two lines of position, course lines, etc., intersect.

crossline glass screen—See **halftone screen.**

crystal clock—A device for keeping accurate time. It consists essentially of a generator of constant frequency controlled by a resonator made of quartz crystal, with suitable methods for producing continuous rotation to operate time-indicating and related mechanisms.

culmination—The position of a heavenly body when at highest apparent altitude (zenith). Known as **upper culmination;** also, for a heavenly body which is continually above the horizon, the position of lowest apparent altitude, known as **lower culmination.** Culmination occurs when the body transits the local meridian. See also **lower transit; transit; upper transit.**

cultural details—See **culture.**

cultural features—See **culture.**

culture—(JCS) Features of the terrain that have been constructed by man. Included are such items as roads, buildings, and canals; boundary lines, and, in a broad sense, all names and legends on a map. Also called **cultural details; cultural features; manmade features.**

currency review—The comparison of an existing MC&G product against source material of later date than that from which it was produced, for the purpose of determining its currency.

current chart—A map of a water area depicting current speeds and directions by current roses, vectors, or other means.

current cycle—A complete set of tidal current conditions, as those occuring during a tidal day, lunar month, or Metonic cycle.

current diagram—A graphic presentation showing the speed of the flood and ebb currents and the times of slack and strength over a considerable stretch of the channel of a tidal waterway, the times being referred to tide or current phases at some reference station.

current meter—A device for determining the velocity of flowing water by ascertaining the speed at which a stream of water rotates a vane or wheel.

current rose—A graphic presentation of currents for specified areas, utilizing arrows at the cardinal and intercardinal compass points to show the direction toward which the prevailing current flows and the percent frequency of set for a given period of time. The arrows on some presentations may be further subdivided (by thickness or pattern) to designate categories of current speeds.

curvature correction—**1.** (astronomy) A correction applied to the mean of a series of observations on a star or planet to take account of the divergence to the apparent path of the star or planet from a straight line. **2.** (geodesy) The correction applied in some geodetic work to take account of the divergence of the surface of the Earth (spheroid) from a plane. In geodetic spirit leveling, the effects of curvature and of atmospheric refraction are considered together, and tables have been prepared from which combined corrections can be taken.

curvature of Earth–1. (obstruction to line of sight) The offset from the tangent to the curve, as a result of the curvature of the Earth and refraction combined. **2.** The divergence of the Earth's surface from a plane.

curvature of field–An aberration affecting the longitudinal position of images off the axis in such a manner that objects in a plane perpendicular to the axis are imaged in a curved or dish-shaped surface.

curve–See **color mixture curve; degree of curve; distortion curve; exterior to a curve; interior to a curve; isoparametric curve; latitudinal curve; middle point; point of compound curvature; point of curvature; point of cusp; point of inflection; point of intersection; point of reverse curvature; point of tangency; point of vertical curve; point of vertical tangent; spiral curve; transition curve; vertical curve.**

curve of alignment–A line connecting two points on the surface of the spheroid and defined by the condition that at every point the azimuths of the two end points of the line differ by exactly 180°. A curve of alignment is a line of double curvature slightly less in length than the normal section lines connecting its two end points.

curve of equal bearing–A curve connecting all points at which the great-circle bearing of a given point is the same.

curve to spiral (PCS)–The common point between the circular arc and the tangent spiral.

curved-path error–The difference between the length of a ray refracted by the atmosphere and the straight-line distance between the ends of the ray.

curvilinear coordinates–Any linear coordinates which are not Cartesian coordinates. Frequently used curvilinear coordinates are **polar coordinates** and **cylindrical coordinates.**

cut–1. An observation between two points, one of which is known. Also a graphic representation of such an observation. See also **intersection; resection. 2.** A printed sheet of specific symbols used in cartography, such as swamp, sand, route markers, etc. **3.** A section of the right-of-way of a line of communication, such as a road or railroad, that has been excavated in order to reduce the grade (vertical inclination) or to allow sufficient lateral clearance for the roadbed, as on the side of a hill. **4.** A notch, passage, or channel worn by natural action, as of water.

cut line–The guide line sketched on a photograph to indicate where it should be cut or torn in order to form the best possible match of detail with the photographs immediately adjacent to it when laying a mosaic.

cut tape–See **subtracting tape.**

cutoff cylinder–An accessory apparatus, used in standardization operations to refer the end of a base tape or bar standard to a ground mark.

cutoff line–A survey line run between two or more stations on a linear traverse for the purpose of producing a closed traverse of that part of the survey.

cutting positive–A printing on glass of the contour drawing used to make the etched zinc plate. A preliminary step in relief model production.

cylindrical coordinates–A system of curvilinear coordinates in which the position of a point in space is determined by (1) its perpendicular distance from a given line, (2) its distance from a selected reference plane perpendicular to this line, and (3) its angular distance from a selected reference line when projected onto this plane. Also called **circular cylindrical coordinates; cylindrical polar coordinates.**

cylindrical equal-area map projection–A cylindrical map projection upon a cylinder tangent to a sphere, showing the geographic meridians as a family of equal-spaced parallel straight lines perpendicular to a second family of parallel straight lines which represent the geographic parallels, and which are so spaced as to produce an equal-area map projection. The equal-area condition preserves a constant ratio between corresponding ground and map areas. This projection must not be confused with the Mercator projection to which it bears some general resemblance.

cylindrical equal-spaced map projection–A cylindrical map projection upon a cylinder tangent to a sphere, showing the geographic meridians as a family of equal-spaced parallel straight lines perpendicular to a second family of equal-spaced parallel straight lines which represent the geographic parallels. The spacing of the parallels need not be the same as that of the meridians.

cylindrical lens–A lens in which the surfaces are segments of cylinders.

cylindrical map projection–A map projection produced by projecting the geographic meridians and parallels onto a cylinder which is tangent to (or intersects) the surface of a sphere, and then developing the cylinder into a plane.

cylindrical polar coordinates–See **cylindrical coordinates.**

D

D log E curve—See **characteristic curve.**

daily aberration—See **diurnal aberration.**

daily inequality—See **diurnal inequality.**

danger line—**1.** A line drawn on a chart to indicate the limits of safe navigation for a vessel of specific draft. **2.** A line of small dots used to draw the navigator's attention to a danger which would not stand out clearly enough if it were represented on the chart solely by specific symbols.

data-acquisition station—A ground station used for performing the various functions necessary to control satellite operations and to obtain data from the satellite.

data chamber—The portion of a cartographic aerial camera where ancillary data is recorded along the film margins. Data usually includes time, altitude, frame number, and other information required for identification and correlation purposes. It may include a character data generator block and extensive binary coded information to include aircraft position and camera orientation data.

data element—(JCS) A basic unit of information having a unique meaning and which has subcategories (data items) of distinct units of values.

data link—(JCS) A communication link suitable for transmission of data.

data reduction—Transformation of observed values into useful, ordered, or simplified information.

date line—See **international date line.**

datum—**1.** (JCS) Any numerical or geometrical quantity or set of such quantities which may serve as a reference or base for other quantities. **2.** (geodesy) A geodetic datum is uniquely defined by five quantities. Latitude (ϕ), longitude (λ), and geoid height (N) are defined at the datum origin. The adoption of specific values for the geodetic latitude and longitude implies specific deflections of the vertical at the origin. A geodetic azimuth is often cited as a datum parameter, but the azimuth and longitude are precisely related by the Laplace condition so there is no need to define both. The other two quantities define the reference ellipsoid: the semimajor axis and flattening or the semimajor axis and semiminor axis. Also called **horizontal datum; horizontal geodetic datum.** See also **horizontal control datum. 3.** (leveling) A level surface to which elevations are referred, usually, mean sea level but may also include mean low water, mean lower low water, or an arbitrary starting elevation(s). Also called **vertical datum.** See also **altitude datum; Cape Canaveral datum; Department of Defense World Geodetic System; European datum; hydrographic datum; Indian datum; international low water; local datum; low water datum; low water springs datum; lower low water datum; Mercury datum; model datum; National Geodetic Vertical datum of 1929; North American datum of 1927; North American datum of 1983; photographic datum; preferred datum; Pulkovo 1932 datum; reference datum; sounding datum; tidal datum; Tokyo datum; vertical-control datum; World Geodetic System.**

datum-centered ellipsoid—The ellipsoid that gives the best fit to the astrogeodetic network of a particular datum, and hence does not necessarily have its center at the center of the Earth.

datum level—(JCS) A surface to which elevations, heights, or depths on a map or chart are related. Also called **datum plane; reference level; reference plane.**

datum line—See **reference line.**

datum plane—See **datum level.**

datum point—(JCS) Any reference point of known or assumed coordinates from which calculations or measurements may be taken. Also called **reference point.**

datum transformation—**1.** The systematic elimination of discrepancies between adjoining or overlapping triangulation networks from different datums by moving the origins, rotating, and stretching the networks to fit each other. **2.** Constants used to transform the coordinates of a station from one datum and/or ellipsoid to another, together with the coordinates on the two systems.

day—The duration of one rotation of the Earth, or occasionally another celestial body, on its axis. It is measured by successive transits of a reference point on the celestial sphere over the meridian, and each type takes its name from the reference used. See also **apparent solar day; astronomical day; calendar day; civil day; constituent day; Julian day; lunar day; mean solar day; modified Julian day; sidereal day; solar day.**

deadbeat compass—See **aperiodic compass.**

Decca—A trade name for a radio phase-comparison system which uses a master and slave stations to establish a hyperbolic lattice and provide accurate position-fixing facilities.

Decca chart—A chart showing Decca lines of position.

December solstice—See **winter solstice.**

declination—**1.** In a system of polar or spherical coordinates, the angle at the origin between a line to a point and the equatorial plane, measured in a plane perpendicular to the equatorial plane. **2.** The arc between the Equator and the point measured on a great circle perpendicular to the Equator. **3.** (JCS) (astronomy) The angular distance to a body on the celestial sphere measured north or south through 90° from the celestial equator along the hour circle of the body. Comparable to latitude on the terrestrial sphere. **4.** Often used as a shortened term for **magnetic declination** although this use is not preferred. See also **grid declination; grid magnetic angle; lunar declination; parallel of declination; solar declination.**

declination difference—The difference between two declinations, particularly between the declination of a celestial body and the value used as an argument for entering a table.

declination of grid north—See **grid declination.**

declination of the Sun—See **solar declination.**

declinatoire—A combined magnetic compass and straightedge, suitable for use on a planetable to mark the magnetic meridian. Also called **box compass; trough compass.**

declinometer—A magnetic instrument similar to a surveyor's compass, but arranged so that the line of sight can be rotated to conform with the needle or to any desired setting on the horizontal circle. Used in determining the magnetic declination.

definition—(JCS) In imagery interpretation, the degree of clarity and sharpness of an image.

deflecting force—See **Coriolis force.**

deflection angle—**1.** (surveying) A horizontal angle measured from the prolongation of the preceding line to the following line. Deflection angles to the right are positive; those to the left are negative. **2.** (photogrammetry) A vertical angle, measured in the vertical plane containing the flight line by which the datum of any model in a stereotriangulated strip departs from the datum of the preceding model.

deflection-angle traverse—A survey, usually an open traverse, in which the measurement is made on the deflection angle of each course or leg from the direction of the preceding leg. See also **deflection angle,** definition 1.

deflection anomaly—The difference between an uncorrected value of the deflection of the vertical as determined by observation and the value after being corrected in accordance with certain assumptions made with reference to the physical condition of the geoid.

deflection of the plumb line—Deflection of the plumb line has the same value as the deflection of the vertical except the sign of the value is reversed. See also **deflection of the vertical.**

deflection of the vertical—The angular difference, at any place, between the upward direction of a plumb line (the vertical) and the perpendicular (the normal) to the reference spheroid. This difference seldom exceeds 30 seconds except in mountainous terrain or great depths of the sea. Often expressed in two components, meridian and prime vertical. Also called **deflection of the plumb line; station error.** See also **astrogeodetic deflection; gravimetric deflection; topographic deflection.**

degaussing range—A station for determining magnetic signatures of ships and other marine craft. Such signatures are used to determine required degaussing coil current settings and other required corrective action.

degenerate amphidromic system—A system of cotidal lines whose center or nodal (no-tide) point appears to be located on land rather than in the open ocean.

degree of curve—The number of degrees of angular measure at the center of a circle subtended by a chord 100 feet in length. In highway surveying, a 100-foot arc is sometimes used instead of a 100-foot chord in defining degree of curve.

delay—(JCS) (radar) **1.** The ground distance from a point directly beneath the aircraft to the beginning of the area of radar scan. **2.** The electronic delay of the start of the time base used to select a particular segment of the total.

densification network—Triangulation stations based on a long-line triangulation scheme but with shorter station-to-station distances established to provide more easily available control of higher accuracy to local users.

densitometer—An instrument which is used for measuring light (either transmitted or reflected) in terms of density tolerances. Transmission densitometers measure the full density range of negatives and the reflection densitometer measures the reflection range (density) of opaque copy. A numbered scale or digital display allows accurate readings of specific areas for comparisons or control purposes.

density—**1.** (photography) A measure of the degree of blackening of an exposed film, plate, or paper after development, or of the direct image (in the case of a printout material). It is defined strictly as the logarithm of the optical opacity, where the opacity is the ratio of the incident to the transmitted (or reflected) light. It varies with the use of scattered or specular light. See also **characteristic curve; contrast.** **2.** (cartography) The amount of detail shown on a map or chart. Density varies with scale and the nature of the area being compiled. **3.** (surveying) The number of control points in a given survey or area.

density altitude—(JCS) An atmospheric density expressed in terms of the altitude which corresponds with that density in the Standard Atmosphere.

density exposure curve—See **characteristic curve.**

Department of Defense libraries—Those Department of Defense libraries designated as responsible for maintaining files and providing library services related to specific MC&G data of common DoD interest, and designated as the primary library of that data with the responsibility for providing service for all authorized agencies.

Department of Defense World Geodetic System (DoD WGS)—A unified world datum based on a combination of all available astrogeodetic, gravimetric, and satellite tracking observations.

Previous World Geodetic Systems were WGS 59, WGS 60, and WGS 66. The current system is WGS 72. The system is revised as new geodetic, gravimetric, and satellite data materials change the currently accepted values.

departure—**1.** (plane surveying) The orthographic projection of a line on an east-west axis of reference. The departure of a line is the difference of the meridian distances or longitudes of the ends of the line. It is east or positive, and sometimes termed the **easting,** for a line whose azimuth or bearing is in the northeast or southeast quadrant; it is west or negative, and sometimes termed the **westing,** for a line whose azimuth or bearing is in the southwest or northwest quadrant. Also called **longitude difference. 2.** (navigation) The distance between two meridians at any given parallel of latitude, expressed in linear units, usually nautical miles; the distance to the east or west made good by a craft in proceeding from one point to another.

dependent resurvey—A resurvey for accomplishing a restoration based on the original conditions according to the records. The dependent resurvey is made, first, by identifying existing corners and other recognized and acceptable points of control of the original survey and, second, by restoring the missing corners by proportionate measurements in harmony with the original survey. This type of resurvey is used where there is a fair agreement between the conditions on the ground and the records of the original survey. Titles, areas, and descriptions should remain unchanged. See also **independent resurvey.**

depressed pole—The celestial pole below the horizon, of opposite name to the latitude. Opposite of **elevated pole.**

depression angle—See **angle of depression.**

depression contour—A closed contour delimiting an area of lower elevation than the surrounding terrain. Directional ticks extend from the contour in a downhill direction.

depth—(JCS) The vertical distance from the plane of the hydrographic datum to the bed of the sea, lake, or river.

depth curve—(JCS) A line on a map or chart connecting points of equal depth below the hydrographic datum. Also called **isobath.**

depth number—A numerical value placed upon a depth contour to denote its depth relative to a given datum. Also a depth sounding as it appears on a chart.

depth of field—The distance between the points nearest and farthest from the camera which are imaged with acceptable sharpness.

depth of focus—The distance that the focal plane can be moved forward or backward from the point of exact focus, and still give an image of acceptable sharpness. Also called **focal range.**

depth of isostatic compensation—The depth below sea level at which the condition of equilibrium known as isostasy is complete.

descending node—The point at which a planet, planetoid, or comet crosses the ecliptic from north to south, or a satellite crosses the Equator of its primary from north to south. Opposite of **ascending node.** Also called **southbound node.**

descending vertical angle—See **angle of depression.**

description—**1.** A term for the formal published data of each triangulation station, bench mark, etc. The data include information of the location and type of mark and enable anyone to go to the immediate locality and identify the mark with certainty. **2.** (cadastral surveying) A document listing the metes and bounds of a property.

descriptive name—(JCS) Written indication on maps or charts used to specify the nature of a feature (natural or artifical) shown by a general symbol. See also **toponym.**

detail points—Selected identified points, especially on oblique photographs, used to assist in correctly positioning features displaced as a result of elevation.

detailing—(surveying) The process of tying topographic details to the control net. Objects to be located in a survey may range from single points to meandering streams and complex geological formations.

detection—In imagery interpretation, the discovering of the existence of an object but without recognition of the object.

develop (development)—In photography, to subject to the action of chemical agents for the purpose of bringing to view the invisible or latent image produced by the action of light on a sensitized surface; also, to produce or render visible in this way.

developable—A surface that can be flattened to form a plane without compressing or stretching any part of it, such as a cone or cylinder.

deviation—(JCS) The angular difference between magnetic and compass headings. Also called **magnetic deviation.** See also **error; residual deviation.**

diagonal check—Measurements made across the opposite corners of the basic frame of a map projection to insure the accuracy of its construction, or to establish and/or check the scale of reproduction.

diagram on the plane of the celestial equator—See **time diagram.**

diagram on the plane of the celestial meridian—A diagram in which the local celestial meridian appears as a circle with the zenith at the top, and the horizon as a horizontal diameter. See also **time diagram.**

diagram on the plane of the equinoctial—See **time diagram.**

diameter—(magnification) See **magnifying power.**

diameter enlargement—A term used to indicate the degree of enlargement of original copy. A one-diameter enlargement of a 4- by 5-inch original would be 8 by 10 inches. See also **times (X) enlargement.**

diapositive—(JCS) (photogrammetry) A positive photograph on a transparent medium. The term is generally used to refer to a transparent positive on a glass plate used in a plotting instrument, a projector, or a comparator.

diapositive printer—A photographic device for producing diapositives from aerial negatives. Also called **reduction printer.** See also **contact printer; fixed-ratio projection printer.**

difference of elevation—The vertical distance between two points, or the vertical distance between the level surfaces that pass through the two points.

difference of latitude—**1.** The shorter arc of any meridian between the parallels of two places, expressed in angular measure. **2.** (plane surveying) The difference of latitude of the two ends of a line is frequently called **latitude of the line,** and defined as the orthographic projection of the line on a reference meridian. The latitude (as above defined) of the middle of a line is also referred to as **latitude of the line.**

difference of longitude—The smaller angle at the pole or the shorter arc of a parallel between the meridians of two places, expressed in angular measure.

differential aberration—The difference between the aberration of stars and that of a moving object (e.g., a satellite). Also called **parallactic aberration.**

differential distortion—The resultant dimensional changes in length and width in any medium. See also **differential shrinkage.**

differential leveling—The process of measuring the difference of elevation between any two points by spirit leveling. See also **direct leveling.**

differential shrinkage—The difference in unit contraction along the grain structure of the material as compared to the unit contraction across the grain structure; frequently refers to photographic film and papers and to map materials in general.

differential temperature—A natural error in surveying whereby temperature variations in the instrument cause reading errors that cannot be detected. It is usually caused by direct sunlight on the instrument which can be minimized by shading the instrument during survey operations.

diffraction—(optics) The bending of light rays around the edges of opaque objects. Due to diffraction, a point of light seen or projected through a circular aperture will always be imaged as a bright center surrounded by light rings of gradually diminishing intensity. Such a pattern is called a **diffraction disk, airy disk,** or **centric.**

diffuse reflection—Any reflection process in which the reflected radiation is sent out in many directions usually bearing no simple relationship to the angle of incidence. See also **diffusion; specular reflection.**

diffusion—(JCS) The scattering of light rays either when reflected from a rough surface or during the transmission of light through a translucent medium.

dihedral angle—The angle between two intersecting planes.

dimensional stability—Ability to maintain size; resistance to dimensional changes caused by changes in moisture content and temperature.

DIN—The German Industrial Standard (Deutsche Industrie-Norm), a European system of standardization for mechanical, engineering, and scientific manufactured products. In MC&G, the most notable application is the measurement of film speed.

diopter—A unit of measurement of the power of a lens, especially a spectacle-type lens. The power in diopters equals the reciprocal of the focal length in meters; thus, a lens whose focal length is 20 cm has a power of 5 diopters.

dioptric system—(optics) An optical system containing only refractive elements (lenses).

dip—**1.** The vertical angle, at the eye of an observer, between the horizontal and the line of sight to the visible horizon. Also called **dip of the horizon. 2.** The angle between the horizontal and the lines of force of the Earth's magnetic field at any point. Also called **inclination; magnetic dip; magnetic latitude; magnetic inclination. 3.** The first detectable decrease in the altitude of a celestial body after reaching its maximum altitude on or near meridian transit.

dip angle—**1.** (surveying) the vertical angle of the observation point between the plane of the true horizon and a sight line to the apparent horizon. **2.** (photogrammetry) The vertical angle, at the air station, between the true and the apparent horizon, which is due to flight height, Earth curvature, and refraction.

dip circle—An instrument for measuring magnetic dip. It consists essentially of a dip needle, or magnetic needle, suspended in such manner as to be free to rotate about a horizontal axis.

dip correction—See **height-of-eye correction.**

dip equator—See **magnetic equator.**

dip needle—See **magnetic dip needle,** definition 1.

dip of the horizon—See **dip,** definition 1.

dip pole—See **magnetic pole.**

direct angle—An angle measured directly between two lines, as distinguished in transit traverse from a **deflection angle.**

direct leveling—The determination of differences of elevation by means of a continuous series of short horizontal lines. Vertical distances from these lines to adjacent ground marks are determined by direct observations on graduated rods with a leveling instrument equipped with a spirit level. See also **differential leveling.**

direct measurement—Any measurement obtained by applying a tape to a line or a protractor to an angle, or by turning an angle with a transit; especially applicable to surveying. See also **indirect measurement.**

direct motion—The apparent motion of a planet or other object eastward among the stars.

direct observation—A measure of the quantity whose value is desired, such as a single measure of a horizontal angle.

direct photography—Photography in which the image of a subject is recorded directly by the camera in the conventional manner.

direct positive—A positive image obtained directly without the use of a negative.

direct problem—The determination of the geodetic position of the end point and the back azimuth at position two when the given information is the latitude and longitude of position one, along with the forward azimuth and the distance between the two points.

direct radial plot—See **direct radial triangulation.**

direct radial triangulation—A graphic radial triangulation made by tracing the directions from successive radial centers directly onto a transparent plotting sheet rather than laying the triangulation by the templet method. Also called **direct radial plot.**

direct-scanning camera—A type of panoramic camera wherein the lenses swing or rotate about the rear nodal point at a given rate.

direct telescope—A telescope is said to be direct when it is in its normal position.

direct vernier—A vernier scale which has spaces or divisions slightly shorter than those of the primary scale. The numbers on the vernier scale coincide with the numbers on the primary scale.

direction—**1.** (JCS) A term used by a spotter or observer in a call for fire to indicate the bearing of the spotting line. **2.** The position of one point relative to another without reference to the distance between them. Direction may be either three-dimensional or two-dimensional, the horizontal being the usual plane of the latter. Direction is usually indicated in terms of its angular distance from a reference direction. See also **great circle direction; grid direction; horizontal direction; magnetic direction; Mercator direction; reference direction; relative direction; true direction; x-direction.**

direction angle—In tracking, the angle between the antenna base line and an imaginary line connecting the center of the base line with the target.

direction instrument theodolite—A theodolite in which the graduated horizontal circle remains fixed during a series of observations, the telescope being pointed on a number of signals or objects in succession, and the direction of each read on the circle, usually by means of micrometer microscopes. Direction instrument theodolites are used almost exclusively in first- and second-order triangulation. Also called **direction theodolite; triangulation theodolite.**

direction method of adjustment—(triangulation and traverse) A method of adjustment of observations which determines corrections to observed directions. The direction method is used in the adjustment of first- and second-order survey observations.

direction method of determining astronomic azimuth—The determination of the astronomic azimuth of a line by measuring, with a direction theodolite, the horizontal angle between a selected star and a suitable mark, and applying that angle to the azimuth of the star computed for the epoch of the observation.

direction method of measuring horizontal angles—See **direction method of observation.**

direction method of observation—A method of observing angular relationships wherein the graduated circle is held in a fixed position, and the directions of the various signals are observed around the horizon. Thus, directions are pointings whereby angles are found by the differences in directions. Also called **direction method of measuring horizontal angles.**

direction of gravity—See **direction of the force of gravity.**

direction of relative movement—The direction of motion relative to a reference point, itself usually in motion.

direction of the force of gravity—The direction indicated by a plumbline. It is perpendicular to the surface of the geoid. Also called **direction of gravity.**

direction of tilt—The direction (azimuth) of the principal plane of a photograph. Also, the direction of the principal line on a photograph.

direction theodolite—See **direction instrument theodolite.**

directional radar prediction—(JCS) A prediction made for a particular heading.

directional reflectance—Reflectance measured for a specific mode of irradiation and collection.

discrepancy—A difference between results of duplicate or comparable measures of a quantity. The difference in computed values of a quantity obtained by different processes using data from the same survey.

dispersion—(optics) The separation of light into its component colors by its passage through a diffraction grating or by refraction such as that provided by a prism.

displacement—**1.** (cartography) The horizontal shift of the plotted position of a topographic feature from its true position, caused by required adherence to prescribed line weights and symbol sizes. **2.** Any shift in the position of an image on a photograph which does not alter the perspective characteristics of the photograph (i.e., shift due to tilt of the photograph, scale change in the photograph, and relief of the objects photographed). See also **refraction displacement; relief displacement; tilt displacement; x-displacement; y-displacement.**

display—The graphic presentation of the output data of any device or system.

distance—The spatial separation of two points, measured by the length of a line joining them. See also **angular distance; double meridian distance; double zenith distance; ecliptic polar distance; electrical distance; external distance; falling; front focal distance; great circle distance; grid length; ground distance; hyperfocal distance; interocular distance; lunar distance; meridian distance; meridional distance; plus distance; polar distance; principal distance; projection distance; relative distance; rhumb-line distance; slant range; Sun-zenith distance; tangent distance; time distance; zenith distance.**

distance angle—An angle in a triangle opposite a side used as a base in the solution of the triangle, or a side whose length is to be computed.

distance-measuring equipment (DME)—See **electronic distance-measuring equipment.**

distance prorate rule—A method of balancing a survey. A rule for holding angles to their recorded values and prorating the lengths of the lines in the traverse; operated by locating the bearings to a convenient meridian, preferably the closing line of the traverse, and prorating the dimensions through a trigonometric process.

distant points—Similar to tie points but which appear only on the obliques facing outward on the perimeter of a compilation. Distant points unite the sets of a strip into a flight unit; but, unlike tie points, do not serve to join several flight strips together.

distortion—Lens aberrations affecting the positions of images from their true relative positions. See also **aberration,** definition 2; **angular distortion,** definition 2; **differential distortion; film distortion; image distortion; image motion compensation distortion; lens distortion; linear distortion; panoramic distortion; radial distortion; scan positional distortion; tangential distortion; tipped panoramic distortion.**

distortion compensation—(photogrammetry) In a double-projection direct-viewing plotter system, that correction applied to offset the effect of radial distortion introduced in an original negative by the objective lens of an aerial camera.

distortion curve—A curve representing the linear distortion characteristics of a lens; it is plotted with image radial distance from the lens axis as abscissas and image radial displacements as ordinates.

distribution map—A map which shows the geographic arrangement of a specific product, commodity, or formation.

disturbing function—See **disturbing potential,** definition 2.

disturbing potential—**1.** (geodesy) The difference between the gravity potential of the actual Earth and the potential function of the normal gravity. Also called **potential disturbance; potential of disturbing masses; potential of random masses. 2.** (astronomy) The difference between the total gravitational potential and the potential pertaining to a spherical mass distribution. Also called **disturbing function.**

diurnal—Having a period of, occurring in, or related to a day.

diurnal aberration—Aberration resulting from the rotation of the Earth on its axis. The value of diurnal aberration varies with the latitude of the observer and ranges from zero at the poles to 0.32 second of arc at the Equator. A correction for diurnal aberration is applied to astronomic observations for longitude and azimuth. Also called **daily aberration.**

diurnal age—See **age of diurnal inequality.**

diurnal arc—See **astronomic arc.**

diurnal circle—The apparent daily path of a celestial body, approximating a parallel of declination.

diurnal constituent—Any tide constituent whose period approximates that of a lunar day (24.84 solar hours). See also **constituent.**

diurnal inequality—The difference in heights and durations of the two successive high waters or of the two successive low waters of each day; also, the difference in speed and direction of the two flood currents or the two ebb currents of each day. Also called **daily inequality; low water inequality.**

diurnal motion—The apparent daily motion of a celestial body.

diurnal parallax—See **geocentric parallax.**

diurnal variation—That component of a determinable magnitude which passes through a complete cycle in one day.

divergence—(leveling) The difference between the numerical values of two runnings over the same section of a line of levels.

diverging lens—See **negative lens.**

dodging—(photography) The process of holding back light from certain areas of sensitized material to avoid overexposure of these areas.

domestic map—A map of an area within the limits of the United States.

dominant wavelength—The wavelength of a spectrally pure energy that if mixed with white light would match a color.

Doppler effect—(JCS) The phenomenon evidenced by the change in the observed frequency of a sound or radio wave caused by a time rate of change in the effective length of the path of travel between the source and the point of observation. Also called **Doppler shift.**

Doppler navigation—1. A system which measures ground speed and drift of an aircraft, based on the Doppler effect of electronically generated signals emitted from the craft and reflected from the terrain. 2. A system which determines positions based on the Doppler effect of satellite signals. See also **broadcast ephemeris; Doppler sonar navigation; Navy Navigation Satellite System; precise ephemeris; point positioning; translocation.**

Doppler satellite survey system—A receiver/antenna combination capable of receiving signals broadcast by satellites of the Navy Navigation Satellite System (NNSS). The position of the antenna is computed by point positioning, translocation, or short arc geodetic adjustment (SAGA), using the collected signals from the satellite passes.

Doppler shift—See **Doppler effect.**

Doppler sonar navigation—A system by which speed and drift of a marine craft are determined from the Doppler effect of sonar-generated signals emitted from the craft and reflected from the ocean bottom or suspended particulate in the seawater. See also **Doppler navigation; sonar.**

dot screen—A photographic negative containing equal-sized dots which are equally spaced in parallel and perpendicular rows. The printed area of coverage determines the screen percent. The percentage of the printed area of coverage is measured by transmission density and/or by dot diameter. Dot screens are used to print tones of a color. Also called **flat tint screen.**

double burn—The intentional exposure of two or more line and/or halftone negatives in succession and register on the same sensitized surface. Not to be confused with **double exposure,** which is usually unintentional. Also called **double shooting.** See also **composite.**

double-center theodolite—See **repeating theodolite.**

double centering—A method of prolonging a line from a fixed point whereby the backsight is taken with the telescope direct. The instrument is plunged, and the foresight is made with the telescope inverted. The point at which the vertical crosshair intersects the hub is then marked. The transit is then rotated to take a backsight on the fixed point with the telescope inverted, the foresight is made with the telescope direct, and a second projected point is marked on the hub. A point midway between the two marked points is the true point on the prolonged line. Also called **double sighting; double reversion; double reversing; reversing in azimuth and altitude; wiggling-in on line; working-in on a line.**

double corners—(USPLS) Normally the two sets of corners along a standard parallel; the standard township, section, and quarter-section corners placed at regular intervals of measurement; additionally, the closing corners established on the line at the points of intersection of the guide meridians, range, and section lines of the surveys brought in from the south. In other cases, not fully in conformity with the rectangular plan, two corners, each common to two townships only, instead of one corner of the four townships. Similarly, two corners, each common to two sections, and two quarter-section corners, each referring to one section only. The term is sometimes used incorrectly to denote two lines established on the ground although the field-note record indicates only one line, thus creating a hiatus or overlap.

double interpolation—Deriving a value from tabulated computations where two or three proportions may be required, as in an hour-angle observation of Polaris for azimuth, where an interpolation is made suitable to the time factor, another to arrive at a result conforming with the latitude of the observing station.

double meridian distance—The algebraic sum of the perpendicular distances from the two ends of any line of a traverse to the initial, or reference, meridian.

double-model stereotemplet—A templet representative of the horizontal plot of two or three adjacent stereoscopic models that have been adjusted to a common, though random, scale.

double-optical projection stereoplotter—See **double-projection direct-viewing stereoplotter.**

double-projection direct-viewing stereoplotter—A class of stereoplotters employing the principle of projecting the images of two correctly oriented overlapping aerial photographs onto a reference datum so the resultant images may be viewed directly without additional optical system support. Also called **double-optical projection stereoplotter.**

double proportionate measurement—A method for restoring a lost corner of four townships or a lost interior corner of four sections. It is based on the principle that monuments north and south should control the latitudinal position of a lost corner, and monuments east and west should control the longitudinal position. In this method the influence of one identified original corner is balanced by the control of a corresponding original corner upon the opposite side of a particular missing corner which is to be restored, each identified original corner being given a controlling weight inversely proportional to its distance from the lost corner.

double reversing—See **double centering.**

double reversion—See **double centering.**

double-rodded line—A line of differential levels wherein two sets of turning points, one high and one low, are used to give independent measures between bench marks.

double shooting—See **double burn.**

double sighting—See **double centering.**

double-target leveling rod—Any target rod having graduations on two opposite faces.

double zenith distance—A value of twice the zenith distance of an object, obtained by observation and not by mathematical process.

doubly azimuthal map projection—An azimuthal map projection having two poles.

dove prism—A prism which reverts the image but does not deviate or displace the beam. A given angular rotation of the prism about its longitudinal axis causes the image to rotate through twice the angle. Also called **rotating prism.**

draconic month—See **nodical month.**

drafting—The art of drawing from given specifications.

drafting guide—See **guide.**

drag—**1.** (theodolite) A slight movement of the graduated circle of a theodolite produced by the rotation of the alidade. See also **atmospheric drag.**

drainage—In mapping, all features associated with water, such as shorelines, rivers, lakes, marshes, etc.

drainage pattern—The pattern or overall appearance made by the network of drainage features on a map or chart.

drift—**1.** (JCS) In ballistics, a shift in projectile direction due to gyroscopic action which results from gravitational and atmospheric induced torques on the spinning projectile. **2.** The lateral shift or displacement of a ship or aircraft from its course, due to the action of wind or other causes. **3.** Aerial photography obtained under this condition produces successive photographs whose edges are parallel but sidestepped. **4.** (precession) See **total drift.**

drift angle—(JCS) The angle measured in degrees between the heading of an aircraft or ship and the track made good. Drift angle is designated right or left to indicate the direction of drift.

drift station—A term sometimes used in shoran operations to designate the ground station about which the aircraft flies during arc navigation. The second ground station is then referred to as the **rate station.**

dummy—**1.** A preliminary drawing or layout showing the position of illustrations and text as they will appear in the final reproduction. **2.** A set of blank pages made up to show the size, shape, and general style of a book, booklet, or pamphlet.

dummy pendulum—A pendulum of similar construction to the working pendulums except that it is equipped with a thermometer and is fastened rigidly in the receiver so that it cannot swing during observations. The dummy pendulum is subject to the same temperature conditions as the working pendulums, and is used in determining their temperature when in use.

dumpy level—A leveling instrument which has its telescope permanently attached to the leveling base, either rigidly or by a hinge that can be manipulated by a micrometer screw.

duplex base-line measuring apparatus—A contact base-line measuring apparatus, composed of two disconnected bars, one of brass and the other of steel, each 5 meters in length and so arranged as to indicate the accumulated difference of length of the measures from the brass and steel components.

duplicate level line—A line of spirit leveling composed of two single lines run over the same route, but in opposite directions, and using different turning points.

duplicate negative—(JCS) A negative reproduced from an original negative or positive. The duplicate negative may be a true reproduction of the original or a reproduction possessing greater or less contrast. With direct positive materials, chemical reversal process, and duplicating film it is not always necessary to make a positive to obtain a duplicate negative.

dynamic correction—The quantity that must be added to the orthometric elevation of a point to obtain its dynamic number.

dynamic elevation—Elevation expressed in length units, but determined by dynamic number.

dynamic gravity meter—A type of gravity instrument in which the period of oscillation is a function of gravity and is the quantity directly observed.

dynamic height—A height derived by dividing the geopotential number by a constant, usually the value of normal gravity at 45° latitude.

dynamic number—The work required to raise a unit mass from sea level to a given point, expressed in absolute units.

dynamic temperature correction—(pendulum) The correction to the observed period of a pendulum for the rate of change of its temperature.

dynamical mean sun—A fictitious sun conceived to move eastward along the ecliptic at the average rate of the apparent Sun.

dyne—A force which, acting on a mass of 1 gram, imparts to that mass an acceleration of 1 centimeter per second per second. The dyne is the unit of force of the c.g.s. system of units. Until about 1930, the dyne was used by the U.S. Coast and Geodetic Survey (now the National Geodetic Survey) in stating values of gravity. Since that time, gravity has been reported in terms of the gal, the c.g.s. unit of acceleration.

E

Earth-centered ellipsoid—A reference ellipsoid whose geometric center coincides with the Earth's center of gravity and whose semiminor axis coincides with the Earth's rotational axis.

Earth-fixed coordinate system—Any coordinate system in which the axes are stationary with respect to the Earth.

Earth inductor—An instrument designed for use in magnetic surveys to determine magnetic dip. In principal, the instrument is a small dynamo by which the electrical flow can be introduced to the coil of the instrument. The presence or absence of current is indicated by a galvanometer which provides for direct reading of magnetic dip.

Earth inductor compass—A compass depending for its indications upon the current generated in a coil revolving in the Earth's magnetic field.

Earth satellite—A body that orbits about the Earth; specifically, an artificial satellite placed in orbit by man.

Earth tide—A periodic movement of the Earth's crust caused by tide-producing forces of the Moon and Sun.

easement curve—See **spiral curve.**

east point—See **prime vertical plane.**

easting—1. (JCS) (grid) Eastward (that is left to right) reading of grid values. See also **false easting. 2.** (plane surveying) See **departure,** definition 1.

ebb tide—The portion of the tide cycle between high water and the following low water. Also called **falling tide.**

eccentric—Not having the same center.

eccentric anomaly—See **anomaly,** definition 3.

eccentric error—Centering error.

eccentric reduction—(triangulation) The correction which must be applied to a direction observed by an instrument with either the instrument or signal (swing), or both, eccentric, to reduce the observed value to what it would have been if there had been no eccentricity. Also called **eccentricity correction.**

eccentric signal—A signal (target) which is not in the same vertical line with the station which it represents.

eccentric station—A survey point over which an instrument is centered and observations made, and which is not in the same vertical line with the station which it represents and to which the observations will be reduced before being combined with observations at other stations. In general, an eccentric station is established and occupied when it is impracticable to occupy the station center, or when it becomes necessary in order to see points which are not visible from the station center.

eccentricity—1. Amount of deviation from a center. See also **eccentric station. 2.** (surveyor's compass) An effect caused by one or a combination of the following conditions: A straight line through the ends of the magnetic needle fails to pass through the center of rotation of the needle; the center of rotation of the needle is not coincident with the center of figure of the graduated circle; the line of sight fails to pass through the vertical axis of the instrument. **3.** The ratio of the distances from any point of a conic section to a focus and the corresponding directrix.

eccentricity correction—(triangulation) See **eccentric reduction.**

eccentricity of alidade—The distance between the center of figure of the index points on an alidade and the center of figure of the graduated circle. See also **eccentricity of instrument.**

eccentricity of circle—The distance between the center of figure of a graduated circle and its center of rotation. See also **eccentricity of instrument.**

eccentricity of ellipse—The ratio of the distance between the center and a focus of an ellipse to the length of its semimajor axis.

eccentricity of instrument—The combination of eccentricity of circle and eccentricity of alidade.

eccentricity of spheroid (ellipsoid) of revolution—The eccentricity of an ellipse forming a meridian section of the spheroid.

echo—See **blip.**

echo sounder—An echo sounding instrument used for depth measurements in water.

echo sounding—A method for measuring depths by recording the time interval required for sound waves to go from a source of sound near the surface to the bottom and back again.

echo timing—The measurement of time required for a short train of energy waves to travel the round-trip path from an originating station to a reflector or transponder.

echogram—A graphic record of depth measurements obtained by echo sounding equipment. Called **fathogram** when obtained from a Fathometer echo sounder.

eclipse—1. The reduction in visibility or disappearance of a non-luminous body by passing into the shadow cast by another non-luminous body. **2.** The apparent cutting off, wholly or partially, of the light from a luminous body by a dark body coming between it and the observer. See also **annular eclipse; lunar eclipse; solar eclipse.**

eclipse year—The interval between two successive conjunctions of the Sun with the same node of the Moon's orbit.

ecliptic—The great circle formed by the intersection of the plane of the Earth's orbit around the Sun (or apparent orbit of the Sun around the Earth) and the celestial sphere.

ecliptic coordinate system—See **ecliptic system of**

coordinates.

ecliptic latitude—See **celestial latitude.**

ecliptic longitude—See **celestial longitude.**

ecliptic meridian—A circle on the celestial sphere containing points of the same celestial longitude.

ecliptic node—See **node.**

ecliptic parallel—A circle on the celestial sphere containing points of the same celestial latitude.

ecliptic polar distance—The complement of the celestial latitude.

ecliptic pole—On the celestial sphere, either of the two points 90° from the ecliptic.

ecliptic system of coordinates—A system of curvilinear celestial coordinates which uses the ecliptic as the primary reference plane and the ecliptic meridian through the vernal equinox as the secondary plane. The points 90° from the ecliptic are the north and south ecliptic poles. Angular distance north or south of the ecliptic analogous to latitude, is celestial latitude. Celestial longitude is measured eastward along the ecliptic from the vernal equinox through 360°.

editing—The process of checking a map or chart in its various stages of preparation to insure accuracy, completeness, and correct preparation from and interpretation of the sources used, and to assure legible and precise reproduction. Edits are usually referred to by a particular production phase, such as compilation edit, scribing edit, etc.

effective area—For any aerial photograph that is one of a series in a flight strip, that central part of the photograph delimited by the bisectors of overlaps with adjacent photographs. On a vertical photograph, all images within the effective area have less displacement than their conjugate images on adjacent photographs.

effective Earth radius—See **effective radius of the Earth.**

effective focal length (EFL)—See **principal distance,** definition 1.

effective radius of the Earth—A fictitious value for the radius of the Earth, used in a place of the geometrical radius to correct for atmospheric refraction when the index of refraction in the atmosphere changes linearly with height.

Egault level—A French instrument having the spirit level attached to a level bar which also carries wyes in which the telescope rests.

electrical distance—Length measured in terms of the distance traveled by radio waves in unit time.

electrical survey-net adjuster (ESNA)—A device used for obtaining least-squares adjustments of level, traverse, and vertical angle nets through the medium of an analogous electrical net.

electromagnetic spectrum—The entire range of wavelengths or frequencies of electromagnetic radiation extending from gamma rays to the longest radio waves and including visible light. Most remote sensing systems are designed to operate within the electromagnetic spectrum.

electronic bearing—A bearing obtained by means of electronic equipment.

electronic control—Control obtained by electronic devices.

electronic distance-measuring (EDM) equipment—Devices that measure the phase difference between transmitted and returned (i.e., reflected or retransmitted) electromagnetic waves, of known frequency and speed, or the round-trip transit time of a pulsed signal, from which distance is computed. A wide range of such equipment is available for surveying and navigational use. Also called **distance-measuring equipment (DME).** See also **electronic position indicator; Electrotape; Geodimeter; laser; sonar; Tellurometer.**

electronic line of position—A line of position established by means of electronic equipment.

electronic-position-indicator (EPI)—A type of electronic distance-measuring equipment used primarily in hydrographic surveying. The offshore range has been extended from the limits of shoran distances to more than 500 miles by use of EPI.

electronic refraction—The refraction due to the effects of the atmosphere and the ionosphere, which introduce appreciable changes in the quantities measured by means of electronic devices, such as in the phase differences measured with interferometers, in the rate of change of phase measured with the Doppler systems, and in the change in phase between the times of transmitting and receiving a signal by the ranging instruments.

electronic sketchmaster—See **universal analog photographic rectification system.**

electronic survey—Any survey utilizing electronic equipment.

electronic telemeter—An electronic device that measures the phase difference or transit time between a transmitted electromagnetic impulse of known frequency and speed and its return.

Electrotape—A trade name for a precise electronic surveying device which transmits a radio-frequency signal to a responder unit which in turn retransmits the signal back to the interrogator unit. The time lapse between original transmission and receipt of return signal is measured and displayed in a direct digital readout for eventual reduction into a precise linear distance. It operates on the same principal as the **Tellurometer.**

elements of a fix—The specific values of the coordinates used to define a position.

elevated pole—The celestial pole above the horizon, agreeing in name with latitude. Opposite of **depressed pole.**

elevation—Vertical distance from a datum, usually mean sea level, to a point or object on the Earth's surface; not to be confused with altitude which refers to points or objects above the Earth's surface. In geodetic formulas, elevations are heights: h is height above ellipsoid, H is height above the geoid or local datum. Occasionally the h and H may be reversed. See also **adjusted elevation; assumed ground elevation; barometric elevation; checked spot elevation; critical elevation; difference of elevation; dynamic elevation; field elevation; fixed elevation; highest elevation; mean ground elevation; optimum ground elevation; orthometric elevation; preliminary elevation; spot elevation; standard elevation; supplemental elevation; unchecked spot elevation.**

elevation angle—See **angle of elevation.**

elevation meter—A mechanical or electromechanical device on wheels that measures slope and distance, and automatically and continuously integrates their product into difference of elevation.

elevation tints—See **hypsometric tinting.**

11/10 peg adjustment—See **peg adjustment.**

ellipsoid—A surface whose plane sections (cross sections) are all ellipses or circles, or the solid enclosed by such a surface. In geodesy, ellipsoid and spheroid are used interchangeably. See also **datum-centered ellipsoid; Earth-centered ellipsoid; Fischer ellipsoid of 1960; triaxial ellipsoid.**

ellipsoid of revolution—See **ellipsoid of rotation.**

ellipsoid of rotation—The surface generated by an ellipse rotating about one of its axes. See also **oblate spheroid; prolate spheroid.** Also called **ellipsoid of revolution.**

ellipsoidal height—The height above the reference ellipsoid, measured along the ellipsoidal outer normal through the point in question. Also called **geodetic height.**

ellipsoidal reflector—A mirror surface which conforms to a portion of an ellipsoid of revolution. Principally employed in several types of stereoplotter projectors such as the ER-55.

ellipticity of an ellipse—The ratio between the difference in length of the semi-axes of an ellipse and its semimajor axis.

ellipticity of the spheroid—See **flattening (of the Earth).**

elongation—**1.** (surveying) The position of a celestial body relative to the observer's meridian, is such that the apparent azimuthal movement is at a minimum. **2.** The angular distance of a body of the solar system from the Sun; the angle at the Earth between lines to the Sun and another celestial body of the solar system. See also **greatest elongation.**

emergency run—See **tide-over run.**

emergent nodal point—See **nodal point,** definition 1.

emissivity—The amount of energy given off by an object relative to the amount given off by a "blackbody" at the same temperature. Normally expressed as a positive number between zero and one.

empirical orientation (rectification)—The composited rectified adjustments of magnification, swing, easel tilt, y-displacement, and x-displacement used to correctly recreate the exact conditions in the projected image that existed in the negative at the instant of exposure.

emulsion—A suspension of either light-sensitive silver salts, Diazos, or photopolymers, in a colloidal medium which is used for coating films, plates, and papers.

emulsion-to-base—A contact exposure in which the base of the copying film is in contact with the emulsion side of the sheet being copied. See also **emulsion-to-emulsion.**

emulsion-to-emulsion—A contact exposure in which the emulsion of the copying film is in contact with the emulsion of the sheet being copied. See also **emulsion-to-base.**

end lap—See **overlap,** definition 1.

engineer's chain—Similar to a Gunter's chain except that it is 100 feet in length and contains 100 links, each 1 foot long.

engineer's level—Any of a group of precision leveling instruments for establishing a horizontal line of sight, used to determine differences of elevation.

engineering map—A map showing information that is essential for planning an engineering project or development and for estimating its cost.

engineering survey—A survey executed for the purpose of obtaining information that is essential for planning an engineering project or development and estimating its cost. The information obtained may, in part, be recorded in the form of an engineering map.

engraver—See **scriber.**

engraver subdivider—A specially designed scribing instrument which permits the selection of uniform tick spacing in subdividing or ticking map projections.

enlargement—(JCS) A negative, diapositive, or paper print made at a larger scale than the original. Also called **blowup.**

enlargement factor—See **scale of reproduction.**

enlargement/reduction diagram—Chart showing the necessary lens extension and copy board extension required for various enlargements and reductions.

enroute chart—A chart of air routes in specific areas that shows the exact location of electronic aids to navigation, such as radio-direction-finder stations, radio and radar marker beacons, and radio-range stations. Also called **radio facility chart.**

entrance pupil—The image of the aperture stop formed by all the lens elements on the object side of the aperture stop.

entrance slit—Slit through which energy enters a spectroscopic instrument.

entrance window—The image of the field stop formed by all the lens elements on the object side of the field stop.

Eötvös correction—Component of uncorrected observed gravity obtained on a moving platform that is attributed to the east-west component of the velocity that either increases or decreases the effect of the centrifugal force caused by the Earth's rotation. The formula for the correction usually includes the term for the secondary velocity correction.

Eötvös effect—A vertical force experienced by a body moving in an east-west direction on the rotating Earth. In gravity measurements, a positive correction is applied if moving eastward, and a negative correction if moving westward.

Eötvös unit—The unit of gravitational gradient 10^{-9} meters per second per second per meter.

ephemeris—(JCS) A publication giving the computed places of the celestial bodies for each day of the year, or for other regular intervals. [A publication giving similar information in a form suitable for use by a navigator is called an *almanac*. An ephemeris is also a statement, not necessarily in a publication, presenting a correlation of time and position of celestial bodies or manmade satellites. A short ephemeris of future location/time of satellite is known as a set of alerts.] See also **broadcast ephemeris; precise ephemeris.**

ephemeris time—The uniform measure of time defined by the laws of dynamics and determined in principal from the orbital motions of the planets, specifically in the orbital motion of the Earth as represented by Newcomb's "Tables of the Sun." Ephemeris time for close-Earth satellites, or more correctly the time associated with satellite ephemerides, is observation dependent. For example, if the observations were made in universal time (UT), the ephemeris time is UT; if the observations were made in international atomic time (IAT), the ephemeris time is in IAT.

epipolar plane—(photogrammetry) Any plane which contains the epipoles; therefore, any plane containing the air base. Also called **basal plane.**

epipolar ray—The line on the plane of a photograph joining the epipole and the image of an object. Also expressed as the trace of an epipolar plane on a photograph.

epipoles—In the perspective setup of two photographs (two perspective projections), the points on the planes of the photographs where they are cut by the air base (extended line joining the two perspective centers). In the case of a pair of truly vertical photographs, the epipoles are infinitely distant from the principal points.

epoch—**1.** A particular instant for which certain data are given. **2.** A given period of time during which a series of related acts or events takes place. **3.** An arbitrary moment in time to which measurements of position for a body or orientation for an orbit are referred.

equal-altitude observations—Observations of celestial objects at a fixed altitude (such as by an astrolabe) taken at more or less uniformly spaced azimuths around the horizon.

equal-altitude observations of the Sun—Azimuth observations that consist of measuring horizontal angles from a southerly reference point to the Sun's limbs at an identical vertical angle, if measured to the right limb in the a.m., then the angle should be measured to the left limb in the p.m. The same limb should be observed in vertical angle. The mean of the two horizontal angles, with small correction for the change in the Sun's declination during the interval from the a.m. to the p.m. readings, gives a resulting horizontal angle to the meridian.

equal-area map projection—A map projection having a constant area scale. Such a projection is not conformal and is not used for navigation. Also called **authalic map projection; equivalent map projection.**

equal energy—Spectral distribution characterized by equal flex per unit wavelength interval.

equation—See **angle equation; azimuth equation; correlate equation; condition equation; error equation; Euler's equation; hydrostatic equation; Laplace equation; latitude equation; length equation; longitude equation; lunar equation; normal equation; observation equation; parametric equations; perpendicular equation; personal equation; side equation; side equation tests.**

equation of the equinox—The difference between the mean and true right ascensions of a body on the Equator, thus the difference between mean and apparent sidereal time. Also called **nutation in right ascension.**

equation of time—The algebraic difference in hour angle between apparent solar time and mean solar time, usually labeled + or − as it is to be applied to mean solar time to obtain apparent solar time.

Equator—The great circle on the Earth midway between the poles and in a plane perpendicular to the Earth's axis of rotation. It is the line of 0° latitude. See also **astronomic equator; celestial equator; fictitious equator; galactic equator; geodetic equator; geomagnetic equator; grid equator; lunar celestial equator; magnetic equator; oblique equator; transverse equator.**

equator system—See **celestial equator system of coordinates.**

equatorial axis—**1.** The diameter of the Earth described between two points on the Equator. **2.** (astronomy) A telescope mounting axis oriented parallel to the Earth's rotational axis.

equatorial bulge—The excess of the Earth's equatorial diameter over the polar diameter.

equatorial chart—**1.** A chart of equatorial areas

2. A chart on an equatorial projection.

equatorial cylindrical orthomorphic chart—See **Mercator chart.**

equatorial cylindrical orthomorphic map projection—See **Mercator map projection.**

equatorial diameter—The diameter of the Earth at the great circle comprising the terrestrial equator.

equatorial gravity value—The mean acceleration of gravity at the Equator, approximately equal to 9.780 3 m/sec^2.

equatorial horizontal parallax—The angle at a celestial object subtended by the equatorial semi-diameter of the Earth used to indicate the distance of the object from the Earth.

equatorial intervals—The angles, expressed in units of time, between the various lines which compose the reticle of an astronomic transit and the mean position of those lines.

equatorial map projection—A map projection centered on the Equator.

equatorial node—Either of the two points where the orbit of the satellite intersects the equatorial plane of its primary.

equatorial radius—The radius assigned to the great circle comprising the terrestrial equator.

equatorial satellite—A satellite whose orbit plane coincides, or almost coincides, with the Earth's equatorial plane.

equatorial stars—Stars having declinations close to zero and whose diurnal path is a parallel of declination close to the Equator. Equatorial stars, because of their apparently greater speed of travel, are preferred for time and longitude determinations.

equatorial system—See **celestial equator system of coordinates.**

equiangular spiral—See **rhumb line.**

equiangulator—An optical instrument, employing a 60° prism, used to determine astronomic latitude and longitude by equal altitudes of heavenly bodies when the time of a prime meridian (Greenwich) is known at the place of observation.

equigeopotential surface—See **equipotential surface.**

equilibrium—A state of balance between forces. A body is said to be in equilibrium when the vector sum of all forces acting upon it is zero.

equilibrium spheroid—The shape that the Earth would attain if it were entirely covered by a tideless ocean of constant depth. See also **geoid.**

equilibrium theory—A hypothesis which assumes an ideal Earth which has no continental barriers and is uniformly covered with water of considerable depth. It also assumes that the water responds instantly to the tide-producing forces of the Moon and Sun to form a surface in equilibrium and moves around the Earth without viscosity or friction. See also **geoid.**

equinoctial—See **celestial equator.**

equinoctial colure—The hour circle through the equinoxes.

equinoctial day—See **sidereal day.**

equinoctial point—See **equinox.**

equinoctial system of coordinates—See **celestial equator system of coordinates.**

equinoctial year—See **tropical year.**

equinox—One of the two points of intersection of the ecliptic and the celestial equator, occupied by the Sun when its declination is 0°. Also called **equinoctial point.** See also **autumnal equinox; mean equinox; vernal equinox.**

equipotential surface—A surface having the same potential of gravity at every point. Also called **level surface.** See also **geoid; geop; geopotential surface.**

equiscalar—A surface along which a scalar quantity has a constant value.

equivalent focal length—(JCS) The distance measured along the optical axis of the lens from the rear nodal point to the plane of best average definition over the entire field used in a camera. See also **focal length.**

equivalent map projection—See **equal-area map projection.**

equivalent scale—The relationship which a small distance on a graphic bears to the corresponding distance on the Earth, expressed as an equivalence, such as 1 inch (on the graphic) equals 1 mile (on the ground). Also called **verbal scale.**

equivalent vertical photograph—A theoretically, truly vertical photograph taken at the same camera station with a camera whose focal length is equal to that of a camera taking a corresponding tilted photograph.

erect image—An image that appears upright, or in the same relative position as the object.

erecting telescope—An observer sees objects right side up when looking through an erecting telescope and upside down when looking through an inverting telescope. The eyepiece in the optical system of an erecting telescope usually has four lenses, and the eyepiece in the optical system of an inverting telescope has two lenses. See also **inverting telescope.**

erratic error—An error caused by an incomplete element in an instrument, such as backlash in a gear train. See also **instrument error.**

error—**1.** The difference between an observed or computed value of a quantity and the ideal or true value of that quantity. **2.** An error is generally classified as one of three types: a blunder (mistake) which can be identified and corrected; a systematic error, either constant or variable, which must be compensated for; and a random error, one of the class of small inaccuracies due to imperfections in equipment, surrounding conditions, or human limitations. See also **absolute error; actual error; accumulative error; blunder; chronometer error; circuit closure; circular error; circular error probable;**

circular near-certainty error; circular standard error; clamping error; collimation error; compass index error; compensating error; constant error; curved-path error; eccentric error; erratic error; external error; graduation error; gross error; index error; inherited error; instrument error; law of propagation of error; linear error; mean-square error; modulation error; natural error; near-certainty error; orthometric error; parallactic error; periodic errors; personal error; pointing errors; positional error; principal-distance error; principal-point error; prismatic error; probable error; random error; residual error; resultant error; scale error; shade error; standard error; systematic error; theoretical error.

error budget—A correlated set of the individual major error sources with the quantified error or uncertainty which each contributes to a total system accuracy or probable error.

error equation—The probability equation which expresses the laws of the occurrence of random errors. This equation expresses the relationship between observed values, plus first-order correction terms, and theoretical values. The error equation is the basis of the method of least squares, used in the adjustment of observations for determining the most probable value of a result from those observations.

error interval—See **confidence interval.**

error of closure—1. (general) The amount by which a quantity obtained by a series of related measurements differs from the true or fixed value of the same quantity. Also called **closing error; closure. 2.** (angles) The amount by which the actual sum of a series of angles fails to equal the theoretically exact value of that sum. Also called **angular error of closure. 3.** (azimuth) The amount by which two values of the azimuth of a line, derived by different surveys or along different routes, fail to be exactly equal to each other. Also called **azimuth error of closure. 4.** (leveling) The amount by which two values of the elevation of the same bench mark, derived by different surveys or through different survey routes or by independent observations, fail to be exactly equal to each other. Also called **leveling error of closure.** See also **circuit closure. 5.** (loop) The error in the closure of a survey on itself. Loops do not protect against systematic errors in distance measurement or blunders in starting position or azimuth. Also called **loop error of closure. 6.** (horizon) The amount by which the sum of a series of adjacent measured horizontal angles around a point fails to equal exactly 360°. Measurement of the last angle of the series is called **closing the horizon;** also called **closure of horizon; horizon closure. 7.** (triangle) The amount by which the sum of the three observed angles of a triangle fails to equal exactly 180° plus the spherical excess of the triangle. Also called **closure of triangle; triangle closure; triangle error of closure. 8.** (traverse) The amount by which a value of the position of a traverse station, as obtained by computation through a traverse, fails to agree with another value of the same station as determined by a different set of observations or routes of survey. Also called **closure of traverse; error of survey; horizontal closure error; traverse error of closure.** See also **linear error of closure; relative error of closure.**

error of collimation—See **collimation error.**

error of observation—The difference between an observed value of a quantity and a value adopted as representing the ideal or true value of that quantity.

error of run—(micrometer) The difference, in seconds of arc, between the intended value of one turn of the micrometer screw and its actual value as determined by measuring the space between two adjacent graduation marks of the circle with the micrometer. Also called **run; run of micrometer.**

error of survey—See **error of closure,** definition 8.

error of the mean—Equal to the standard error of the sample divided by the square root of the number of items sampled. Also called **standard error of the mean.**

escape and evasion graphic—A map, chart or other graphic, usually produced on a lightweight durable material, specifically designed to guide personnel to safety from enemy-held territory.

establishment—See **lunitidal interval.**

establishment of the port—The average interval between upper and lower lunar transit near time of new and full Moon and the next high water. See also **lunitidal interval.** Also called **common establishment; high water full and change; vulgar establishment.**

etch—1. To remove selected areas of the emulsion either chemically or manually. **2.** Chemical treatment of a lithographic plate to make non-printing areas grease-repellent and water-receptive or to produce the image on deep-etch plates. **3.** An acid solution mixed with the dampening fountain water on an offset press to help control ink on the pressplate.

etch slip—A pencil-shaped abrasive used in removing unwanted marks on a metal pressplate. Also called **snakeslip.**

etched zinc plate—An etched copy of the contour drawing of the base map, used as the guide in cutting the stepped terrain base of a model for making relief models.

Euler's equation— A relation in a parabolic orbit involving two radius vectors, their chord, and the time interval between them.

Euler's theorem—A mathematical expression to obtain the radius of curvature of a normal section in any azimuth on the reference ellipsoid. The azimuth angle, and the radius of curvature in the meridian and in the prime vertical must be known.

Eulerian angles—A system of three angles which uniquely define with reference to one coordinate system (e.g., Earth axes) the orientation of

European datum—The initial point of this sytem is located at Potsdam, Germany. Numerous national systems have been joined into a large datum based upon the International ellipsoid which was oriented by the astrogeodetic method. The European and African triangulation chains have been connected and the gap of the African arc measurement from Cairo to Cape Town has been filled. Thus, all of Europe, South Africa, and North Africa are molded into one system. Through common survey stations, it was also possible to convert data from the Russian Pulkovo 1932 system to the European datum and, as a result, the European datum includes triangulation as far east as the 84th meridian. Additional ties across the Middle East have permitted connection of the Indian and European datums. See also **preferred datum.**

evection—A perturbation of the Moon in its orbit due to the attraction of the Sun. This results in an increase in the eccentricity of the Moon's orbit when the Sun passes the Moon's line of apsides and a decrease when perpendicular to it. See also **lunar inequality,** definition 1.

Everest spheroid (ellipsoid)—A reference ellipsoid having the following approximate dimensions: semimajor axis—6,377,276.3 meters; flattening or ellipticity—1/300.80. Used in India, Burma, Pakistan, Laos, Cambodia, Thailand, and Vietnam.

exaggerated stereo—See **hyperstereoscopy.**

existent corner—(USPLS) A corner whose position can be identified by verifying the evidence of the monument or its accessories, by reference to the description that is contained in the field notes; or where the point can be located by an acceptable supplemental survey record, some physical evidence, or testimony.

existing data—Source material and/or information assumed or known to be in the possession of a given source and subject to "off-shelf" collection, as in contrast to data obtained by operational field surveys.

exit pupil—The image of the aperture stop formed by all the lens elements on the image side of the aperture stop.

exit window—The image of the field stop formed by all the lens elements on the image side of the field stop.

ex-meridian altitude—An altitude of a celestial body near the celestial meridian of the observer to which a correction is to be applied to determine the meridian altitude.

ex-meridian observation—Measurement of the altitude of a celestial body near the celestial meridian of the observer, for conversion to a meridian altitude; or the altitude so measured.

experience radar prediction—The determination of size, shape, and relative intensity of radar returns and a determination of radar shadow and no-return areas based primarily on the radar knowledge and experience of the individual making the prediction rather than on proven formulas, power tables, or graphs. Also called **art-work prediction.**

experimental map—A sample of a new map product prepared either to obtain user approval of the adequacy of content and symbolization or to disclose any problems which may occur in the various production stages. Also called **prototype.** See also **pilot sheet.**

explement—The difference between an angle and 360°.

exploratory survey—A survey executed for the purpose of obtaining general information concerning areas about which such information is not a matter of record.

exposure—1. The total quantity of light received per unit area on a sensitized plate or film; may be expressed as the product of the light intensity and the exposure time. **2.** The act of exposing a light-sensitive material to a light source. **3.** One individual picture of a strip of photographs, usually called **frame.**

exposure interval—The time required between successive exposures of a series of photographs for the purpose of obtaining desired forward lap.

exposure station—See **air station.**

exposure time—The time during which a light-sensitive material is subjected to the action of light.

extended color—See **bleed.**

extension—1. (surveying) See **prolongation. 2.** (photogrammetry) Extending existing control from a controlled area into an area without control. The term is usually further qualified as horizontal or vertical according to the primary purpose. See also **cantilever extension.** Also called **horizontal extension; horizontal/vertical extension; vertical extension.**

extension of control—Execution of additional control from existing control by any method.

exterior orientation—The determining (analytically or in a photogrammetric instrument) of the position of the camera station and the attitude of the taking camera at the instant of exposure. In stereoscopic instrument practice, exterior orientation is divided into two parts, relative and absolute orientation. Also called **outer orientation.** See also **resection,** definition 3.

exterior perspective center—See **perspective center.**

exterior to a curve—Any area adjacent to a curve lying toward its convex side; the area not included within the circle of which the curve is part of the circumference.

external distance—The distance from the vertex

of a circular curve to the middle point of the curve.

external error—The repeatability of a measurement with any condition extraneous to the measuring method itself changed; contrasted to internal error. See also **standard error.**

extra foresight—(leveling) The rod reading made at an instrument station in a line of levels and on a leveling rod standing on a bench mark or another point not in the continuous line of levels. In spirit leveling there may be one or more extra foresights from a single instrument station or setup, but there can be only one backsight and one foresight from any one instrument station.

extrapolation—The process of estimating the value of a quantity beyond the limits of known values by assuming that the rate or system of change between the last few known values continues.

eye base—See **interocular distance.**

eyepiece—In an optical device, the lens group which is nearest the eye and with which the image formed by the preceding elements is viewed.

eyepiece micrometer—See **ocular micrometer**.

F

f-number—See **relative aperture.**

face—The emulsion side of a negative or layout plate, or the printing surface of a plate.

factored transparency—A system of radar simulation which utilizes a pair of photographic images on a glass plate or plates to store topographic and radar reflection data. The data are scanned by a flying-spot scanner cathode-ray tube, and the density of the images is read by two photo-multiplier tubes. The two planar dimensions of the two images are the *x* and *y* dimensions of the topographic and reflectance data respectively. The densities of the reflectance images are used to store the instrinsic strength of radar target reflectance. The images are identical in their *x* and *y* values but separated in one dimension by the optical spacing of the dual readout system. Also called **land mass simulator plate.**

falling—(USPLS) The distance by which a random line falls to the right or left of a corner on which the true line is too close. Usually the direction of falling is expressed as cardinal.

falling tide—See **ebb tide.**

false bearing—The difference between the true bearing and the back bearing caused by the convergence of meridians.

false color—(photogrammetry) See **infrared.**

false easting—A value assigned the central meridian of a coordinate system to avoid the inconvenience of using negative departures. See also **grid coordinates.**

false-fix probability—A statistical value or ratio which reflects the likelihood of a false match occurring between prestored digital cartographic data and data or imagery acquired by electronic aerial sensor systems. See also **image correlation; terrain contour matching (TERCOM); terrain correlation.**

false horizon—A line resembling the visible horizon, but above or below it.

false northing—A value assigned the origin of northings, or grid coordinates, to avoid the inconvenience of using negative coordinates. See also **grid coordinates.**

false origin—(JCS) A fixed point to the south and west of a grid zone from which grid distances are measured eastward and northward. See also **grid origin.**

false parallax—(JCS) The apparent vertical displacement of an object from its true position when viewed stereoscopically, due to movement of the object itself as well as to change in the point of observation.

false stereo—(JCS) An imaginary impression of stereoscopic relief. See also **pseudoscopic stereo.**

fan-camera photography—(JCS) Photography taken simultaneously by an assembly of three or more cameras, systematically installed at fixed angles relative to each other so as to provide wide lateral coverage with overlapping images.

fan cameras—(JCS) An assembly of three or more cameras systematically disposed at fixed angles relative to each other so as to provide wide lateral coverage with overlapping images.

fathogram—A graphic record of depth measurements obtained by echo-sounding equipment. Also called **echogram.**

fathom curve—See **depth curve.**

fathom line—See **depth curve.**

Fathometer—A trade name for an echo sounder.

Faye anomaly—See **free-air anomaly.**

Faye correction—See **free-air correction.**

featheredging—**1.** (cartography) The technique of progressively dropping contours, to avoid congestion on steep slopes, and tapering the line weight near the end of the contour to be dropped. Also called **feathering. 2.** (photomosaicking) The thinning of overlapping edges of photographs before assembling into a mosaic in order to make match lines less noticeable. When overlapping edges are feathered, shadows and sharp changes in contrast are reduced or eliminated. Also called **feathering.**

feathering—See **featheredging.**

feature extraction—The art of extracting and classifying features contained in an image.

felt side—The top or smooth side of paper that is contacted by the felt belt for extraction of moisture during manufacture. This is the correct side of the paper for printing.

fence—**1.** A line of readout or tracking stations for pickup of signals from an orbiting satellite. **2.** A line or network of radar or radio stations for detection of a satellite in orbit.

fermenting dough theory—See **Pratt-Hayford theory of isostasy.**

ferrotype—To burnish photographic prints by squeegeeing wet upon a japanned sheet of iron or stainless plate and allowing to dry. This produces a harder, glossier surface on the photographic print.

fiber optics—A device for relaying an image by means of a large number of transparent fibers (filaments) by multiple total internal reflection. The fibers are most commonly glass and less often a highly transparent plastic. Each fiber carries only one element of the image, so that the image is a mosaic in which the cell size is the fiber cross section rather than a continuous picture.

fictitious—In cartography, pertaining to or measured from an arbitrary reference line.

fictitious equator—A reference line serving as the origin for measurement of fictitious latitude.

fictitious graticule—The network of lines representing fictitious parallels and fictitious meridians on a map or chart. See also **oblique**

graticule; transverse graticule.

fictitious latitude—Angular distance from a fictitious equator. It may be called transverse, oblique, or grid latitude depending upon the type of fictitious equator.

fictitious longitude—The arc of fictitious equator between the prime fictitious meridian and any given fictitious meridian. It may be called transverse, oblique, or grid longitude depending upon the type of fictitious meridian.

fictitious loxodrome—See **fictitious rhumb line.**

fictitious loxodromic curve—See **fictitious rhumb line.**

fictitious meridian—One of a series of great circles or lines used in place of a meridian for certain purposes. It may be called transverse, oblique, or grid meridian depending upon the type of fictitious meridian. See also **prime fictitious meridian.**

fictitious parallel—A circle or line parallel to a fictitious equator, connecting all points of equal fictitious latitude. It may be called transverse, oblique, or grid parallel depending upon the type of fictitious equator.

fictitious pole—One of the two points 90° from a fictitious equator. It may be called transverse or oblique pole depending upon the type of fictitious equator.

fictitious rhumb line—A line making the same oblique angle with all fictitious meridians. It may be called transverse, oblique, or grid rhumb line depending upon the type of fictitious meridian. Also called **fictitious loxodrome; fictitious loxodromic curve.**

fictitious sun—A fictitious point termed the mean Sun, which is imagined to move at a uniform rate along the Equator, its rate of motion being such that it makes one apparent revolution around the Earth in the same time as the actual Sun—that is, in 1 year.

fictitious year—The period between successive returns of the Sun to a sidereal hour angle of 80° (about January 1). The length of the fictitious year is the same as that of the tropical year, since both are based upon the position of the Sun with respect to the vernal equinox. Also called **Besselian year.**

fiducial axes—The lines joining opposite fiducial marks on a photograph. The *x*-axis is generally considered to be the one nearly parallel with the line of flight.

fiducial mark(s)—**1.** (surveying) An index line or point. A line or point used as a basis of reference. **2.** (JCS) (photogrammetry) See **collimating marks. 3.** Also, markers in any instrument which define the axes whose intersection fixes the principal point of a photograph and fulfills the requirements of interior orientation.

field calibration—A term generally applied where only a combination of field and office computer techniques are available to check instrument accuracy. Adjustments, other than normal operator adjustments, cannot be made during field calibration.

field check—The operation of checking a map compilation manuscript on the ground. See also **field classification.**

field classification—Field inspection and identification of features which a map compiler is unable to delineate; identification and delineation of political boundary lines, place names, road classifications, buildings hidden by trees, and so forth. Field classification may be included as part of the control survey effort and normally is completed prior to the actual stereocompilation phase. See also **field inspection.**

field comparator—A short line whose length is measured with accuracy and precision, and is used to check the lengths of apparatus (tapes) used in the actual field operations. Also called **calibration course; comparator base.**

field completion—A combination of field inspections or surveys, either before or after compilation, to classify and complete the map content, correct erroneous data, and add informatio such as names, civil boundaries, and simila classification data. Its purpose is to fill in o confirm that portion of a map manuscrip prepared by stereocompilation.

field control—(JCS) A series of points whos relative positions and elevations are known These positions are used in basic data i mapping and charting. Normally, these posi tions are established by survey methods and ar sometimes referred to as **trig control.** See als **ground control.**

field contouring—Contouring a topographic ma by field methods accomplished by planetabl surveys on a prepared base or by stadia survey Generally, this operation applies to terrai unsuitable for contouring by photogrammetri methods. Also used in limited areas whe engineering design (drainage) requires 1-foo contours. See also **contour sketching.**

field correction—Adjustments made to fiel measurements, such as angles or distances, t correct for geometric or length discrepancies

field correction copy—A map or tracing prepare in the field, delineating corrections for subs quent reproduction of a map.

field elevation—An elevation taken from the fiel computation of a line of levels.

field inspection—The process of comparing aeri photographs with conditions as they exist o the ground, and of obtaining information t supplement or clarify that which is not readil discernible on the photographs themselve Also called **classification survey.**

field intensity—See **field strength.**

field of view—(JCS) The angle between two ra passing through the perspective center (re nodal point) of a camera lens to the tw opposite sides of the format. Not to be confused with **angle of view.**

field position—A position computed while field work is in progress to determine the acceptability of the observations or to provide a preliminary position for other purposes.

field sheet—The hydrographer's or topographer's work sheet; it presents a graphic display of all surface and subsurface features in the area being surveyed. See also **boat sheet.**

field standardization of tape—The comparison of the length of a tape to be used for survey measurements with the length of a standard tape, to determine the true length of the former.

field stop—The physical element (such as a stop, diaphragm, or lens periphery) of an optical system which limits the field of view covered by the system. See also **aperture stop.**

field strength—For any physical field, the flux density, intensity, or gradient of the field at the point in question. Also called **field intensity.**

figure adjustment—(surveying) The adjustment of a single chain of triangles made to satisfy the requirement that the sum of the angles in each triangle equals 180°, and in the case of a quadrilateral that the sum of the angles equals 360°. An office computation.

figure of the Earth—See **geoid.**

filar micrometer—A device attached to a telescope or microscope, consisting of a wire thread (filament) connected with a screw in such manner that as the screw is turned, the wire moves through a continuous succession of parallel positions, all in the focal plane of the instrument.

film—A film base which is coated with a light sensitive emulsion for use in a camera or printing frame. See also **aerial film; autoscreen film; cartographic film; infrared film; stable-base film; strip film; topographic base film.**

film base—A thin, flexible, transparent sheet of stable plastic material to which a light sensitive emulsion may be applied.

film distortion—The dimensional changes which occur in photographic film with changes in humidity or temperature, or from aging, handling, or other causes.

film mosaic—See **panel base.**

film negative—See **negative,** definition 1.

film positive—See **positive,** definition 1.

film titling—See **titling.**

filter—Any transparent material which, by absorption, selectively modifies the light transmitted through an optical system.

final composite—A composite of the principal color separations made after all corrections have been made.

firing chart—(JCS) Map, photomap, or grid sheet showing the relative horizontal and vertical positions of batteries, base points, base point lines, check points, targets, and other details needed in preparing firing data.

first approximation chart—See **historical chart.**

first of Aries—See **first point of Aries; vernal equinox.**

first-order bench mark—A bench mark connected to the datum (usually mean sea level) by continuous first-order leveling.

first-order level—A leveling instrument which meets the following criteria: the sensitivity of the level bubble vial must be 10″ of arc or less per division of 2 mm; the instrument must be constructed of low expansion metal to minimize the effect of unequal heating; the objective lens must have an effective opening of at least 40 mm and a magnification of 40X.

first-order leveling—Spirit leveling conforming to the specifications of the current "Classification, Standards of Accuracy and General Specifications of Geodetic Control Surveys." Formerly known as precise leveling and leveling of high precision. Recommended for primary National Networks, as a basis for all subordinate elevation determinations, scientific studies such as crustal movement over large regions, extensive engineering projects such as hydroelectric dams. Such leveling generally includes the determination of geopotential values through simultaneous gravity measurements.

first-order traverse—A survey traverse which extends between adjusted positions of other first-order control surveys and conforms to the current specifications of first-order traverse, per "Classification, Standards of Accuracy and General Specifications of Geodetic Control Surveys."

first-order triangulation—First-order triangulation was at one time known as primary triangulation; changed in 1921 to precise triangulation; and in 1925 to first-order triangulation. These surveys conform to the current "Classification, Standards of Accuracy and General Specifications of Geodetic Control Surveys." Recommended for: primary National Networks, as a basis for all subordinate surveys; metropolitan area surveys, where high value is attached to land and its line of communication frontage; and in scientific studies, such as crustal movement and space exploration.

first-order work—The designation given survey work of the highest prescribed order of precision and accuracy. Such surveys were formerly called primary.

first point of Aries—See **vernal equinox.** Also called **first of Aries.**

first point of Cancer—See **summer solstice,** definition 1.

first point of Capricornus—See **winter solstice,** definition 1.

first point of Libra—See **autumnal equinox.**

Fischer ellipsoid of 1960—A reference ellipsoid with two primary uses. In the Mercury datum it has the following approximate dimensions:

semimajor axis—6,378,166.0 meters; the flattening or ellipticity—1/298.3. In the South Asia datum the semimajor axis is 6,378,155.0 meters, and the flattening or ellipticity is 1/298.3.

Fischer level—A dumpy level capable of first-order leveling.

fix—A relatively accurate position determined without reference to any former position, from terrestrial, electronic, or astronomic data. Also, the point thus established.

fixed elevation—An elevation which has been adopted, either as a result of tide observations or previous adjustment of spirit leveling, and which is held at its accepted value in any subsequent adjustment.

fixed position—See **adjusted position.**

fixed-ratio pantograph—See **pantograph.**

fixed-ratio projection printer—A diapositive printer having an optical system in which a lens is placed between the negative and the diapositive plate, the object and image distances being set at nominal values according to the laws of optics, but with freedom of adjustment within narrow limits, to produce diapositives whose scale is at a predetermined ratio to the negative scale.

fixed satellite—See **synchronous satellite.**

fixer network—(JCS) A combination of radio or radar direction-finding installations which, operating in conjunction, are capable of plotting the position relative to the ground of an aircraft in flight.

fixing—The process of rendering a developed photographic image permanent by removing the unaffected light-sensitive material.

flare triangulation—A method of triangulation in which simultaneous observations are made on parachute flares. This method is used for extending triangulation over lines too long to be observed by ordinary methods.

flash apparatus—An auxiliary apparatus used in timing a pendulum during observations for intensity of gravity.

flash plate—See **calibration plate.**

flat—**1.** (lithography) An assembly of photographic negatives or positives on goldenrod paper or vinyl acetate for contact exposure with a sensitized metal pressplate. May contain illustrations as well as text. See also **key flat; layout. 2.** (photography) Lacking in contrast. **3.** (optics) See **optical flat.**

flat model—Any spatial model which is capable of being leveled. See also **warped model.**

flat stock—**1.** Charts or maps which are not folded and kept for filling official and sales orders. **2.** Flat sheets of map paper as opposed to roll paper.

flat tint screen—See **dot screen.**

flattening (of the Earth)—The ratio of the difference between the equatorial and polar radii of the Earth (semimajor and semiminor axes of the spheroid) to its equatorial radius (semimajor axis). Also called **compression; ellipticity of the spheroid.** See also **eccentricity of ellipse; eccentricity of spheroid of revolution; ellipticity of ellipse.**

flexure—**1.** (pendulum) The bending of a swinging pendulum, due to its lack of perfect rigidity. **2.** (pendulum support) The forced movement of a pendulum support caused by the motion of the swinging pendulum.

flicker method—**1.** The alternate projection of corresponding photographic images onto a tracing table platen or projection screen, or into the optical train of a photogrammetric instrument. **2.** (stereoscopy) The alternate blinking of the eyes and mentally comparing the appearance of images in a stereoscopic pair to determine differences between the two photos.

flight altitude—The vertical distance above a given datum, usually mean sea level, of an aircraft in flight.

flight block—An adjustable unit of photographic coverage consisting of overlapping strips of photography. A minimum size block consists of at least three overlapping flight strips.

flight chart—See **route chart,** definition 2.

flight information and air facilities data—Data concerning airfields and seaplane stations and related information required for the operation of aircraft.

flight line—(JCS) In air photographic reconnaissance, the prescribed ground path over which an air vehicle moves during the execution of its photo mission.

flight line spacing—The distance between adjacent tracks in a series of parallel aerial photographic flight strips.

flight map—A map on which are indicated the proposed lines of flight and/or positions of exposure stations. Flight data are plotted on the best available map of the area. Generally used for planning purposes.

flight strip—A succession of overlapping aerial photographs taken along a single course. Also called **strip.**

flipping—The act of superimposing and comparing identical areas of two overlapping vertical photographs as an aid in laying an uncontrolled mosaic.

float gage—Any of the tide or stream gages which permit direct reading of changes of water height by the action of a float, contained within a restricted pipe or channel, attached to a graduated tape or chain.

floating—(cartography) The technique of making minor adjustments of detail in order to maintain their proper relative position.

floating lines—(JCS) In photogrammetry, lines connecting the same two points of detail on each print of a stereo pair, used to determine whether or not the points are intervisible. The lines may be drawn directly onto the prints or superimposed by means of strips of transparent material.

floating mark—(JCS) (photogrammetry) A mark seen as occupying a position in the three-dimensional space formed by the stereoscopic fusion of a pair of photographs and used as a reference mark in examining or measuring the stereoscopic model.

flood control map—A special map, or set of maps, designed for study and planning the control of areas subject to inundation.

flood tide—The portion of the tide cycle between low water and the following high water. Also called **rising tide.**

flowline—The slope extending from the heights along the neatline to the model datum, at an angle no greater than 45°, to preclude forming the plastic sheet at a 90° angle at the neatline of a plastic relief map.

fluorescent map—A map reproduced with fluorescent ink or on fluorescent paper, which enables the user to read the map in darkness under ultraviolet light.

flux-gate magnetometer—An instrument designed to measure the Earth's magnetic field. Also called **saturable reactor.**

fluxmeter—An instrument for measuring the intensity of a magnetic field.

fly leveling—See **flying levels.**

fly-by method—(surveying) A technique of determining approximate elevations where extremely rugged terrain is encountered. The principle is identical to the two-base method except the roving barometers are air transported and read in the aircraft as it passes on a level with the topographic feature whose elevation is required.

flying levels—**1.** A level line run at the close of a working day to check the results of an extended line run in one direction only. Longer sights and fewer setups are used as the purpose is to detect large mistakes. Also called **fly leveling. 2.** Level lines run with the engineer's ordinary leveling equipment but with a distinctly low order of accuracy. Error of closure may be perhaps one foot (or more) times the square root of the distance in miles.

focal length—A general term for the distance between the center, vertex, or rear node of a lens (or the vertex of a mirror) and the point at which the image of an infinitely distant object comes into critical focus. The term must be preceded by an adjective such as "equivalent" or "calibrated" to have a precise meaning. See also **back focal length; calibrated focal length; effective focal length; equivalent focal length; nominal focal length.**

focal plane—(photography) The plane, perpendicular to the axis of the lens in which images of points in the object field of the lens are focused.

focal-plane plate—A glass plate set in the camera so that the surface away from the lens coincides with the focal plane. Its purpose is to position the emulsion of the film in the focal plane when the film is physically pressed into contact with the glass plate. Also called **contact glass; contact plate.**

focal point—See **focus.**

focal range—See **depth of focus.**

focus—The point toward which rays of light converge to form an image after passing through a lens. Also defined as the condition of sharpest imagery. Also called **focal point; principal focus.** See also **hyperfocal distance; sidereal focus.**

folded optics—(photogrammetry) Any optical or lens system containing reflecting components which reduces the physical length of a photographic or sensing system, or changes the path of an optical axis.

folding vernier—A single vernier so constructed and numbered that it may be read in either direction.

foot-meter rod—A stadia rod, marked in feet and tenths on one side, and meters and hundredths on the other side, used to determine distances and elevations in one unit of measurement and to check them by readings in a different unit.

force function—See **potential.**

foresight—**1.** An observation of the distance and direction to the next instrument station. **2.** (transit traverse) A point set ahead to be used for reference when resetting the transit on line or when verifying the alignment. **3.** (leveling) The reading on a rod that is held at a point whose elevation is to be determined. Also called **minus sight.** See also **backsight.**

forestry map—A map prepared principally to show the size, density, kind, and value of trees on a given area.

form lines—(JCS) Lines resembling contours, but representing no actual elevations, which have been sketched from visual observation or from inadequate or unreliable map sources, to show collectively the configuration of the terrain.

forming machine—The equipment for forming, by heat and vacuum, preprinted plastic maps over a mold representing the terrain of the area.

formula for theoretical gravity—A formula expressing gravity on the spheroid of reference in terms of geographic position, it being assumed that the spheroid of reference is a level surface.

forward azimuth—See **azimuth,** definition 1.

forward lap—See **overlap,** definition 1.

found corner—A term adopted by the U.S. Geological Survey to designate an existent corner of the public-land surveys which has been recovered by field investigation.

four-pole chain—See **Gunter's chain.**

four-rod chain—See **Gunter's chain.**

fourth-order traverse—A survey traverse of an

accuracy less than third-order traverse or which fails to meet third-order official standards. In fourth-order traverse, angles are observed with a transit or sextant or are determined graphically, and distances are measured with tape or stadia.

fractional scale—See **representative fraction.**

fractional section—(USPLS) A section containing an area appreciably different from 640 acres, usually as a result of an invasion by a segregated body of water, or by other land which cannot properly be surveyed or disposed of as part of that section. See also **section.**

fractional township—(USPLS) A township containing less than 36 normal sections, usually because of invasion by a segregated body of water, or by other land which cannot properly be surveyed as part of that township, or by closing the public-land surveys on State boundaries, or other limiting lines. Half ranges and half townships are fractional townships by definition. See also **township.**

frame—(JCS) In photography, any single exposure contained within a continuous sequence of photographs.

frame camera—A camera in which an entire frame or format is exposed through a lens that is fixed relative to the focal plane. See also **panoramic camera.**

framework of control—See **survey net.**

free-air anomaly— (JCS) The difference between observed gravity and theoretical gravity which has been computed for latitude and corrected for elevation of the station above or below the geoid, by application of the normal rate of change of gravity for change of elevation, as in free-air. Also called **Faye anomaly.**

free-air correction—Correction factor, usually expressed as milligals per meter, which is applied to observed gravity to reduce the value to sea level. Also called **Faye correction.**

free-swinging pendulum—A pendulum moving wholly under the influence of gravity and an initial momentum imparted to it by mechanical or other means. In gravity work, the initial momentum may be imparted by drawing the pendulum slightly out of plumb and then releasing it.

Fresnel lens—A lens which consists of a thin stepped disc with each step having the curvature of a much thicker lens. A similar design would be the roof of a factory that looks like the teeth of a saw. This pattern, embossed in plastic, is used to distribute image (or light) brightness over a given area.

frilling—The separation, along the edges, of the photographic emulsion from its base.

front element—See **lens element.**

front focal distance—The distance measured from the vertex of the front surface of the lens to the front focal point.

front nodal point—See **nodal point,** definition 1.

front surface mirror—An optical mirror on which the reflecting surface is applied to the front surface of the mirror instead of to the back; i.e., to the first surface of incidence.

fundamental circle—See **primary great circle.**

fundamental star places—The apparent right ascensions and declinations of standard comparison stars obtained by leading observatories and published annually.

fundamental tables, deformation of the geoid and its effect on gravity—Tables giving the deformation of the geoid and its effect on gravity, computed for masses of unit density extending to various distances above and below the surface of the geoid. Fundamental tables serve as the basis for the preparation of special tables corresponding to particular assumptions respecting density, isostasy, etc.

gain-riding technique—A method of controlling the hiran signal rise time by maintaining the amplitude of the receiver output at the same reference level when zeroing and measuring distance.

gal—A unit of acceleration equal to 1 centimeter per second per second, or 1000 milligals, used in measuring the acceleration of gravity.

galactic circle—See **galactic equator.**

galactic equator—A great circle of the celestial sphere, inclined 62° to the celestial equator and coinciding approximately with the center line of the Milky Way, constituting the primary great circle for the galactic system of coordinates. It is everywhere 90° from the galactic poles. Also called **galactic circle.**

galactic latitude—Angular distance north or south of the galactic equator; the arc of a great circle through the galactic poles, between the galactic equator and a point on the celestial sphere, measured northward or southward from the galactic equator through 90° and labeled N or S to indicate the direction of measurement.

galactic longitude—Angular distance east of sidereal hour angle (SHA) 94°.4 along the galactic equator; the arc of the galactic equator or the angle at the galactic pole between the great circle through the intersection of the galactic equator and the celestial equator in Sagittarius (SHA 94°.4) and a great circle through the galactic poles measured eastward from the great circle through SHA 94°.4 through 360°.

galactic pole—On the celestial sphere, either of the two points 90° from the galactic equator.

galactic system of coordinates—An astronomic coordinate system using latitude measured north and south from the galactic equator and longitude measured in the sense of increasing right ascension from 0° to 360°. The system was originally defined such that the pole was at RA = 12^h40^m, Dec = + 28°; however, in 1958 the International Astronomical Union (IAU) introduced the IAU galactic system which defined the pole at RA = 12^h49^m, Dec = 27°24′.

galley proof—A proof from type on a galley before it is made up in pages; also, such proofs.

gamma—**1.** (photography) The tangent of the angle which the straight-line portion of the characteristic curve makes with the log-exposure axis. It indicates the slope of the straight-line portion of the curve and is a measure of the extent of development and the contrast of the photographic material. **2.** (geomagnetism) A small unit of magnetic field intensity sometimes used in describing the Earth's magnetic field. It is defined as being equal to 10^{-5} oersted. See also **nanotesla.**

gap—(JCS) (imagery) Any space where imagery fails to meet minimum coverage requirements. This might be a space not covered by imagery or a space where the minimum specified overlap was not obtained. See also **holiday.**

gauss—A centimeter-gram-second electromagnetic unit of magnetic induction equal to 10^4 tesla. See also **tesla.**

Gauss-Kruger grid—See **transverse Mercator grid.**

gazetteer—An alphabetical list of place names giving feature identification and geographic and/or grid coordinates. See also **Index to Names.**

general chart—A nautical chart intended for offshore coastwise navigation. A general chart is of smaller scale than a coast chart, but of larger scale than a sailing chart.

general map—(JCS) A map of small scale used for general planning purposes.

general precession—The motion of the equinoxes westward along the ecliptic at the rate of about 50″.3 per year. See also **lunisolar precession; planetary precession; precession in declination; precession in right ascension; precession of the equinoxes.**

general-purpose map—A map which provides a broad range of information and which satisfies the needs of a broad range of users.

generalization—Smoothing the character of features without destroying their visible shape. Generalization increases as map scale decreases. See also **character.**

generation—(photography) The preparation of successive positive/negative reproductions from an original negative (first generation). The first positive produced is a second-generation product; the negative made from this positive is a third-generation product, and the next positive or print from that negative is a fourth-generation product. With each successive generation, quality deteriorates.

generic term—That part of a name which describes the kind of feature to which the name is applied, and which has the same meaning in current local usage. For example, the generic term "wan" in "Tokyo-wan" means "bay."

Geoceiver—Trade name for an antenna-receiver capable of receiving signals from the Navy navigation satellites, from which three-dimensional positions can be computed for the antenna location. See also **Doppler navigation,** definition 2.

geocentric—Relative to the Earth as a center; measured from the center of the Earth.

geocentric coordinate system—See **geocentric coordinates.**

geocentric coordinates—(terrestrial) Coordinates that define the position of a point with respect to the center of the Earth. Geocentric coordinates can be either Cartesian (x,y,z) or spherical (geocentric latitude and longitude, and radial distance). Also called **geocentric coordi-**

nate system; geocentric position.

geocentric diameter—The diameter of a celestial body measured in seconds of arc as viewed from the Earth's center.

geocentric geodetic coordinates—Geodetic coordinates referred to a geocentric reference ellipsoid.

geocentric gravitational constant—The product of the Earth's mass and the gravitational constant. This product is known to a far greater precision than either factor.

geocentric horizon—The plane through the center of the Earth, parallel to the topocentric horizon.

geocentric latitude—The angle at the center of the Earth between the plane of the celestial equator and a line to a point on the surface of the Earth. Geocentric latitude is used as an auxiliary latitude in some computations in astronomy, geodesy, and cartography, in which connection it is defined as the angle formed with the major axis of the ellipse (meridional section of the spheroid) by the radius vector from the center of the ellipse to the given point. In astronomic work, geocentric latitude is also called **reduced latitude**, a term that is sometimes applied to **parametric latitude** in geodesy and cartography. The geocentric and isometric latitudes are approximately equal.

geocentric longitude—See **geodetic longitude.**

geocentric parallax—The difference in the apparent direction or position of a celestial body as observed from the center of the Earth and a point on its surface. This varies with the body's altitude and distance from the Earth. Also called **diurnal parallax.**

geocentric position—See **geocentric coordinates.**

geocentric radius vector—The vector from the center of the Earth to the point in question. See also **geocentric coordinates.**

geocentric station position—The location of a station defined in terms of geocentric coordinates.

geocentric zenith—The point where a line from the center of the Earth through a point on its surface meets the celestial sphere.

geodesic—See **geodesic line.**

geodesic line—A line of shortest distance between any two points on any mathematically defined surface. A geodesic line is a line of double curvature, and usually lies between the two normal section lines which the two points determine.If the two terminal points are in nearly the same latitude, the geodesic line may cross one of the normal section lines. It should be noted that, except along the Equator and along the meridians, the geodesic line is not a plane curve and cannot be sighted over directly. However, for conventional triangulation the lengths and directions of geodesic lines differ inappreciably from corresponding pairs of normal section lines. Also called **geodesic; geodetic line.**

geodesy—The science which deals with the determination of the size and figure of the Earth; which determines the external gravitational field of the Earth and, to a limited degree, the internal structure; and which derives three-dimensional positions for points above, on, and below the surface of the Earth.

geodetic and geophysical data reduction—The process of enhancing the value of geodetic and geophysical data by analysis, evaluation, computation, and adjustment. The process includes (1) transforming unadjusted survey data and observations into an adjusted form with reliability statements; (2) establishing basic frameworks of horizontal and vertical control in advance of map, chart, and target materials production, and publishing of trig lists; and (3) analysis, evaluation, and computation of geodetic and geophysical data obtained by surface, airborne, or satellite techniques to establish, extend, connect, and transform datums and to relate datums to the Department of Defense World Geodetic System.

geodetic anomaly—See **anomaly,** definition 2.

geodetic astronomy—The branch of geodesy which utilizes astronomic observations to extract geodetic information.

geodetic azimuth—The angle between the geodetic meridian and the tangent to the geodesic line at the observer, measured in the plane perpendicular to the ellipsoid normal of the observer; preferably clockwise from north. Although older surveys, particularly by the Coast and Geodetic Survey (now National Geodetic Survey), used south, the Department of Defense now uses north.

geodetic azimuth mark—A marked point established in connection with a triangulation (or traverse) station to provide a starting azimuth for dependent surveys.

geodetic control—A system of horizontal and/or vertical control stations that have been established and adjusted by geodetic methods and in which the shape and size of the Earth (geoid) have been considered in position computations.

geodetic control data—Information concerning the precise horizontal and vertical geodetic location of points on the surface of the Earth and celestial bodies, including points obtained by photogrammetric techniques.

geodetic coordinates—The quantities of latitude, longitude, and height (ellipsoid), which define the position of a point on the surface of the Earth with respect to the reference spheroid. Also imprecisely called geographic coordinates.

geodetic data sheet—See **control data card.**

geodetic datum—See **datum,** definition 2.

geodetic equator—The line of zero geodetic latitude; the great circle described by the semimajor axis of the reference ellipsoid as it is rotated about the minor axis. See also **astro-**

nomic equator.

geodetic height—See **ellipsoidal height.**

geodetic latitude—The angle which the normal at a point on the reference spheroid makes with the plane of the geodetic equator. Geodetic latitudes are reckoned from the Equator, but in the horizontal control survey of the United States they are computed from the latitude of station Meades Ranch as prescribed in the North American datum of 1927. The new North American datum of 1983 will be Earth-mass centered. A geodetic latitude differs from the corresponding astronomic latitude by the amount of the meridian component of the deflection of the vertical. Also called **topographical latitude.**

geodetic leveling—Spirit leveling of a high order of accuracy, usually extended over large areas, to furnish accurate vertical control as a basis for the control in the vertical dimension for all surveying and mapping operations. Spirit leveling follows the geoid and its associated level surfaces which are irregular, rather than any mathematically determined spheroid or ellipsoid and associated regular level surfaces.

geodetic line—See **geodesic line.**

geodetic longitude—The angle between the plane of the geodetic meridian and the plane of an initial meridian, arbitrarily chosen. A geodetic longitude can be measured by the angle at the pole of rotation of the reference spheroid between the local and initial meridians, or by the arc of the geodetic equator intercepted by those meridians. In the United States, geodetic longitudes are numbered from the meridian of Greenwich, but are computed from the meridian of station Meades Ranch as prescribed in the North American datum of 1927. The new North American datum of 1983 will be Earth-mass centered. A geodetic longitude differs from the corresponding astronomic longitude by the amount of the prime vertical component of the local deflection of the vertical divided by the cosine of the latitude. Also called **geocentric longitude.**

geodetic meridian—A line on a reference ellipsoid which has the same geodetic longitude at every point. Also called **geographic meridian.**

geodetic meridian plane—A plane that contains the normal to the reference ellipsoid at a given point and the rotation axis of the reference ellipsoid.

geodetic parallel—A line on the reference spheroid which has the same geodetic latitude at every point. A geodetic parallel, other than the Equator, is not a geodesic (geodetic) line. In form, it is a small circle whose plane is parallel with the plane of the geodetic equator.

geodetic position—A position of a point on the surface of the Earth expressed in terms of geodetic latitude, geodetic longitude, and geodetic height. A geodetic position implies an adopted geodetic datum.

Geodetic Reference System 1967 (GRS 67)—A reference system defined by the four constants: semimajor axis—6,378,160 m; geocentric gravitational constant of the Earth (including the atmosphere)—398603×10^9 m^3 s^{-2}; dynamical form factor of the Earth—0.0010827; angular velocity—$7.2921151467 \times 10^{-5}$ s^{-1}. Derived parameters for this system (adopted by the International Association of Geodesy in 1971) are: semiminor axis—6,356,774.5161 m; flattening—1/298.247167427; equatorial gravity—978.03184558 gal.

geodetic satellite—Any satellite whose orbit and payload render it useful for geodetic purposes.

geodetic stellar camera—A precision terrestrial camera, usually employing glass plates, used to photograph elevated illuminated objects against a star background.

geodetic survey—A survey in which the figure and size of the Earth is considered. It is applicable for large areas and long lines and is used for the precise location of basic points suitable for controlling other surveys.

geodetic zenith—The point where the normal (to the reference spheroid) extended upward, meets the celestial sphere.

Geodimeter—A trade name for an instrument that measures distance by precise electronic phase-comparison of modulated lightwaves which travel to a reflector and return.

geoelectric survey—A survey to determine the electricity potential of the Earth rocks, or resistivity.

geographic (geographical)—Signifying basic relationship to the Earth considered as a globe-shaped body. The term **geographic** is applied alike to data based on the geoid and on other spheroids.

geographic coordinates—(JCS) The quantities of latitude and longitude which define the position of a point on the surface of the Earth with respect to the reference spheroid. Also called **astronomic coordinates; gravimetric coordinates; terrestrial coordinates.** See also **coordinates; geodetic coordinates.**

geographic latitude—A general term, applying alike to astronomic latitudes and geodetic latitudes.

geographic limits—The lines having latitude and longitude values bounding the area of a map or chart; that area exclusive of overlap areas. See also **neatlines.**

geographic location—See **geographic position.**

geographic longitude—A general term, applying alike to astronomic and to geodetic longitudes.

geographic meridian—A general term, applying alike to an astronomic or geodetic meridian.

geographic parallel—A general term, applying alike to an astronomic parallel or a geodetic parallel.

geographic position—The position of a point on

the surface of the Earth expressed in terms of latitude and longitude, either geodetic or astronomic.

geographic survey—A general term, not susceptible of defined limitation, covering a wide range of surveys lying between and merging into exploratory surveys on the one hand and basic topographic surveys on the other. Geographic surveys usually cover large areas, are based on coordinated control, and are used to record physical and statistical characteristics of the area surveyed.

geographic vertical—See **vertical.**

geographical area classification system—A logical and orderly geographical division of the world using numbers, letters, and combinations of the same for the designation of areas and subareas.

geographical exploration traverse—A route followed across some parts of the Earth, approximate positions along which are determined by surveying or navigational methods.

geographical mile—The length of 1 minute of arc of the Equator, or 6,087.08 feet (on the Clarke spheroid of 1866).

geographical pole—Either of the two points of intersection of the surface of the Earth with its axis, where all meridians meet.

geoid—The equipotential surface in the gravity field of the Earth which coincides with the undisturbed mean sea level extended continuously through the continents. The direction of gravity is perpendicular to the geoid at every point. The geoid is the surface of reference for astronomic observations and for geodetic leveling. See also **compensated geoid; equilibrium theory; equipotential surface; geoidal horizon; gravimetric geoid; isostatic geoid; reference spheroid.**

geoidal contour—A line on the surface of the geoid of constant elevation with reference to the surface of the spheroid of reference. Geoidal contours depend on the surface of reference as well as on the shape of the geoid. The same geoid referred to different surfaces of reference will give different sets of geoidal contours.

geoidal height—The distance of the geoid above (positive) or below (negative) the mathematical reference spheroid. Also called **geoid separation; undulation of the geoid.** See also **astrogeodetic undulations.**

geoidal height profile—See **astrogeodetic leveling.**

geoidal horizon—That circle of the celestial sphere formed by the intersection of the celestial sphere and a plane tangent to the sea-level surface of the Earth at the zenith-nadir line.

geoidal separation—See **geoidal height.**

geokinetics—Local and global motion of the Earth or sea, its measurement, isolation from, and effect upon precision equipment and measuring instruments. Usually applied to the design and test of inertial instruments and systems and stable platforms.

geologic survey—A survey or investigation of the Earth, of the physical changes which the Earth's crust has undergone or is undergoing, and of the causes producing those changes.

geological map—A map showing the structure and composition of the Earth's crust.

geomagnetic coordinates—A system of spherical coordinates based on the best fit of a centered dipole to the actual magnetic field of the Earth.

geomagnetic equator—The terrestrial great circle everywhere 90° from the geomagnetic poles. **Geomagnetic equator** should not be confused with **magnetic equator**, the line connecting all points of zero magnetic dip.

geomagnetic latitude—Angular distance from the geomagnetic equator, measured northward or southward through 90° and labeled N or S to indicate the direction of measurement. **Geomagnetic latitude** should not be confused with **magnetic latitude.** See also **dip.**

geomagnetic meridian—The meridional lines of a geomagnetic coordinate system. Not to be confused with **magnetic meridian.**

geomagnetic pole—Either of two antipodal points marking the intersection of the Earth's surface with the extended axis of a powerful bar magnet assumed to be located at the center of the Earth and approximating the source of the actual magnetic field of the Earth. The expression **geomagnetic pole** should not be confused with **magnetic pole,** which relates to the actual magnetic field of the Earth.

geomagnetism—**1.** The magnetic phenomenon, collectively considered, exhibited by the Earth and its atmosphere, and by extension the magnetic phenomena in interplanetary space. **2.** The study of the magnetic field of the Earth. Also called **terrestrial magnetism.**

geometric latitude—See **parametric latitude.**

geometric map projection—See **perspective map projection.**

geometrical dip—The vertical angle, at the eye of an observer, between the horizontal and a straight line tangent to the surface of the Earth. It is larger than dip by the amount of terrestrial refraction.

geometrical horizon—Originally, the celestial horizon; now more commonly the intersection of the celestial sphere and an infinite number of straight lines tangent to the Earth's surface, and radiating from the eye of the observer. If there were no terrestrial refraction, geometrical and visible horizons would coincide.

geop—An equipotential surface in the gravity field of the Earth. Also called **geopotential surface.**

geophysics—The science of the Earth with respect to its structure, composition, and development. Geophysics is a branch of experimental physics dealing with the Earth, including its atmosphere and hydrosphere. It includes the sciences of dynamical geology and physical geography, and makes use of geodesy, geology, seismology, meteorology, oceanography, magnetism, and other Earth sciences in collecting and interpreting Earth data.

geopotential—The gravity potential of the actual Earth. The sum of the gravitational (attraction) potential and the potential of the centrifugal force. A function describing the variation of the geopotential in space. The function whose partial derivative in any direction gives the gravity component in that direction.

geopotential number—The difference between the geopotential on the geoid and the geopotential at a point.

geopotential surface—See **geop.**

GEOREF—(JCS) A worldwide position reference system that may be applied to any map or chart graduated in latitude and longitude [with Greenwich as prime meridian] regardless of projection. It is a method of expressing latitude and longitude in a form suitable for rapid reporting and plotting. This term is derived from the words "The World Geographic Reference System."

geosphere—The solid and liquid portions of the Earth; the lithosphere plus the hydrosphere.

giant planets—See **major planets.**

gisement—See **grid declination.**

Global Navigation Chart (GNC)—A 1:5,000,000 scale series of multicolored charts designed for general planning purposes for operations involving long distances or large areas of in-flight navigation in long range, high altitude, high speed aircraft.

Global Positioning System (GPS)—See **NAVSTAR Global Positioning System (GPS).**

globular map projection—A map projection representing a hemisphere, on which the Equator and a central geographic meridian are represented by straight lines intersecting at right angles; these lines are divided into equal parts. All meridians, except the central one, are represented by circular arcs connecting points of equal division on the Equator with the poles. Excepting the Equator, the parallels are circular arcs dividing the central and extreme outer meridians into equal parts. The extreme outer meridian limits the projection and is a full circle.

gnomonic chart—A chart on the gnomonic projection. Also called **great circle chart.**

gnomonic map projection—A perspective map projection on a plane tangent to the surface of a sphere having the point of projection at the center of the sphere. The projection is neither conformal nor equal-area. It is the only projection on which great circles on the sphere are represented as straight lines.

goldenrod paper—A paper, usually a shade of yellow or red, for blocking-out nonprinting areas of negatives or film layouts. Also called **masking paper.**

goniometer—An instrument for measuring angles. See also **photogoniometer.**

Goode's Interrupted Homolosine projection—An equal-area projection, based on the Mollweide and sinusoidal projections, using the sinusoidal from the Equator to 40°N and 40°S, and the Mollweide in higher latitudes. The oceans are "interrupted" to allow the continents to be recentered on several meridians, so as to attain good overall shape. It is used widely for maps of economic distributions.

gore—**1.** (surveying) An irregularly shaped tract of land, generally triangular, left between two adjoining surveyed tracts, because of inaccuracies in the boundary surveys or as a remnant of a systematic survey. **2.** (globe) A lune-shaped map which may be fitted to the surface of a globe with a negligible amount of distortion.

gradation—The range of tones from the brightest highlights to the deepest shadows.

grade—The rate of slope or degree of inclination. See also **gradient.**

grade correction—(land surveying) A correction applied to a distance measured on a slope to reduce it to a horizontal distance between the vertical lines through its end points. Also called **correction for inclination of tape; inclination correction; slope correction.**

gradient—**1.** A rate of rise or fall of a quantity against horizontal distance expressed as a ratio, decimal, fraction, percentage, or the tangent of the angle of inclination. Also called **percent of slope; slope. 2.** The rate of increase or decrease of one quantity with respect to another.

gradient speed—The speed of a photographic material determined on the basis of the exposure corresponding to a particular gradient of the characteristic curve.

gradient tints—See **hypsometric tinting.**

gradienter—An attachment to an engineer's transit with which an angle of inclination is measured in terms of the tangent of the angle rather than in degrees and minutes. It may be used as a telemeter in measuring horizontal distances.

gradiometer—An instrument used to measure gravity gradients.

graduation error—Inaccuracy in the graduations of the scale of an instrument.

Graf sea gravimeter—A balance-type gravity meter designed for ocean surveys which consists of a mass at the end of a horizontal arm that is supported by a torsion spring rotational axis.

grain—**1.** (photography) One of the discrete silver particles resulting from the development of an exposed light-sensitive material. The random distribution of these particles in an area of uniform exposure gives rise to the appearance known as "graininess." **2.** (paper) See **grain direction.** **3.** (lithography) See **grained surface.**

grain direction—The alignment of paper fibers parallel to the movement on the paper machine during manufacture.

grained surface—The roughened or irregular surface of an offset printing plate.

graining—The mechanical roughening or grinding of an abrasive into the surface of a metal pressplate to increase the surface area and improve the water receptiveness of the surface.

granularity—The graininess of a developed photographic image, evident particularly on enlargements, that is due either to agglomerations of developed grains or to an overlapping pattern of grains.

graphic—(JCS) Any and all products of the cartographic and photogrammetric art. A graphic may be either a map, chart, mosaic, or even a film strip that was produced using cartographic techniques.

graphic scale—(JCS) A graduated line by means of which distances on the map, chart, or photograph may be measured in terms of ground distances. Also called **bar scale.**

graphical radial triangulation—A radial triangulation performed by other than analytical means. A radial triangulation is assumed to be made with principal points as radial centers unless the definitive term designates otherwise (as, for example, nadir-point triangulation or nadir-point plot and isocenter triangulation or isocenter plot, and nadir-point slotted-templet plot, etc.).

graphical rectification—Any rectification technique employing a graphic method for determining the solution as contrasted with mechanical techniques. See also **paper-strip method.**

graticule—**1.** A network of lines representing the Earth's parallels of latitude and meridians of longitude. See also **fictitious graticule.** **2.** A scale at the focal plane of an optical instrument to aid in the measurement of objects. See also **reticle.**

Gravatt level—A dumpy level with the spirit level mounted on top of a short telescope tube having a large object glass. Later made with wyes.

Gravatt leveling rod—A speaking rod, marked with rectangles each 0.01 foot high, the rectangles at the tenths of foot being longer, and those at the half-tenths being identified by dots.

graver—See **scriber.**

gravimeter (gravity meter)—An accelerometer designed to measure relative differences in the acceleration due to gravity at different locations.

gravimetric coordinates—See **astronomic coordinates.**

gravimetric datum orientation—Adjustment of the ellipsoid of reference for a particular geodetic datum so that the differences between the gravimetric and astrogeodetic deflection components and geoidal undulations are minimized.

gravimetric deflection—A deflection of the vertical determined by methods of gravimetric geodesy.

gravimetric geodesy—The science that utilizes measurements and characteristics of the Earth's gravity field as well as theories regarding this field to deduce the shape of the Earth and in combination with arc measurements, the Earth's size. Also called **physical geodesy.**

gravimetric geoid—An approximation to the geoid as determined from gravity observation.

gravimetric map—A map on which contour lines are used to represent points at which the acceleration of gravity is equal.

gravimetric survey—A survey made to determine the acceleration of gravity at various places on the Earth's surface.

gravimetric undulations—Separations between a gravimetrically determined geoid and a reference ellipsoid of specified flattening.

gravitation—The mutual interaction of two masses producing a force between them acting along the line joining their centers of mass. The force is proportional to the product of the two masses divided by the square of the distance between the two centers of mass.

gravitational constant—See **constant of gravitation.**

gravitational disturbance—See **gravity disturbance.**

gravitational flattening—The ratio of the difference between the polar and equatorial normal gravities to the equatorial normal gravity. Also called **gravity flattening.**

gravitational gradient—The change in the gravity per unit distance.

gravitational harmonics—The spherical harmonics used in approximating the gravitational field of the Earth. See also **gravity field of the Earth; sectorial harmonics; spherical harmonics; tesseral harmonics; zonal harmonics.**

gravitational perturbations—Perturbations caused by body forces due to nonspherical terrestrial effects, lunisolar effect, tides, and the effect of relativity.

gravitational potential—**1.** The potential associated with the force of gravitation arising from the attraction between mass points, e.g., the Earth's center and a particle in space. **2.** At any point, the work needed to remove an object from that point to infinity.

gravity—Viewed from a frame of reference fixed in the Earth, acceleration imparted by the Earth to a mass which is rotating with the

Earth. Since the Earth is rotating, the acceleration observed as gravity is the resultant of the acceleration of gravitation and the centrifugal acceleration arising from this rotation and the use of an earthbound rotating frame of reference. It is directed normal to sea level and to its geopotential surfaces. See also **absolute gravity; Clairaut's theorem; center of gravity; constant of gravitation; direction of the force of gravity; equatorial gravity value; equipotential surface; formula for theoretical gravity; gravitation; Hayford deflection templets; Hayford effect; Helmert's gravity formula of 1901; Helmert's gravity formula of 1915; international gravity formula; intensity of gravity; isostasy; longitude term gravity formula; normal gravity; observed gravity; reduced gravity; regional gravity; relative gravity; residual gravity; resolution; resolution limit; standard gravity; subgravity; theoretical gravity; virtual gravity.**

gravity anomaly—The difference between the observed gravity value properly reduced to sea level, and the theoretical gravity obtained from gravity formula. Also called **observed gravity anomaly.** See also **Bouguer anomaly; free-air anomaly; gravity disturbance; Hayford gravity anomalies/Hayford anomalies; isostatic anomaly.**

gravity anomaly map—A map showing the positions and magnitudes of gravity anomalies. Also, a map on which contour lines are used to represent points at which the gravity anomalies are equal.

gravity corer—Any type of corer that achieves bottom penetration solely as a result of gravity.

gravity data—Information concerning that acceleration which attracts bodies and is expressed as observations or in the form of gravity anomaly charts or spherical harmonics for spatial representation of the Earth and other celestial bodies.

gravity disturbance—The difference between the observed gravity and the normal gravity at the same point (the vertical gradient of the disturbing potential) as opposed to **gravity anomaly** which uses corresponding points on two different surfaces. Because the centrifugal force is the same when both are taken at the same point, it can also be called **gravitational disturbance.**

gravity field of the Earth—The field of force arising from a combination of the mass attraction and rotation of the Earth. The field is normally expressed in terms of point values, mean area values, and/or series expansion for the potential of the field.

gravity flattening—See **gravitational flattening.**

gravity instrument—A device for measuring the acceleration due to gravity (absolute) or gravity differences between two or more points (relative). See also **astatized gravimeter; Brown gravity apparatus; dynamic gravity meter; Graf sea gravimeter; gravimeter; La Coste-Romberg gravimeter; stable gravimeter; stable-type gravimeter; static gravity meter; torsion balance; unstable-type gravimeter.**

gravity network—A network of gravity stations.

gravity reduction—A combination of gravity corrections to obtain reduced gravity on the geoid. See also **Bouguer correction; free-air correction; isostatic correction; terrain correction.**

gravity reference stations—Stations which serve as reference values for a gravity survey, i.e., with respect to which the differences at the other stations are determined in a relative survey. The absolute value of gravity may or may not be known at the reference stations.

gravity station—A station at which observations are made to determine the value of gravity.

gray scale—See **step wedge.**

great circle—(JCS) A circle on the surface of the Earth, the plane of which passes through the center of the Earth. Also called **orthodrome.**

great circle bearing—The initial direction of a great circle through two terrestrial points, expressed as angular distance from a reference direction. It is usually measured from 000° at the reference direction clockwise to 360°.

great circle chart—A chart on which a great circle appears as a straight line; a chart on the gnomonic projection.

great circle direction—Horizontal direction of a great circle, expressed as angular distance from a reference direction.

great circle distance—The length of the shorter arc of the great circle joining two points. It is usually expressed in nautical miles.

great circle line—In land surveying, the line of intersection of the surface of the Earth and the plane of a great circle of the celestial sphere.

great circle route—(JCS) The route which follows the shortest arc of a great circle between two points.

great elliptic arc—An arc defined by a plane which contains the two points and the center of the reference spheroid.

great year—The period of one complete cycle of the equinoxes around the ecliptic, about 25,800 years. Also called **platonic year.**

greatest elongation—The maximum angular distance of a body of the solar system from the Sun, as observed from the Earth. The direction of the body east or west of the Sun is usually specified, as the greatest elongation west.

Greenwich apparent time (GAT)—Local apparent time at the Greenwich meridian.

Greenwich civil time (GCT)—See **universal time.**

Greenwich hour angle (GHA)—Angular distance west of the Greenwich celestial meridian; the arc of the celestial equator, or the angle at the celestial pole, between the upper branch of the

Greenwich celestial meridian and the hour circle of a point on the celestial sphere, measured westward from the Greenwich celestial meridian through 360°; local hour angle at the Greenwich meridian.

Greenwich interval—An interval based on the Moon's transit of the Greenwich celestial meridian, as distinguished from a local interval based on the Moon's transit of the local celestial meridian.

Greenwich lunar time—Local lunar time at the Greenwich meridian; the arc of the celestial equator or the angle, at the celestial pole, between the lower branch of the Greenwich celestial meridian and the hour circle of the Moon, measured westward from the lower branch of the Greenwich celestial meridian through 24 hours; Greenwich hour angle of the Moon, expressed in time units, plus 12 hours.

Greenwich mean time (GMT)—(JCS) Mean solar time at the meridian of Greenwich, England, used as a basis for standard time throughout the world. Normally expressed in four numerals 0001 through 2400. Also called **Greenwich civil time; universal time; z-time; zulu time.**

Greenwich meridian—The meridian through Greenwich, England, serving as the reference for Greenwich time, in contrast with local meridians. It is accepted almost universally as the prime meridian, or the origin of measurement of longitude.

Greenwich sidereal date—The number of mean sidereal days that have elapsed on the Greenwich meridian since the beginning of the sidereal day that was in progress at Greenwich mean noon on January 1, 4713 B.C. See also **Greenwich sidereal day number.**

Greenwich sidereal day number—The integral part of the Greenwich sidereal date. It is a means of numbering consecutively successive sidereal days beginning at the instants of upper transit of the mean vernal equinox over the Greenwich meridian. See also **Greenwich sidereal date.**

Greenwich sidereal time (GST)—Local sidereal time at the Greenwich meridian. The arc of the celestial equator, or the angle at the celestial pole, between the upper branch of the Greenwich celestial meridian and the hour circle of the vernal equinox, measured westward from the upper branch of the Greenwich celestial meridian through 24 hours; Greenwich hour angle of the vernal equinox, expressed in time units.

Greenwich time—Time based upon the Greenwich meridian as reference, as contrasted with that based upon a local or zone meridian.

grid—(JCS) **1.** Two sets of parallel lines intersecting at right angles and forming squares; the grid that is superimposed on maps, charts, and other similar representations of the Earth's surface in an accurate and consistent manner to permit identification of ground locations with respect to other locations and the computation of direction and distance to other points. Also called **reference grid. 2.** A term used in giving the location of a geographic point by grid coordinates. See also **arbitrary grid; atlas grid; British grid reference system; GEOREF; Lambert grid; major grid; military grid; military grid reference system; National grid; overlapping grid; parallactic grid; perspective grid; point-designation grid; polar grid; secondary grid; tangent plane grid system; transverse Mercator grid; Universal Polar Stereographic (UPS) grid; Universal Transverse Mercator (UTM) grid; world polyconic grid.**

grid amplitude—Amplitude relative to grid east or west. See also **amplitude.**

grid azimuth—The angle in the plane of projection measured clockwise between a straight line and the central meridian of a plane-rectangular coordinate system.

grid bearing—(JCS) Bearing measured from grid north.

grid computation—The determination, from a set of tables derived from formulas, of the true shape and dimensions of a grid, for the purpose of constructing such a grid. The grid is mathematically coordinated with its related map projection; they are usually computed concurrently.

grid convergence—(JCS) The horizontal angle at a place between true north and grid north. It is proportional to the longitude difference between the place and the central meridian.

grid coordinates—(JCS) Numbers and letters of a coordinate system which designate a point on a gridded map, photograph, or chart.

grid coordinate system—(JCS) A plane-rectangular coordinate system usually based on, and mathematically adjusted to, a map projection in order that geographic positions (latitudes and longitudes) may be readily transformed into plane coordinates and the computations relating to them may be made by the ordinary methods of plane surveying.

grid declination—The angular difference in direction between grid north and true north. It is measured east or west from true north. Also called **declination of grid north; gisement.**

grid direction—Horizontal direction expressed as angular distance from grid north.

grid distance—See **grid length.**

grid equator—A line perpendicular to a prime grid meridian, at the origin.

grid interval—(JCS) The distance represented between the lines of a grid.

grid inverse—The computation of grid length and grid azimuths from grid coordinates.

grid junctions—Those lines delineating the joining of two or more grid systems on a map or chart

grid latitude—Angular distance from a grid equator. See also **fictitious latitude.**

grid length—The distance between two points obtained by computation from grid coordinates of the points. It differs from the geodetic length by the amount of a small correction based on the scale factor for the line. Also called **grid distance.**

grid line—One of the lines of a grid.

grid longitude—Angular distance between a prime grid meridian and any given grid meridian. See also **fictitious longitude.**

grid magnetic angle—(JCS) Angular difference in direction between grid north and magnetic north. It is measured east or west from grid north. Also called **grivation; grid variation.**

grid meridian—One of the grid lines extending in a grid north-south direction. The reference grid meridian is called prime grid meridian. In polar regions the prime grid meridian is usually the 180°–0° geographic meridian. See also **fictitious meridian.**

grid method—(photogrammetry) A method of plotting detail from oblique photographs by superimposing a perspective of a map grid on a photograph and transferring the detail by eye, that is, by using the corresponding lines of the map grid and its perspective as placement guides. See also **perspective grid.**

grid north—(JCS) The northerly or zero direction indicated by the grid datum of directional reference.

grid number—The numerical value of a grid line indicating the distance of that line from the false origin of the grid. See also **grid coordinates.**

grid origin—The point, usually near the center of a grid zone, where a parallel intersects a north-south grid line coincident to a meridian. See also **false origin.**

grid parallel—A line parallel to a grid equator, connecting all points of equal grid latitude. See also **fictitious parallel.**

grid plate—**1.** (cartography) See **color-separation drawing. 2.** (photogrammetry) See **reseau.**

grid prime vertical—The vertical through the grid east and west points of the horizon.

grid rhumb line—A line making the same oblique angle with all grid meridians. Grid parallels and meridians may be considered special cases of the grid rhumb line. See also **fictitious rhumb line.**

grid ticks—(JCS) Small marks on the neatline of a map or chart indicating additional grid reference systems included on that sheet. Grid ticks are sometimes shown on the interior grid lines of some maps for ease of referencing.

grid variation—See **grid magnetic angle.**

grid zone—An arbitrary division of the Earth's surface designated for identification without reference to latitude or longitude.

gridded oblique—An oblique aerial photograph printed with a superimposed grid to assist in the identification of a particular area within the photograph; used chiefly for artillery spotting.

grivation—See **grid magnetic angle.**

gross error—The result of carelessness or a mistake. May be detected through repetition of the measurements.

gross model—The total overlap area of a pair of aerial photographs. See also **neat model.**

ground camera—See **terrestrial camera.**

ground control—(JCS) A system of accurate measurements used to determine the distances and directions or differences in elevation between points on the Earth. See also **field control.**

ground control point—See **control station.**

ground distance— The great circle distance between two ground positions, as contrasted with **slant range,** the straight-line distance between two points. Also called **ground range.**

ground gained forward (GGF)—(aerial photography) The net gain per photograph in the direction of flight for a specified overlap. The GGF is used to compute the number of exposures in a strip of aerial photography.

ground gained sideways (GGS)—(aerial photography) The net lateral gain per flight for a specified sidelap. The GGS is used to compute the number of flight lines for an area to be photographed.

ground nadir—(JCS) The point on the ground vertically beneath the perspective center of the camera lens. On a true vertical photograph this coincides with the principal point. Also called **ground plumb point.**

ground parallel—The intersection of the plane of the photograph with the plane of reference of the ground. See also **axis of homology.**

ground photogrammetry—See **terrestrial photogrammetry.**

ground photograph—See **terrestrial photograph.**

ground plane—The horizontal plane passing through the ground nadir of a camera station.

ground plumb point—See **ground nadir.**

ground pyramid—A component of an analytical method for determining the precise degree of photographic tilt, representing a specific spatial configuration from three ground control points (forming a triangle) on the ground to the exposure station of the photograph containing the identical points. When used with the photo pyramid, the ground pyramid permits the exact analytical determination of tilt in the photograph. See also **photo pyramid.**

ground range—See **ground distance.**

ground resolution—The minimum distance which can be detected between two adjacent features, or the minimum size of a feature expressed in size of objects or distances on the ground.

ground return—(JCS) The reflection from the terrain as displayed and/or recorded as an image.

ground-space coordinate system—A scheme by which positions of triangulation stations, control points, and other ground features are related by distance and azimuth or by *x*- and *y*-coordinates.

ground station—A monumented station, established by field survey methods, which is used as a base for ground station equipment for the procurement of shoran- or shiran-controlled photography or control data.

ground survey—A survey made by ground methods, as distinguished from an aerial survey. A ground survey may or may not include the use of photographs.

ground swing—An error-causing condition in electronic distance measuring which is brought about by the reflection of the microwave beam from the ground or water surface. The reflected beam mixes with the direct beam at the receiving antenna, thereby changing the phase of the direct beam and causing an error in the distance measured. By varying the carrier frequency, the error becomes cyclic, making possible mean instrument readings that are substantially accurate.

ground trace—See **ground parallel.**

guard stake—(surveying) A stake driven near a hub, usually sloped with the top of the guard stake over the hub. The guard stake protects, and its markings identify, the hub.

guide—A drafting or scribing surface bearing a map image to be traced by drafting or scribing for reproduction. Also called **color-separation guide; drafting guide; scribing guide.**

guide meridian—(USPLS) An auxiliary governing line projected north along an astronomic meridian, from points established on the base line or a standard parallel, usually at intervals of 24 miles east or west of the principal meridian, on which township, section, and quarter-section corners are established. See also **auxiliary guide meridian; principal meridian.**

Gunter's chain—A measuring device used in land surveying, composed of 100 metal links fastened together with rings, the length of the chain being 66 feet. Also called **four-pole chain; four-rod chain.**

gyro-azimuth theodolite—See **gyro theodolite.**

gyro-magnetic compass—(JCS) A directional gyroscope whose aximuth scale datum is maintained in alignment with the magnetic meridian by applying precession torques derived from a magnetic detector unit.

gyro-meridian indicating instrument—See **gyro theodolite.**

gyro theodolite—A theodolite with a gyrocompass attached or built in, whereby a true azimuth reference can be established in any weather, day or night, without the aid of stars, landmarks, or other visible stations. The azimuth obtained from the gyro or inertial theodolite is essentially the astronomic azimuth at the point of observation. This azimuth will differ from the corresponding geodetic azimuth by the amount of the Laplace correction.

gyrocompass—A compass which functions by virtue of the couples generated in a rotor when the latter's axis is displaced from parallelism with that of the Earth. A gyrocompass is independent of magnetism and will automatically align itself in the celestial meridian. However, it requires a steady source of motive power and is subject to dynamic error under certain conditions. Certain aircraft compasses also use gyroscopes to gain stability, while relying basically on the magnetic meridian; these are to be distinguished from the true gyrocompass.

gyroscope—A device consisting of a spinning rotor and associated supporting readouts which makes use of Newton's Law of Rotation to give an indication of the angular velocity of the instrument's case with respect to an inertial reference frame. This instrument is used as the basic sensor in many direction-seeking, direction-keeping, and attitude stabilization systems.

gyroscopic stabilization—Equilibrium in the attitude and/or course of a ship or airborne vehicle maintained by the use of gyroscopes. Also, the maintenance (by the use of gyroscopes) of a camera in a desired attitude within an airborne vehicle.

H

H and D curve—See **characteristic curve.**

hachuring—(JCS) A method of representing relief upon a map or chart by shading in short disconnected lines drawn in the direction of the slopes.

hack chronometer—A chronometer used for visual reference, and not usually for record purposes.

halation—(photography) A spreading of a photographic image beyond its proper boundaries, due especially to reflection from the side of the film or plate support opposite to that on which the emulsion is coated. Particularly noticeable in photographs of bright objects against a darker background.

half mark—See **index mark.**

half model—The stereoscopic model formed by the overlap of two adjacent right-hand or left-hand exposures of convergent photographs.

half section—(USPLS) Any two quarter sections within a section which have a common boundary; usually identified as the north half, south half, east half, or west half of a particular section.

half-tide level—See **mean tide level.**

halftone—(JCS) Any photomechanical printing surface or the impression therefrom in which detail and tone values are represented by a series of evenly spaced dots of varying size and shape, varying in direct proportion to the intensity of the tones they represent. See also **middletone.**

halftone screen—(JCS) A series of regularly spaced opaque lines on glass, crossing at right angles, producing transparent apertures between intersections. Used in a process camera to break up a solid or continuous-tone image into a pattern of small dots. Also called **crossline glass screen.** See also **contact screen; halftone.**

Hammer projection—A variation, by E. Hammer in 1892, on the zenithal equal area (Lambert) projection, made by doubling the horizontal distances along each parallel from the central meridian. This transforms the circular shape of the Lambert into an ellipse, similar in appearance to a Mollweide projection, but with all parallels curved except the Equator, which is a straight line.

hand level—A hand-held instrument for approximate leveling. It consists of a sighting tube with a split field of view; a horizontal crosshair in one-half of the field bisects the image of a spirit level in the other half when the instrument is held level. See also **Abney level.**

hand proof—In offset lithography, a proof of a plate made on a hand-proof press where operations are manual for inking, dampening, and taking the impression.

hand templet—A templet made by tracing the radials from the photograph onto a transparent plastic medium. Hand templets are laid out and adjusted by hand to form the radial triangulation.

hand-templet plot—See **hand-templet triangulation.**

hand-templet triangulation—A graphical radial triangulation using any form of hand templet. Also called **hand-templet plot.**

hanging level—A spirit level so mounted that, when in use, its level tube is lower in elevation than its points of support.

harbor chart—A nautical chart intended for navigation and anchorage in harbors and smaller waterways.

harmonic—A sinusoidal quantity having a frequency that is an integral multiple of the frequency of a periodic quantity to which it is related. See also **compound harmonic motion; gravitational harmonics; sectorial harmonics; simple harmonic motion; spherical harmonics; tesseral harmonics; zonal harmonics.**

harmonic coefficients—The coefficients of trigonometric terms of an infinite series used to approximate an irregular closed surface. See also **spherical harmonics.**

harmonic component—Any of the simple sinusoidal components into which a periodic quantity may be resolved.

harmonic constants—The amplitude and epochs of the harmonic constituents of the tide or tidal current at any place.

harmonic constituent—See **constituent.**

harmonic expressions—Trigonometric terms of an infinite series used to approximate irregular curves in two or three dimensions.

harmonic function—Any real function that satisfies Laplace's equation.

harmonic motion—The projection of circular motion on a diameter of the circle of such motion.

Hassler base-line measuring apparatus—An optical base-line measuring apparatus consisting of four rectangular iron bars mounted end to end in a wooden box. Each bar is 2 meters long, the combined length of the apparatus being 8 meters.

Hayford-Bowie method of isostatic reduction—A method of computing the effect of topography and isostatic compensation on gravity by which the effect of topography is computed directly and then corrected for the effect of isostatic compensation. The mechanics of this method involve the use of the Hayford gravity templets.

Hayford-Bullard (or Bullard) method of isostatic reduction—A method by which the topographic effect of an infinite slab of density 2.67 and a thickness equal to the elevation of the gravity station is first computed, and then corrected for curvature of the sea level surface and for difference of elevation between the station and the topography.

Hayford deflection templets—Templets used in connection with studies for the figure of the Earth and isostasy. In obtaining elevation readings from maps in connection with topographic and isostatic reductions, templets of plastic are used. The templets used in connection with deflection of the vertical studies have circles and radial lines drawn upon them, so proportioned with reference to scale of map and azimuth that land elevations and ocean depths within each compartment formed by adjacent arcs and radii can be easily averaged and the effect of the mass therein on a plumb line at the station (center of circles), under various hypotheses, can be computed. See also **Hayford gravity templets.**

Hayford effect—The direct effect on gravity of masses of unit density extending to various distances above and below sea level; it neglects the differences of elevation between the reference spheroid and the geoid.

Hayford gravity anomalies/Hayford anomalies—Isostatic anomalies obtained by computing the isostatic compensation according to the Pratt theory of isostasy as developed by Hayford, using various depths for purposes of comparative analysis.

Hayford gravity templets—Templets used in connection with gravity studies which are similar to Hayford deflection templets except that no account is taken of azimuth, all compartments bounded by a given pair of circles being of the same size and shape. A given templet can be used only on maps of the scale and projection for which it is constructed. See also **Hayford deflection templets.**

Hayford spheroid (ellipsoid)—A reference ellipsoid having the following approximate dimensions: semimajor axis—6,378,388.0 meters; semiminor axis—6,356,909.0 meters; and the flattening or ellipticity 1/297.00.

heading—(JCS) (navigation) The direction in which the longitudinal axis of an aircraft or ship is pointed, usually expressed in degrees clockwise from north (true, magnetic, compass, or grid).

height—(JCS) The vertical distance of an object, point, or level above the ground or other established reference plane. Height may be indicated as follows: *very low*—below 500 feet (above ground level); *low*—500 to 2,000 feet (above ground level); *medium*—2,000 to 25,000 feet; *high*—25,000 to 50,000 feet; *very high*—above 50,000 feet. See also **altitude; elevation; ellipsoidal height; geoidal height.**

height anomaly—The difference between the height of a terrain point above the reference spheroid and the corresponding normal height, measured along the normal plumb line.

height differential—The difference in height between predominant height groupings in a homogeneous surface area.

height displacement—See **relief displacement.**

height finder—A stereoscopic range finder so constructed as to indicate vertical heights rather than slant range. See also **stereometer.**

height-of-eye correction—That correction to sextant altitude due to dip of the horizon. Also called **dip correction.**

height of instrument—1. (spirit leveling) The height of the line of sight of a leveling instrument above the adopted datum. **2.** (stadia surveying) The height of the center of the telescope (horizontal axis) of transit or telescopic alidade above the ground or station mark. **3.** (trigonometric leveling) The height of the center of the theodolite (horizontal axis) above the ground or station mark.

height of the tide—The vertical distance from chart datum to the surface water level at any stage of the tide usually measured in feet.

heliocentric—Relative to the center of the Sun as origin.

heliocentric parallax—See **annual parallax.**

heliotrope—A device used in geodetic surveying for reflecting the Sun's rays to a distant point, to aid in long-distance observations. See also **selenotrope.**

helipad—(JCS) A prepared area designated and used for takeoff and landing of helicopters. (Includes touchdown or hoverpoint.)

heliport—(JCS) A facility designated for operating, basing, servicing, and maintaining helicopters.

Helmert's gravity formula of 1901—A formula for theoretical gravity developed from the gravity observations available at the time (1901), but not fitted to any pre-assigned value of the Earth's ellipticity.

Helmert's gravity formula of 1915—A formula for theoretical gravity based on a triaxial ellipsoid and therefore includes a longitude term. See also **longitude term gravity formula.**

hemispherical map—A map of one-half of the Earth's surface, bounded by the Equator, or by meridians.

high—(JCS) A height between 25,000 and 50,000 feet.

high altitude—(JCS) Conventionally, an altitude above 10,000 meters (33,000 feet). See also **altitude.**

high-oblique photograph—See **oblique air photograph.**

high rod—See **long rod.**

high tide—See **high water.**

high water (HW)—The highest limit of the surface water level reached by the rising tide. High water is caused by the astronomic tide-producing forces and/or the effects of meteorological conditions. Also called **high tide.**

high water full and change (HWF&C)—See **establishment of the port.**

high water interval—See **lunitidal interval.**

high water line—The line on the bank or shore to which the waters normally rise at high water. In tidal waters, the high water line is, in strictness, the intersection of the plane of the mean high water with the shore. The high water line is the boundary line between the bed and the bank of a stream.

high water lunitidal interval—See **lunitidal interval.**

high water springs—See **mean high water springs.**

higher high water (HHW)—The higher of two high waters occurring during a tidal day where the tide exhibits mixed characteristics.

higher high water interval (HHWI)—The interval of time between the transit (upper or lower) of the Moon over the local or Greenwich meridian and the next higher high water. This expression is used when there is considerable diurnal inequality. See also **lunitidal interval.**

higher low water (HLW)—The higher of two low waters of a tidal day where the tide exhibits mixed characteristics.

higher low water interval (HLWI)—The interval of time between the transit (upper or lower) of the Moon over the local or Greenwich meridian and the next higher low water. This expression is used when there is considerable diurnal inequality. See also **lunitidal interval.**

highest elevation—That elevation which is the highest point of relief within the area of a map or chart. See also **critical elevation.**

hill plane—The plane containing the positions of three ground marks constituting control points. This may be, but rarely is, a horizontal plane.

hill shading—(JCS) A method of representing relief on a map by depicting the shadows that would be cast by high ground if light were shining from a certain direction. Also called **hillwork.** See also **shaded relief.**

hillwork—See **hill shading.**

historical chart—A chart based on data from previous years to determine probable oceanographic patterns for a specified time. Also called **first approximation chart.**

history overlay—A specially prepared matte plastic material which shows the sources of all sounding data used in a bathymetric compilation.

holiday—(JCS) An unintentional omission in imagery coverage of an area. See also **gap.**

hologram—The hologram offers a reconstruction of the external appearance of an object with unique three-dimensional properties. This is accomplished by illuminating the object with coherent light.

hologrammetry—The art or science of interpreting the three-dimensional holographic image and obtaining reliable measurements by means of holography.

homogeneous area—(JCS) An area which has uniform radar reflecting power at all points.

homogeneous surface area—A grouping of features having the same general surface composition.

homologous—The condition where an image of a given object point or series of such points is common to two or more projections having different perspective centers.

homologous images—The images of a single object point that appears on each of two or more overlapping photographs having different perspective centers.

homologous photographs—Two or more overlapping photographs having different camera stations.

homologous rays—The two perspective rays corresponding to a pair of homologous image points.

homolographic (homalographic) map projection—An equal-area map projection. This term is found in the designations given some particular map projections, such as the Mollweide homalographic projection.

horizon—(JCS) In general, the apparent or visible junction of Earth and sky, as seen from any specific position. See also **apparent horizon; artificial horizon; celestial horizon; false horizon; geocentric horizon; geoidal horizon; geometrical horizon; radar horizon; sensible horizon; true horizon.**

horizon camera—A camera used in conjunction with another aerial camera to photograph the horizon simultaneously with the other photographs. The horizon photographs indicate the tilts of the other photographs.

horizon closure—See **error of closure,** definition 6.

horizon coordinate system--See **horizon system of coordinates.**

horizon photograph—**1.** A photograph of the horizon taken simultaneously with another photograph for the sole purpose of obtaining an indication of the orientation of the other photograph at the time of exposure. **2.** (surveying) A continuous matched set of horizon photographs defining obstructions 360° around a given station.

horizon prism—A prism which can be inserted in the optical path of an instrument, such as a bubble sextant, to permit observation of the visible horizon.

horizon profile—A plot of vertical angles against the horizontal angles taken 360° around a point with annotation. See also **horizon photograph; horizon sweep.**

horizon sweep—(surveying) A preliminary reconnaissance technique where the instrument is pointed initially at the farthest visible known point and recording clockwise angles to tanks, spires, buildings, signals, etc., for purposes of identification and subsequent use. See also

horizon profile.

horizon system of coordinates—A set of celestial coordinates, usually altitude and azimuth or azimuth angle, based on the celestial horizon as the primary great circle. Also called **horizon coordinate system.**

horizon trace—An imaginary line, in the plane of a photograph, which represents the image of the true horizon; it corresponds to the intersection of the plane of a photograph and the horizontal plane containing the internal perspective center or rear nodal point of the lens. See also **true horizon.**

horizontal angle—Angle in a horizontal plane.

horizontal axis—The axis about which the telescope of a theodolite or transit rotates when moved vertically.

horizontal bridging—See **bridging.**

horizontal circle—A graduated circle affixed to the lower plate of a transit by which horizontal angles can be measured.

horizontal closure error—See **error of closure,** definition 8.

horizontal control—A network of stations of known geographic or grid positions referred to a common horizontal datum, which control the horizontal positions of mapped features with respect to parallels and meridians, or northing and easting grid lines shown on the map. Horizontal control includes basic (marked) and supplementary (unmarked) stations.

horizontal control datum—A geodetic reference point which is the basis for horizontal control surveys, and of which five quantities are known: latitude, longitude, azimuth of a line from this point, and two constants which are the parameters of the reference ellipsoid. The horizontal control datum may extend over a continent or be limited to a small area. Also called **horizontal datum; horizontal geodetic datum.**

horizontal control point—See **horizontal control station.**

horizontal control station—A station whose position has been accurately determined in x- and y-grid coordinates, or latitude and longitude. Also called **horizontal control point.**

horizontal control survey net—See **survey net,** definition 1.

horizontal coplane—See **basal coplane.**

horizontal datum—See **datum,** definition 2.

horizontal deformation—In relative orientation, the cumulative model warpage affecting the horizontal datum from z-motion error, bridging error, and swing error.

horizontal direction—Observed horizontal angles at a triangulation station reduced to a common initial direction.

horizontal extension—See **extension,** definition 2.

horizontal geodetic datum—See **horizontal-control datum.**

horizontal intensity—The intensity of the horizontal component of the magnetic field in the plane of the magnetic meridian.

horizontal line—A line perpendicular to the vertical.

horizontal parallax—**1.** (astronomy) The geocentric parallax of a body on the observer's horizon. This is equal to the angular semidiameter of the Earth as seen from the body. **2.** (photogrammetry) See **absolute stereoscopic parallax.**

horizontal pass point—See **supplemental position.**

horizontal photograph—A photograph taken with the axis of the camera horizontal.

horizontal plane—**1.** A plane perpendicular to the direction of gravity; and plane tangent to the geoid or parallel to such a plane. **2.** (surveying) A plane perpendicular to the plumb line within which, or on which, angles and distances are observed. For any planimetric survey it is assumed that all plumb lines therein are parallel, and all horizontal planes therein are parallel.

horizontal refraction—A natural error in surveying which is the result of the horizontal bending of light rays between a target and an observing instrument. Usually caused by the differences in density of the air along the path of the light rays, resulting from temperature variations. See also **terrestrial refraction.**

horizontal stadia—A method of measuring distances wherein the stadia rod is held in a horizontal position and the stadia hairs of the instrument are vertical during observations.

horizontal taping and plumbing—A method whereby the tape is held horizontally, and the positions of its pertinent graduations are projected to the ground with plumb bobs.

horizontal/vertical bridging—See **bridging.**

horizontal/vertical extension—See **extension,** definition 2.

horizontalizing the model—See **leveling,** definition 2.

horizontally controlled photography—Cartographic aerial photography obtained simultaneously with recording of distance measurements between the taking aircraft and each of two or more geodetically positioned ground stations. Shoran or shiran are normally used as the distance-measuring equipment. The result is precise relative horizontal positioning of each photograph which has associated recorded distances. This positioning information is used as horizontal control data in the mapping process.

Horrebow level attachment—A level used in conjunction with a micrometer in a telescope whereby latitude observations by the Horrebow-Talcott method can be made.

Horrebow-Talcott method of latitude determination—See **latitude determination; zenith-telescope method.**

hot spot—(photography) A small area of undesired brilliancy of illumination in the image projected by a printer or a reader.

hour angle—The hour angle of a celestial body is the time elapsed since its upper transit. It is the angle between the observer's (astronomic) meridian and the declination circle of the body, measured positive westward from the meridian. See also **Greenwich hour angle; local hour angle; sidereal hour angle.**

hour angle difference—See **meridian angle difference.**

hour angle system (of coordinates)—An equatorial system of curvilinear celestial coordinates which has the Equator and the local meridian as primary and secondary reference planes, respectively. The position of a celestial body is given by its hour angle and declination.

hour circle—Any great circle on the celestial sphere whose plane is perpendicular to the plane of the celestial equator. Also called **circle of declination; circle of right ascension.** See also **celestial meridian; colures.**

hub—A temporary traverse-station marker, usually of wood. The stake is driven flush with the ground with a tack or small nail on top to mark the exact point of reference for angular and linear measurements.

Huygen's principle—A very general principle applying to all forms of wave motion which states that every point on the instantaneous position of an advancing phase front (wave front) may be regarded as a source of secondary spherical wavelets. The position of the phase front a moment later is then determined as the envelope of all of the secondary wavelets (ad infinitum). This principle is extremely useful in understanding effects due to refraction, reflection, diffraction, and scattering of all types of radiation, including sonic radiation as well as electromagnetic radiation and applying even to ocean-wave propagation.

hydrographic chart—(JCS) A nautical chart showing depths of water, nature of bottom, contours of bottom and coastline, and tides and currents in a given sea or sea and land area. Also called **marine map; nautical chart.**

hydrographic datum—The plane of reference of soundings, depth curves, and elevations of foreshore and offshore features. Also called **chart datum.** See also **international low water; low water datum; National Geodetic Vertical datum of 1929; sounding datum; tidal datum.**

hydrographic detail—The features along the shore and the submerged parts of bodies of water. Also called **' ydrographic feature.**

hydrographic feature—See **hydrographic detail.**

hydrographic reconnaissance—A reconnaissance of an area of water to determine depths, beach gradients, the nature of the bottom, and the location of coral reefs, rocks, shoals, and manmade obstacles.

hydrographic sextant—See **surveying sextant.**

hydrographic sounding—See **sounding,** definitions 1 and 2.

hydrographic survey—A survey made in relation to any considerable body of water, such as a bay, harbor, lake, or river for the purposes of determination of channel depths for navigation, location of rocks, sand bars, lights, and buoys; and in the case of rivers, made for flood control, power development, navigation, water supply, and water storage.

hydrography—**1.** (JCS) The science which deals with the measurements and description of the physical features of the oceans, seas, lakes, rivers, and their adjoining coastal areas, with particular reference to their use for navigational purposes. **2.** That part of topography pertaining to water and drainage features.

hydrology—Hydrology in its broadest extent deals with the properties, laws, and phenomena of water; of its physical, chemical, and physiological relations; of its distribution throughout the habitable Earth; and of the effect of this circulation on human lives and interests.

hydrophone—An electro-acoustic transducer that converts sound energy into electrical energy. See also **transducer.**

hydrosphere—That part of the Earth that consists of the oceans, seas, lakes, and rivers; a similar part of any other spatial body if such a body exists. Distinguished from the atmosphere and lithosphere.

hydrostatic equation—In numerical equations, the form assumed by the vertical component of the vector equation of motion when all Coriolis force, Earth curvature, frictional, and vertical acceleration terms are considered negligible compared with those involving the vertical pressure force and the force of gravity.

hydrostatic equilibrium—The state of a fluid whose surfaces of constant pressure and constant mass (or density) coincide and are horizontal throughout. Complete balance exists between the force of gravity and the pressure force. The relation between the pressure and the geometric height is given by the hydrostatic equation.

hygrometric—Relating to the relative humidity or comparative amount of moisture in the atmosphere. Since the atmosphere penetrates the pores or cells of material bodies in varying degrees depending upon the substances of which they are composed, the amount of moisture which it contains will affect the shapes and dimensions of certain instruments and equipment used in surveying and mapping. For this reason it is necessary to select materials which are not sensitive to hygrometric conditions for the construction of leveling rods, planetable sheets, etc., and for the construction and printing of maps.

hygroscopic—The property of materials such as

paper and films to absorb or release moisture and, in so doing, to expand or contract.

hyperbolic line of position—A line of position in the shape of a hyperbola, determined by measuring the difference in distance to two fixed points. Loran lines of position are an example.

hyperfocal distance—The distance from the camera lens to the nearest object in focus when the camera lens is focused at infinity.

hyperstereoscopy—(JCS) Stereoscopic viewing in which the relief effect is noticeably exaggerated, caused by the extension of the camera base. Also called **appearance ratio; exaggerated stereo; relief stretching; stereoscopic exaggeration.** See also **vertical exaggeration,** definition 1.

hypsograph—An instrument of the slide-rule type used to compute elevations from vertical angles and horizontal distances.

hypsographic detail—The features pertaining to relief or elevation of terrain.

hypsographic map (or chart)—A map or chart showing land or submarine bottom relief in terms of height above, or below, a datum by any method, such as contours, hachures, shading, or tinting. Also called **hypsometric map (or chart).**

hypsography—**1.** The science or art of describing elevations of land surfaces with reference to a datum, usually sea level. **2.** That part of topography dealing with relief or elevation of terrain.

hypsometer—An instrument used in determining elevations of points on the Earth's surface in relation to sea level by determining atmospheric pressure through observation of the boiling point (temperature) of water at each point.

hypsometric map (or chart)—See **hypsographic map.**

hypsometric tinting—(JCS) A method of showing relief on maps and charts by coloring, in different shades, those parts which lie between selected levels. Also called **altitude tints; color gradients; elevation tints; gradient tints; layer tints.**

hypsometric tint scale—A graphic scale in the margin of maps and charts which indicates heights or depths by graduated shades of colors. See also **hypsometric tinting.**

hypsometry—The art of determining, by any method, surface elevations on the Earth with reference to sea level.

ice chart—A chart showing prevalence of ice, usually with reference to navigable waterways.

iced-bar apparatus—An apparatus for measuring linear distance with great precision and accuracy, and consisting essentially of a steel bar which is maintained at a constant temperature by being surrounded with melting ice. The bar is rectangular in cross section, and is carried in a Y-shaped trough which is filled with melting ice and mounted on a car which moves on a track. Bar lengths are observed with micrometer microscopes mounted on stable supports. Also called **Woodward base-line measuring apparatus.**

ideal Earth—See **equilibrium theory.**

ideal sea level—The theoretical sea surface which is everywhere normal to the plumb line. Reference of all depth soundings to this level would make them all comparable.

identification posts—Posts of wood or other suitable material, appropriately marked and inscribed, and placed near survey stations to aid in their recovery and identification. Also called **supplemental posts for survey monuments.**

idle pendulum—A working pendulum placed in the receiver in advance of its being used, so that it may assume the same temperature as the dummy pendulum.

image—**1.** The permanent record of the likeness of any natural or manmade features, objects, and activities reproduced on photographic materials. This image can be acquired through the sensing of visual or any other segment of the electromagnetic spectrum by sensors, such as thermal infrared, and high resolution radar. See also **erect image; homologous image; inverted image; latent image; real image; reverted image; thermal imagery; virtual image.** **2.** A visual representation, as on a radarscope.

image correlation—The matching of position and physical characteristics between imagery of the same geographic area from different types of sensors, between sensor imagery and a data base, or between two images from the same sensor.

image degradation—(photometry) The reduction of the inherent optimum imaging potential of individual sensor systems caused by error in sensor operations or processing procedures. Reductions in quality caused by unavoidable factors; i.e., atmospherics, snow cover, etc., are not associated with the term.

image direction—A term used to describe the image orientation of a photographic negative or positive relative to the position of the emulsion. See also **emulsion-to-base; emulsion-to-emulsion; right-reading; wrong-reading.**

image distortion—Any shift in the position of an image on a photograph which alters the perspective characteristics of the photograph. Causes of image distortion include lens aberration, differential shrinkage of film or paper, and motion of the film or camera.

image enhancement—Any of several processes that might improve the interpretation quality of an image. Such processes include contrast improvement, greater resolution, special filtering, etc.

image motion—The smearing or blurring of imagery on an aerial photograph because of the relative movement of the camera with respect to the ground.

image motion compensation (IMC)—(JCS) Movement intentionally imparted to film at such a rate as to compensate for the forward motion of an air or space vehicle when photographing ground objects.

image motion compensation distortion—In a panoramic camera system, the displacement of images of ground points from their expected cylindrical position caused by the translation of the lens or negative surface (a motion used to compensate for image motion during exposure time).

image motion factors—Those factors wherein the image motion varies directly with the aircraft ground speed and lens focal length and inversely with the altitude.

image plane—See **photograph plane.**

image point—(photogrammetry) Image on a photograph corresponding to a definite object on the ground.

image ray—Straight line from a ground object, through the camera lens, to the image on the photograph. See also **perspective ray.**

imagery—(JCS) Collectively, the representations of objects reproduced electronically or by optical means on film, electronic display devices, or other media.

imagery interpretation—(JCS) The process of location, recognition, identification, and description of objects, activities, and terrain represented on imagery.

imagery interpretation key—(JCS) Any diagrams, charts, tables, lists, or sets of examples, etc., which are used to aid imagery interpreters in the rapid identification of objects visible on imagery.

imagery sortie—(JCS) One flight by one aircraft for the purpose of recording air imagery.

impersonal micrometer—See **transit micrometer.**

imposition—Positioning and assembling negatives or positives into printing location on a flat.

impression—The inked image received by a sheet in a press. Commonly used as a measure of printing production or capacity.

in-and-out station—A recoverable but unoccupied station incorporated into a traverse by recording a fictitious deflection angle of 180° to reverse the azimuth of the course leading into it, so that the next station coincides with the preceding station and the in-and-out station is used as the backsight for continuing the traverse. In the computations it is treated as an ordinary station in the traverse.

incident nodal point—See **nodal point,** definition 1.

inclination—**1.** The angle which a line or surface makes with the vertical, horizontal, or with another line or surface. **2.** The angle between orbit plane and reference plane; for example, the Equator is the reference plane for geocentric, and the ecliptic for heliocentric orbits. **3.** See **dip,** definition 2.

inclination correction—See **grade correction.**

inclination of the horizontal axis—The vertical angle between the horizontal axis of a surveying or astronomic instrument and the plane of the horizon.

incline sight—A sight made with a surveying instrument at an angle above or below the horizon.

inclinometer—An instrument for measuring inclination to the horizontal of a ship or aircraft, or of the lines of force of the Earth's magnetic field.

independent resurvey—A resurvey which is not dependent on the records of the original survey but is intended to supersede them in establishing new land boundaries and subdivisions. See also **dependent resurvey.**

index chart—An outline chart showing the limits and identifying designations of navigational charts, volumes of sailing directions, etc.

index contour line—(JCS) A contour line accentuated by a heavier line weight to distinguish it from intermediate contours. Index contours are usually shown as every fifth contour with their assigned values, to facilitate reading elevations.

index correction—**1.** A correction applied to the reading from any graduated measuring device to compensate for a constant error such as would be caused by misplacement of the scale; the reverse of the index error. **2.** (leveling) That correction which must be applied to an observed difference of elevation to eliminate the error introduced into the observations when the zero of the graduations on one or both leveling rods does not coincide exactly with the actual bottom surface of the rod.

index error—The instrumental error which is constant and attributable to displacement of a vernier or some analogous effect.

index mark—A real mark, such as a cross or dot, lying in the plane or the object space of a photograph and used singly as a reference mark in certain types of monocular instruments, or as one of a pair to form a floating mark as in certain types of stereoscopes. In stereoscopic map plotting instruments which utilize a stereo pair of index marks, each mark is called a **half mark.** See also **floating mark.**

index of refraction—See **Snell's law of refraction.**

index prism—A sextant prism which can be rotated to any angle corresponding to altitudes between established limits. It is the bubble or pendulum sextant counterpart of the index mirror of a marine sextant.

Index to Names—An alphabetical list of geographic names keyed to a map series or to maps covering a specific country giving feature designation, geographic and grid coordinates, and sheet number for each name appearing on the series. Essentially the same as **gazetteer,** the term **Index to Names** is an official DoD designation, established by international agreements.

index to photography—See **photo index,** definition 1.

Indian datum—The Indian datum is accepted as the preferred datum for India and several adjacent countries in Southeast Asia. It is computed on the Everest ellipsoid with its origin at Kalianpur in Central India. Derived in 1830, the Everest ellipsoid is the oldest of the ellipsoids in common use and is much too small. As a result of the latter, the datum cannot be extended too far from the origin or very large geoid separations will occur. For this reason and the fact that the ties between local triangulation in Southeast Asia are typically weak, the Indian datum is probably the least satisfactory of the preferred datums.

Indian spring low water—The approximate mean water level determined from all lower low waters at spring tides.

Indian tide plane—The datum of Indian spring low water.

indicated corner—A term adopted by the U.S. Geological Survey to designate a corner of the public land surveys whose location cannot be verified by the criteria necessary to class it as a found or existent corner, but which is accepted locally as the correct corner and whose location is perpetuated by such marks as fence-line intersections, piles of rocks, and stakes or pipes driven into the ground, which have been recovered by field investigation.

indicated principal point—See **principal point.**

indirect effect on the deflections—See **topographic deflection.**

indirect leveling—See **barometric leveling; thermometric leveling; trigonometric leveling.**

indirect measurement—Any measurement secured by determining its quantity from its relation to some measured quantity. A technique used in surveying when it is impossible to actually tape a distance across a river or other such obstruction. See also **direct measurement.**

indirect observation—A measure of a quantity which is a function of the quantity or quantities whose value is desired, such as an observed difference in elevation with a spirit level, used to obtain the elevation of a bench mark.

indirect photography—Photography in which the camera records an image cast upon a screen or similar display surface by electronic (television, radar, etc.) or other means.

inequality—A systematic departure from the mean value of a quantity. See also **annual inequality; diurnal inequality; lunar inequality; parallactic inequality; parallax inequality; phase inequality; variational inequality.**

inertial azimuth—An azimuth which approximates the value which could be obtained from astronomic observations, but which is derived from direct observations along the line of sight with an inertial azimuth measuring device consisting of a north-seeking gyroscope combined with a theodolite. See also **gyro theodolite.**

inertial coordinate system—A coordinate system in which the axes do not rotate with respect to the "fixed stars" and in which dynamic behavior can be described using Newton's laws of motion.

inertial guidance system—A system in which guidance is permitted by means of the measurement and integration of acceleration from within the craft.

inertial measurement unit (IMU)—(missile guidance) A compact component of an inertial guidance system which has three accelerometers mounted on a gyro-stabilized platform. See also **accelerometer.**

inertial navigation—The process of measuring a craft's velocity, attitude (in the submarine missile-launching applications), and displacement (including changes in altitude in the aircraft application) from a known starting point through sensing the accelerations acting on it in known directions by means of devices that mechanize Newton's laws of motion.

inertial navigation system—A system which is not dependent on manmade electromagnetic signals. Newton's second law of motion is utilized with a system consisting of accelerometers mounted on gyro-stabilized platforms, each for measuring longitudinal, lateral, and vertical accelerations. The double integration of all accelerations in three mutually perpendicular directions provides distance traveled (from a known starting point) in three mutually perpendicular directions. Navigation is by a highly refined form of dead reckoning with system position being updated from other navigational references in the more sophisticated systems.

inertial reference photography—Cartographic aerial photography obtained simultaneously with magnetic tape recorded inertial reference positioning data.

inferior conjunction—The conjunction of an inferior planet and the Sun when the planet is between the Earth and the Sun.

inferior planets—The planets with orbits smaller than that of the Earth, i.e., Mercury and Venus.

inferior transit—See **lower transit.**

infinity—The point, line, or region beyond measurable limits. An unaltered source of light is regarded as at infinity if it is at such a great distance that rays from it can be considered parallel.

infrared—Pertaining to or designating the portion of the electromagnetic spectrum with wavelengths just beyond the red end of the visible spectrum, such as radiation emitted by a hot body. Invisible to the eye, infrared rays are detected by their thermal and photographic effects. Their wavelengths are longer than those of visible light and shorter than those of radio waves.

infrared distance measurement—A distance determined by measuring the phase delay of modulation signals on a light beam (infrared) traveling at a known velocity between a distance meter and reflector.

infrared film—(JCS) Film carrying an emulsion especially sensitive to "near-infrared." Used to photograph through haze, because of the penetrating power of infrared light, and in camouflage detection to distinguish between living vegetation and dead vegetation or artificial green pigment.

inherited error—The error in initial values used in a computation; especially the error introduced from the previous steps in a step-by-step integration.

initial monument—(USPLS) A physical structure which marks the location of an initial point in the rectangular system of surveys. See also **initial point.**

initial point—**1.** That point from which any survey is initiated. Also called **point of origin. 2.** (USPLS) A point which is established under the rectangular system of surveys and from which is initiated the cadastral survey of the principal meridian and base line that controls the cadastral survey of the public lands within a given area. See also **base line; initial monument; principal meridian.**

inner orientation—See **interior orientation.**

inner planets—The four planets nearest the Sun: Mercury, Venus, Earth, and Mars. See also **major planets.**

inserted grouping (radar)—(JCS) The inclusion of one area of homogeneous surface material in an area of different material.

inset—(JCS) (cartography) A separate map positioned within the neatline of a larger map. Three forms are recognized: (1) an area geographically outside a sheet but included therein for convenience of publication, usually at the same scale; (2) a portion of the map or chart at an enlarged scale; (3) a smaller scale map or chart of surrounding areas included for location purposes.

instantaneous field of view (IFOV)—The smallest solid angle resolvable by a scanner when expressed in radians. When expressed in feet, it is the projected area of the detector image on the ground and is a measure of the resolution of a scanner or similar remote sensor with discrete samples.

instantaneous-reading tape—A survey tape on which the foot mark is repeated at each subdivision. Thus, a tape divided into tenths of a foot would have the foot mark imprinted at each tenth of a foot division.

instrument adjustment—The process of mechanical manipulation of the relation of component parts of an instrument in order to obtain the highest practicable precision and facility in the designed use of the instrument.

instrument approach chart—An aeronautical chart designed for use under instrument flight conditions, for making instrument approach and letdown to contact flight conditions in the vicinity of an airfield.

instrument error—A systematic error resulting from imperfections in, or faulty adjustment of, instruments or devices used. Also called **calibration error.**

instrument parallax—**1.** A change in the apparent position of an object with respect to the reference marks of an instrument which is caused by imperfect adjustment of the instrument. Also called **optical parallax.** **2.** Parallax caused by a change in the position of the observer. Also called **personal parallax.**

instrument phototriangulation—See **stereotriangulation.**

instrument station—See **set-up,** definition 1.

integrated station instrument—An instrument combining horizontal and vertical angles with electronic distance measurement and programmed computer capability in a single piece of hardware.

integration—(JCS) **1.** (intelligence) The process of forming an intelligence pattern through selection and combination of evaluated information. **2.** (photography) A process by which the average radar picture seen on several scans of the time base may be obtained on a print, or the process by which several photographic images are combined into a single image.

intensity of gravity—The magnitude with which gravity acts, expressed in suitable units, usually as an acceleration, in gals; as a force, in dynes.

intercardinal point—Any of the four directions midway between the cardinal points: northeast, southeast, southwest, or northwest.

interferometer—An apparatus used to produce and measure interference from two or more coherent wave trains from the same source. Used to measure wavelengths, to measure angular width of sources, to determine the angular position of sources (as in satellite tracking), and for other purposes. See also **radio interferometer.**

interior angle traverse—In surveying, a closed traverse wherein distances are measured and only interior angle measurements are used.

interior orientation—The determining (analytically or in a photogrammetric instrument) of the interior perspective of the photograph as it was at the instant of exposure. Elements of interior orientation are the calibrated focal length, location of the calibrated principal point, and the calibrated lens distortion. Also called **inner orientation.**

interior perspective center—See **perspective center.**

interior to a curve—That area lying toward the concave side of a curve and included within the area of the circle of which the curve is a part of the circumference.

interlocking angle—In tilt analysis of oblique photographs, the angle between the optical axes of the vertical and oblique cameras. The dihedral angle between the planes of the vertical and oblique photographs.

intermediate contour line—(JCS) A contour line drawn between index contours. Depending on the contour interval, there are three or four intermediate contours between the index contours.

intermediate orbit—A central force orbit that is tangent to the real orbit at some point. A fictitious satellite travelling in the intermediate orbit would have the same position, but not the same velocity, as the real satellite at the point of tangency.

international date line—(JCS) The line coinciding approximately with the antimeridian of Greenwich, modified to avoid certain habitable land. In crossing this line there is a date change of one day. Also called **date line.**

international gravity formula—A development of the formula for theoretical gravity, based on the assumptions that the spheroid of reference is an exact ellipsoid of revolution having the dimensions of the International ellipsoid of reference (Madrid, 1924), rotating about its minor axis once in a sidereal day; that the surface of the ellipsoid is a level surface; and that gravity at the Equator equals 978.049 gals.

International Gravity Standardization Net 1971 (IGSN 71)—An adjusted worldwide network of gravity measurements consisting of absolute, pendulum, and gravimeter observations. The IGSN 71 was approved and adopted by the International Union of Geodesy and Geophysics in 1971, and replaces the Potsdam datum as the international gravity standard.

international low water (ILW)—A plane of reference below mean sea level by the following amount: half the range between mean lower low water and mean higher high water multiplied by 1.5.

International spheroid (ellipsoid)—A reference ellipsoid having the following approximate dimensions: semimajor axis—6,378,388.0

meters; semiminor axis—6,356,911.9 meters; and the flattening or ellipticity—1/297. This spheroid was based on the Hayford spheroid (ellipsoid).

international system of units (SI)—The practical international metric system of units adopted by the Eleventh General Conference of Weights and Measures in 1960. It consists of seven basic units with additional supplementary and derived units.

interocular distance—(JCS) The distance between the centers of rotation of the eyeballs of an individual or between the oculars of optical instruments. Also called **eye base; interpupillary distance.**

interpolate—To determine intermediate values between given fixed values. As applied to logical contouring, to interpolate is to ratio vertical distances between given spot elevations.

interpupillary distance—See **interocular distance.**

interrogation—Transmission of a radio signal or combination of signals intended to trigger a transponder or group of transponders.

interrogator—**1.** A radar set or other electronic device that transmits an interrogation. **2.** An interrogator-responsor or the transmitting component of an interrogator-responsor.

interrogator-responsor—A radio transmitter and receiver combined to interrogate a transponder and display the resulting replies. Also called **challenger; interrogator.**

interrupted map projection—A projection having several standard meridians, each centered over a continent, and with lobate-shaped sections of the projection plotted from each standard meridian. The projection is broken in the ocean areas between the continents, thus reducing the linear scale discrepancies and the overall shape distortion, especially toward the margins, while retaining the equal area property.

intersected point—See **intersection station.**

intersection—**1.** (surveying) The procedure of determining the horizontal position of an unoccupied point (intersection station) by direction observations from two or more known positions. **2.** (photogrammetry) The procedure of determining the horizontal position of a point by intersecting lines of direction obtained photogrammetrically. The lines of direction may be obtained directly from vertical photographs or by graphic or mathematical rectification of tilted photographs. See also **resection.**

intersection station—An object whose horizontal position is determined by observations from other survey stations, no observations being made at the object itself. Where the object is observed from only two stations, the position is termed a **no-check position,** as there is no proof that such observations are free from blunders. Intersection stations are either objects which would be difficult to occupy with an instrument, or survey signals whose positions can be determined with sufficient accuracy without being occupied. Also called **intersected point.**

intervalometer—A timing device for automatically operating at specified intervals certain equipment such as a camera shutter for the purpose of obtaining a desired end lap between successive photographs.

intervisibility test—Any of the several tests used to determine the possible visibility along a sightline in a proposed survey net. Its purpose is to determine the existence of obstructions along a proposed line of sight from which tower and signal requirements may be developed.

Invar—An alloy of nickel and steel having a very low coefficient of thermal expansion. Invar is used in the construction of Jaderin wires (base-line measuring apparatus), subtense bars, precise leveling rods, tapes, and pendulums. See also **Lovar.**

Invar leveling rod—See **precise leveling rod.**

Invar pendulum—A quarter-meter pendulum made of Invar.

Invar scale—A measuring bar made from Invar. Normally, one side is graduated in the metric system and the other side in the English system.

Invar tape—Any survey tape made of Invar.

invariable pendulum—A pendulum so designed and equipped with means of support that it can be used in only one position.

inventory survey—A survey for the purpose of collecting and correlating engineering data of a particular type, or types, over a given area. An inventory survey may be recorded on a base map.

inverse—See **transverse; inverse position computation.**

inverse chart—See **transverse chart.**

inverse computation—See **inverse position computation.**

inverse cylindrical orthomorphic chart—See **transverse Mercator chart.**

inverse cylindrical orthomorphic map projection—See **transverse Mercator map projection.**

inverse equator—See **transverse equator.**

inverse latitude—See **transverse latitude.**

inverse longitude—See **transverse longitude.**

inverse Mercator chart—See **transverse Mercator chart.**

inverse Mercator map projection—See **transverse Mercator map projection.**

inverse meridian—See **transverse meridian.**

inverse parallel—See **transverse parallel.**

inverse position computation—The derivation of the length, and the forward and back azimuths of a line by computation based on the known positions of the ends of the line. Also called **inverse; inverse computation; inverse position problem.**

inverse position problem—See **inverse position**

computation.

inverse rhumb line—See **transverse rhumb line.**

inversors—(photography) Mechanical devices used to maintain correct conjugate distances and collinearity of negative, lens, and easel planes in autofocusing optical instruments, such as copy cameras and rectifiers. See also **Carpentier inversor; Peaucellier inversor; Peaucellier-Carpentier inversor; Pythagorean right-angle inversor.**

inverted image—An image that appears upside down in relation to the object.

inverted stereo—See **pseudoscopic stereo.**

inverting—See **transit,** definition 3.

inverting telescope—An instrument with the optics so arranged that the light rays entering the objective of the lens meet at the crosshairs and appear inverted when viewed through the eyepiece without altering the orientation of the image. See also **erecting telescope.**

ionosphere—(JCS) The region of the atmosphere, extending from roughly 40 to 250 miles altitude, in which there is appreciable ionization. The presence of charged particles in this region profoundly affects the propagation of electromagnetic radiations of long wavelengths (radio and radar waves). See also **atmosphere.**

ionospheric correction—The correction made to electromagnetic measurements between satellites and ground stations to compensate for the effect of the ionosphere.

irradiance—Radiant flux incident per unit area.

irregular error—See **random error.**

isanomal—A line connecting points of equal variations from a normal value.

isentropic—Of equal or constant entropy with respect to either space or time.

isobar—(JCS) A line along which the atmospheric pressure is, or is assumed to be, the same or constant.

isobaric chart—A chart showing isobars. Also called **constant pressure chart.**

isobath—See **depth curve.**

isocenter—**1.** (JCS) The point on a photograph intersected by the bisector of the angle between the plumb line and the photograph perpendicular. **2.** The unique point common to the plane of a photograph, its principal plane, and the plane of an assumed truly vertical photograph taken from the same camera station and having an equal principal distance. **3.** The point of intersection on a photograph of the principal line and the isometric parallel. The isocenter is significant because it is the center of the radiation for displacements of images due to tilt.

isocenter plot—See **isocenter triangulation.**

isocenter triangulation—Radial triangulation utilizing isocenters as radial centers. Also called **isocenter plot.**

isochrone—A line on a chart connecting all points having the same time of occurrence of a particular phenomenon or of a particular value of a quantity.

isoclinal—(JCS) A line drawn on a map or chart joining points of equal magnetic dip. Also called **isoclinic line.**

isoclinic chart—A chart of which the chief feature is a system of isoclinic lines, each for a different value of the magnetic inclination.

isoclinic line—See **isoclinal.**

isodiff—One of a series of lines on a map or chart connecting points of equal correction or difference in datum, especially useful in readjustment of surveys from one datum to another. See also **isolat; isolong.**

isodynamic line—A line connecting points of equal magnitude of any force.

isogal—A contour line of equal gravity values on the surface of the Earth.

isogonal—(JCS) A line drawn on a map or chart joining points of equal magnetic declination for a given time. Also called **isogonic line.** See also **agonic line.**

isogonic chart—A chart of which the chief feature is a system of isogonic lines, each for a different value of the magnetic declination.

isogonic line—See **isogonal.**

isogram—See **isopleth.**

isogriv—A line on a map or chart which joins points of equal angular difference between grid north and magnetic north.

isogriv chart—A chart with lines connecting points of equal grivation.

isolat—An isodiff connecting points of equal latitude corrections.

isoline—**1.** A line representing the intersection of the plane of a vertical photograph with the plane of an overlapping oblique photograph. If the vertical photograph were tilt-free, the isoline would be the isometric parallel of the oblique photograph. **2.** A line along which values are, or are assumed to be, constant.

isolong—An isodiff connecting points of equal longitude corrections.

isomagnetic chart—A chart showing the configuration of the Earth's magnetic field by isogonic, isoclinic, or isodynamic lines.

isometric (conformal) latitude—An auxiliary latitude used in the conformal mapping of the spheroid on a sphere. By transforming geographic latitudes on the spheroid into isometric latitudes on a sphere, a conformal map projection (the Mercator) may be calculated, using spherical formulas, for the plotting of geographic data.

isometric parallel—The intersecting line between the plane of a photograph and a horizontal plane having an equal perpendicular distance from the same perspective center.

isoperimetric curve—A line on a map projection along which there is no variation from exact scale. There are two isoperimetric curves passing through every point on an equal-area map projection. This characteristic gives that class of projections some preference for engineering maps.

isopleth—A line of equal or constant value of a given quantity, with respect to either space or time. Also called **isogram.**

isopor—A line found on magnetic charts showing points of equal annual change. Also called **magnetic isoporic line.**

isoporic chart—A chart with lines connecting points of equal magnetic annual change.

isopycnic—A line connecting points of equal density, particularly of ocean water and atmosphere.

isoradial—A radial from the isocenter.

isostasy—A condition of approximate equilibrium in the outer part of the Earth, such that the gravitational effect of masses extending above the surface of the geoid in continental areas is approximately counterbalanced by a deficiency of density in the material beneath those masses, while the effect of deficiency of density in ocean waters is counterbalanced by an excess of density in the material under the oceans. See also **depth of isostatic compensation; Hayford-Bowie method of isostatic reduction; Hayford-Bullard (or Bullard) method of isostatic reduction; isostatic adjustment; isostatic compensation; isostatic correction; Pratt-Hayford theory of isostasy.**

isostatic adjustment—The natural process by which the crust of the Earth adjusts to restore or maintain its state of equilibrium. See also **isostasy.**

isostatic anomaly—The difference between an observed value of gravity and a theoretical value at the point of observation which has been corrected for elevation of the station above the geoid, and for the effect of topography over the whole Earth, and for its isostatic compensation.

isostatic compensation—The departure from normal density of material in the lower part of a column of the Earth's crust which balances (compensates) land masses (topography) above sea level and deficiency of mass in ocean waters, and which produces the condition of approximate equilibrium of Earth's crust. See also **isostasy; topographic deflection.**

isostatic correction—The adjustment made to values of gravity or to deflections of the vertical observed at a point to take account of the assumed mass deficiency under topographic features for which a topographic correction is also made.

isostatic geoid—An ideal geoid derived from the spheroid of reference by the application of computed values of the deflection of the vertical which depend upon the topography and isostatic compensation.

isostere—A line connecting points of equal atmospheric density.

isotimic—Pertaining to a quality which has equal value in space at a particular time.

J

Jacob's staff—A single staff or pole used for mounting a surveyor's compass or other instrument. Sometimes used in place of a tripod.

Jäderin wires (base apparatus)—An apparatus used for base-line measurement. It consists of separate steel and brass wires, extended under constant tension over reference tripods in the line of the base. The coefficients of expansion and lengths at a certain temperature of the two wires having been found, the temperatures of the wires themselves may be deduced from the difference of the measurement of the same distance by the two wires. With these temperatures known, the length of the base may be accurately obtained.

Jet Navigation Chart (JNC)—A 1:2,000,000 scale, coordinated series of multi-colored charts, designed to satisfy long-range navigation of high-altitude, high-speed aircraft.

Joint Operations Graphic (JOG)—The standard 1:250,000 scale Department of Defense cartographic product which may be produced in any of the following three versions to meet the validated Unified and Specified Commands and Military Departments area requirements: the JOG/G (Series 1501) is designed to meet ground use requirements; JOG/A (Series 1501 Air) is designed to meet air use requirements; and JOG/R (Series 1501 Radar) is the Air Target Material version in support of radar/intelligence planning and operations requirements.

Julian calendar—The calendar established by Julius Caesar in 46 B.C., and based upon the assumption that the true length of the tropical year was exactly 365.25 mean solar days.

Julian day—The number of each day, as reckoned consecutively since the beginning of the present Julian period on January 1, 4713 B.C. The Julian day number denotes the number of days that has elapsed at Greenwich noon on the day designated, since this epoch day. See also **modified Julian day.**

junction—(leveling) The place where two or more lines of levels are connected.

junction bench mark—A bench mark selected as the common meeting point for lines of levels or links of levels.

junction detail—A sketch or working diagram showing the details of the various levelings at a junction.

junction figure—A triangulation figure in which three or more triangulation arcs meet, or two or more arcs intersect.

K

K-factor—See **base-altitude ratio.**

Kalman filtering—The recursive minimum variance estimation of an unbiased stochastic variable. An a priori estimate and covariance are linearly combined with new data to form an updated estimate and covariance.

Kepler's laws—The three laws governing the motions of planets in their orbits. These are: (1) The orbits of the planets are ellipses, with the Sun at a common focus; (2) As a planet moves in its orbit, the line adjoining the planet and Sun sweeps over equal areas in equal intervals of times; (3) The squares of the periods of revolution of any two planets are proportional to the cubes of their mean distances from the Sun. Also called **Kepler's planetary laws.**

Kepler's planetary laws—See **Kepler's laws.**

key flat—The principal or master layout or flat used as a positioning guide for stripping up other flats. Also called **layout guide.** See also **layout; flat,** definition 1.

kiss plate—A pressplate used to make an addition or correction to a previously printed sheet. Also called **touch plate.**

Klimsch-Variomat compilation method (relief)—A process by which reproduction material of large-scale sheets is photographically filtered to retain only index contours. The index contours are then used as the relief compilation for a medium-scale sheet.

Krasovsky spheroid (ellipsoid)—A reference ellipsoid having the following approximate dimensions: semimajor axis—6,378,245.0 meters; flattening or ellipticity—1/298.3.

L

Laborde map projection—Similar to the transverse Mercator projection, except that the Laborde projects a spheroid rather than a sphere onto a plane. This conformal projection is best suited for regions which are elongated in a direction which is at a considerable angle to the meridian. USAF Special Flight Charts are based on this projection.

LaCoste-Romberg gravimeter—A long-period spring suspended cantilevered weight system adapted to the measurement of gravity differences.

ladder grid numbers—Those grid numbers which identify the grid lines within the neatline.

Lambert azimuthal equal-area map projection—See **Lambert zenithal equal-area map projection.**

Lambert azimuthal polar map projection—A Lambert equal-area map projection with the pole of projection at the pole of the sphere, and the radii of the circles which represent the geographic parallels corresponding to the chords of those parallels.

Lambert bearing—A bearing as measured on a Lambert conformal chart or plotting sheet. This approximates a great-circle bearing.

Lambert central equivalent map projection upon the plane of the meridian—An azimuthal map projection having the pole of the projection on the Equator. Also called **Lambert equal-area meridional map projection.**

Lambert conformal chart—A chart on the Lambert conformal projection.

Lambert conformal conic map projection—A conformal map projection of the so-called conical type, on which all geographic meridians are represented by straight lines which meet in a common point outside the limits of the map, and the geographic parallels are represented by a series of arcs of circles having this common point for a center. Meridians and parallels intersect at right angles, and angles on the Earth are correctly represented on the projection. This projection may have one standard parallel along which the scale is held exact; or there may be two such standard parallels, both maintaining exact scale. At any point on the map, the scale is the same in every direction. It changes along the meridians and is constant along each parallel. Where there are two standard parallels, the scale between those parallels is too small; beyond them, too large. Also called **Lambert conformal map projection.**

Lambert conformal map projection—See **Lambert conformal conic map projection.**

Lambert equal-area meridional map projection—See **Lambert central equivalent map projection upon the plane of the meridian.**

Lambert grid—An informal designation for a coordinate system based on a Lambert conformal map projection.

Lambert zenithal equal-area map projection—An azimuthal map projection having the pole of the projection at the center of the area mapped. The azimuths of great circles radiating from this center (pole) are truly represented on the map: equal distances on those great circles are represented by equal linear distances on the map, but the scale along those great circle lines so varies with distance from the pole of the projection, that an equal-area projection is produced. Also called **Lambert azimuthal equal-area map projection.**

laminate—**1.** The process of preserving a map sheet or other graphic by sandwiching between two sheets of clear synthetic material (polyethylene-polyester plastic). The laminating equipment uses heat and pressure but no adhesive. **2.** (relief model) See **plastic block.**

land boundary—A line of demarcation between two parcels of land.

land effect—See **coastal refraction.**

land-line adjustment—Positioning the public-land lines on a topographic map to indicate their true, theoretical, or approximate location relative to the adjacent terrain and culture.

land mass simulator plate—See **factored transparency.**

land survey—The process of determining boundaries and areas of tracts of land. The term **cadastral survey** is sometimes used to designate a land survey, but in this country its use should be restricted to the surveys of public lands of the United States (USPLS). Also called **boundary survey; property survey.** See also **cadastral survey.**

landmark—**1.** An object of enough interest or prominence in relation to its surroundings to make it outstanding or to make it useful in determining a location or a direction. **2.** Any monument, material mark, or fixed object used to designate the location of a land boundary on the ground.

Landsat—A series of NASA satellites designed to acquire information about the Earth's resources.

landscape map—A topographic map made to a relatively large scale and showing all details. Such maps are required by architects and landscape gardeners for use in planning buildings to fit the natural topographic features and for landscaping parks, playgrounds, and private estates. These are generally maps of small areas, and scales vary from 1 inch = 20 feet to 1 foot = 50 feet, depending on the amount of detail.

Laplace azimuth—A geodetic azimuth derived from an astronomic azimuth by use of the Laplace equation.

Laplace azimuth mark—An astronomic azimuth mark at a Laplace station.

Laplace condition—The Laplace condition, expressed by the Laplace equation, arises from the fact that a deflection of the vertical in the plane of the prime vertical will give a difference between astronomic and geodetic longitude and between astronomic and geodetic azimuth; or, conversely, that the observed differences between astronomic and geodetic values of the longitude and of the azimuth may both be used to determine the deflection in the plane of the prime vertical.

Laplace control—Control and correction of astronomic azimuths through observations of the deflection of the plumb line in the prime vertical (comparison of astronomic and geodetic longitude).

Laplace equation—**1.** The equation which expresses the relationship between astronomic and geodetic azimuths in terms of astronomic and geodetic longitudes and geodetic latitude. **2.** (potential) A partial differential equation of the second order which is satisfied by the Newtonian potential of every finite body at all exterior points.

Laplace station—A triangulation or traverse station at which a Laplace azimuth is determined. At a Laplace station both astronomic longitude and astronomic azimuth are determined.

large-scale map—(JCS) A map having a scale of 1:75,000 or larger.

laser—A device producing coherent-energy beams in the spectrum of light-or-near-light frequencies. A laser-equipped Geodimeter makes it possible to measure greater distances.

laser terrain profile recorder—An electronic instrument that emits a continuous wave laser beam from an aircraft to measure vertical distances between the aircraft and the Earth's surface.

latent image—The invisible image produced in radiation-sensitive materials which becomes visible upon processing.

lateral chromatic aberration—An aberration which affects the sharpness of images off the lens axis because different colors undergo different magnifications.

lateral gain—(JCS) The amount of new ground covered laterally by successive photographic runs over an area.

lateral magnification—The ratio of a length in the image, perpendicular to the lens axis, to a corresponding length in the object.

lateral refraction—The horizontal component of the refraction of light through the atmosphere.

lateral shift—The offset of the position of the peak of an anomaly with the mass of magnetization (or gravitation).

lateral tilt—See **roll,** definition 2.

latitude—**1.** (general) A linear or angular distance measured north or south of the Equator on a sphere or spheroid. **2.** (plane surveying) The perpendicular distance in a horizontal plane of a point from an east-west axis of reference. See also **difference of latitude,** definition 2. **3.** (on a sphere) The angle at the center of a sphere between the plane of the Equator and the line to the point on the surface of the sphere. **4.** (traverse) The north-south component of a traverse course. **5.** (photography) The ability of an emulsion to record a range of brightness values. See also **argument of latitude; assumed latitude; astronomic latitude; authalic (equal-area) latitude; celestial latitude; circle of latitude; difference of latitude; fictitious latitude; galactic latitude; geocentric latitude; geodetic latitude; geographic latitude; geomagnetic latitude; grid latitude; isometric (conformal) latitude; latitude correction; middle latitude; oblique latitude; parallel; parametric latitude; rectifying latitude; terrestrial latitude; transverse latitude; variation of latitude.**

latitude band—(JCS) Any latitudinal strip designated by accepted units of linear or angular measurement, which circumscribes the Earth. Also called **band; latitudinal band.**

latitude correction—The amount of the adjustment of observed gravity values to an arbitrarily chosen base latitude. Also, correction to latitude in a traverse course.

latitude determination, zenith-telescope method—A precise method of determining astronomic latitude by measuring the difference of the meridional zenith distances of two stars of known declination, one north and the other south of zenith. Also called **Horrebow-Talcott method of latitude determination.**

latitude difference—(plane surveying) Length of the projection of a traverse course onto a meridian. Also called **northing; southing.**

latitude equation—A condition equation which expresses the relationship between the fixed latitudes of two points which are connected by triangulation or traverse.

latitude factor—The change in latitude along a celestial line of position per one minute change in longitude.

latitude level—A sensitive spirit level attached to the telescope of an instrument employed for observing astronomic latitude, in such manner that when the telescope is clamped in position, the level measures, in a vertical plane, variations in the direction of the line of collimation.

latitude of the line—See **difference of latitude,** definition 2.

latitudinal band—See **latitude band.**

latitudinal curve—This term denotes an easterly and westerly property line adjusted to the same mean bearing from each monument to the next one in regular order, as distinguished from the long chord or great circle that would connect the initial and terminal points.

lattice—(JCS) A network of intersecting positional lines printed on a map or chart from which a fix may be obtained.

law of propagation of error—The probable (standard) error of the sum of two or more quantities is equal to the square root of the sum of the squares of their probable (standard) errors.

law of universal gravitation—See **constant of gravitation.**

lay—**1.** See **layout. 2.** To assemble a photo mosaic. Often referred to by the method used in assembly, such as **wet lay, staple lay,** etc.

laydown—A mosaic. Often used to designate a mosaic temporarily assembled from uncropped prints.

layer tints—See **hypsometric tinting.**

layout—The planned positioning of reproduction material to fit the requirements and limitations of lithographic plates, paper, and finishing. Also called **lay.** See also **flat,** definition 1.

layout guide—See **key flat.**

lead line—A long, graduated chain or line at the end of which is attached a lead weight, used to measure depths of water. The lead line is usually used when making soundings by hands in water less than 25 fathoms deep.

leap-frog method—A rapid means of obtaining elevations of stations along a route between two base stations, or to obtain a closed loop of altimeter elevations. The system uses four barometers operating in pairs. One pair of barometers remains at the base station while the other pair is advanced to the first station at which time barometer and weather conditions are read and recorded simultaneously. The original base station pair are advanced to the second station and the process repeated. This method does not produce reliable elevations. The two-base method is considered better.

leap second—The step adjustment made to UTC to compensate for the approximately 1 second that the time transmitted by UTC signals gains on UT1 or UT2 each year. Normally, UTC is decreased by exactly 1 second (i.e., the leap second) at 24^h on the last day of December and/or June. See also **international atomic time; UT0; UT1; UT2; universal time coordinated.**

least count—(micrometer or vernier) The finest reading that can be made directly (without estimation) from a vernier or micrometer.

least squares—A method of adjusting observations in which the sum of the squares of all the deviations or residuals derived in fitting the observations to a mathematical model is made a minimum. Such an adjustment is based on the assumption that blunders and systematic errors have been removed from the data, and that only random errors remain. See also **adjustment of observations.**

left bank—That bank of a stream or river on the left of the observer when he is facing in the direction of flow, or downstream.

Legendre polynomial—A special case of the associated Legendre function in which the function becomes a polynomial.

Legendre's theorem—The lengths of the sides of a spherical triangle (very short by comparison with the radius of the sphere) are equal to the lengths of the corresponding sides of a plane triangle in which the plane angles are derived by reducing each of the spherical angles by approximately one-third of the spherical excess.

Lehmann's method—See **triangle-of-error method.**

length correction—(taping) The difference between the nominal length of a tape and its effective length under conditions of standardization. The standard length of a tape is usually expressed by a number of whole units (the nominal length) plus or minus a small distance which is the length correction defined above.

length equation—A condition equation which expresses the relationship between the fixed lengths of two lines which are connected by triangulation.

length of degree—The length of a degree of latitude measured along a meridian of longitude. The length varies somewhat with the degrees of latitude. Those near the pole are longer and those near the Equator are shorter. The length also varies with different selections of spheroids.

Lenoir level—An instrument which has the telescope passing through steel blocks, one near each end, whose upper and lower faces are plane and closely parallel; the lower faces rest upon a brass circle; the upper faces support a spirit level, which is reversed in leveling the instrument.

lens—A disk of optical glass, or plastic, or a combination of two or more such disks, by which rays of light may be made to converge or to diverge. Such disks have two surfaces, which may both be spherical, one plane and one spherical, or various other combinations (cylindrical, paraboloid, or hyperboloid). See also **achromatic lens; anastigmatic lens; aplanatic lens; apochromatic lens; astigmatizer; aspherical lens; coated lens; compensating lens; convertible lens; cylindrical lens; eyepiece; Fresnel lens; Metrogon lens; narrow-angle lens; negative lens; normal-angle lens; objective lens; positive lens; process lens; spherical lens; superwide-angle lens; thick lens; thin lens; wide-angle lens.**

lens axis—See **optical axis.**

lens calibration—See **camera calibration.**

lens component—See **lens element.**

lens distortion—(JCS) Image displacement caused by lens irregularities and aberrations.

lens element—One lens of a complex lens system. In a photographic lens, the terms front element and rear element are often used. Also called **lens component.**

lens speed—See **relative aperture.**

lensatic compass—A type of compass equipped with a lens which permits the observer to read the far side of the movable dial.

level constant—The amount by which the actual line of sight through a leveling instrument (when the bubble is centered in its vial) departs from the truly horizontal line through the center of the instrument, computed in millimeters per millimeter of stadia interval. When leveling rods graduated in yards instead of meters are used, the level constant, C, would be expressed in milliyards per milliyard of stadia interval. Also called **C-constant.**

level control—A series of bench marks or other points of known elevation, established throughout a project.

level correction—That correction which is applied to an observed difference of elevation to correct for the error introduced by the fact that the line of sight through the leveling instrument is not absolutely horizontal when the bubble is centered in its vial. See also **level constant.**

level line—**1.** A line on a level surface; therefore a curved line. **2.** A line over which leveling operations are accomplished. See also **duplicate level line; line of levels; multiple level line; simultaneous level line; spur line of levels.**

level net—See **survey net,** definition 2.

level rod—See **leveling rod.**

level surface—See **equipotential surface.**

level trier—An apparatus for use in measuring the angular value of the divisions of a spirit level.

leveling—**1.** (surveying) The operations of measuring vertical distances, directly or indirectly, to determine elevations. See also **astrogeodetic leveling; astrogravimetric leveling; barometric leveling; differential leveling; direct leveling; first-order leveling; flying levels; geodetic leveling; profile leveling; reciprocal leveling; second-order leveling; spirit leveling; stadia trigonometric leveling; thermometric leveling; third-order leveling; three-wire leveling; trigonometric leveling; vertical angulation; water leveling.** **2.** (photogrammetry) In absolute orientation, the operation of bringing the model datum parallel to a reference plane, usually the tabletop of the stereoplotting instrument. Also called **horizontalizing the model; leveling the model.** See also **aeroleveling; orientation,** definition 7.

leveling error of closure—See **error of closure,** definition 4.

leveling instrument—An instrument used for determining differences of elevations between points. See also **Abney level; dumpy level; Egault level; engineer's level; first-order level; Fischer level; Gravatt level; hand level; hanging level; latitude level; Lenoir level; locator's hand level; military level; pendulum level; plate level; precise level; reversible level; self-leveling level; spirit level; Stampfer level; striding level; telescope level; tilting level; Troughton level; U.S. Geological Survey level; wye (Y) level.**

leveling rod—A straight rod or bar, designed for use in measuring a vertical distance between a point on the ground and the line of collimation of a leveling instrument which has been adjusted to a horizontal position. Also called **level rod; rod.** See also **Barlow leveling rod; Boston leveling rod; double-target leveling rod; foot-meter rod; Gravatt leveling rod; Invar leveling rod; long rod; Molitor precise leveling rod; New York leveling rod; Pemberton leveling rod; Philadelphia leveling rod; precise leveling rod; range rod; self-reading leveling rod; short rod; single-target leveling rod; Stephenson leveling rod; tape rod; target leveling rod; U.S. Engineer precise leveling rod; U.S. Geological Survey precise leveling rod.**

leveling the model—See **leveling,** definition 2.

library negative mold—A negative mold which has been extended to a size compatible with the printed plastic map and forming equipment, and which is kept in file for subsequent castings.

libration—A real or apparent oscillatory motion, particularly the apparent oscillation of the Moon, which results in more than half of the Moon's surface being revealed to an observer on the Earth, even though the same side of the Moon is always toward the Earth because the Moon's periods of rotation and revolution are the same.

lift—See **selection overlay.**

limb—**1.** The graduated curved part of an instrument for measuring angles, as that part of a marine sextant carrying the altitude scale, or arc. **2.** The circular outer edge of a celestial body. See also **lower limb; upper limb.**

limit of reliable photo coverage—A label placed along a dashed line separating reliable photo compilation from map compilation on a chart. The label is always placed on the photo compilation side of the limit line.

line copy—Any copy suitable for reproduction without using a screen; copy composed of lines as distinguished from continuous-tone copy.

line map—See **planimetric map.**

line of apsides—The major axis of an elliptical orbit extended indefinitely in both directions. Also called **apse line.**

line of collimation—(optics) The line through the second nodal point of the objective lens of a telescope and the center of the reticle. Also called **aiming line; line of sight; pointing line; sight line.**

line of constant scale—Any line on a photograph which is parallel to the true horizon or to the isometric parallel. Also called **line of equal scale.**

line of equal scale—See **line of constant scale.**

line of force—A line indicating the direction in which a force acts, as in a magnetic field.

line of levels—A continuous series of measured differences of elevation. The individual measured differences may be single observations in the case of single-run leveling or the means of repeated observations in the case of double-run leveling.

line of nodes—The straight line connecting the two points of intersection of the ecliptic with the orbit of a planet, planetoid, or comet; or the line of intersection of the planes of the orbit of a satellite and the equator of its primary.

line of position—A line indicating a series of possible positions, determined by observation or measurement.

line of sight—**1.** The straight line between two points. This line is in the direction of a great circle, but does not follow the curvature of the Earth. **2.** The line extending from an instrument along which distant objects are seen, when viewed with a telescope or other sighting device. Also called **aiming line. 3.** (optics) See **line of collimation.**

line of soundings—A series of soundings obtained by a vessel underway, usually at regular intervals.

line pattern—A photographic negative containing parallel lines of equal-sized widths, which are equally spaced. Line patterns are used for printing tones of a color or to present a pattern of coverage for a chart feature. See also **area pattern screen.**

line rod—See **range rod.**

line-route map—A map or overlay for signal communication operations that shows the actual routes and types of construction of wire circuits in the field. It also gives the locations of switchboards and telegraph stations.

line tree—(USPLS) A tree intersected by a surveyed line, reported in the field notes of the survey, and marked with two hacks or notches cut on each of the sides facing the line. Also called **sight tree.**

lineal convergency—The length by which meridians approach one another when extended from one parallel to another.

linear building frontage—(CENTO, NATO) In air photographic interpretation, the side elevation of structures of homogeneous area.

linear distortion—The failure of a lens to reproduce accurately to scale all distances in the object.

linear error—A one-dimensional error (such as an error in elevation) defined by the normal distribution function.

linear error of closure—The straight-line distance by which a traverse fails to close.

linear magnification—The ratio of a linear quantity in the image to a corresponding linear quantity in the object. It may be lateral magnification or longitudinal magnification.

linear parallax—See **absolute stereoscopic parallax.**

lines of communications (LOC)—(JCS) All the routes, land, water, and air, which connect an operating military force with a base of operations, and along which supplies and reinforcements move.

lines on a spheroid—Any direct line between two positions on a spheroid, represented by two points on the Earth. Such a line may be one of mathematical definition, or it may result from a direct survey between the points on the Earth. See also **curve of alignment; geodesic line; normal section line.**

lining pole—See **range rod.**

link—**1.** (leveling) A line, a part of a line, or a combination of lines or parts of lines of levels, which, taken as a unit, make a continuous piece of leveling directly from one junction bench mark to another junction bench mark without passing through or over any other junction bench marks. Also called **link of levels. 2.** A unit of linear measure, one one-hundredth of a chain, and equivalent to 7.92 inches. See also **chain.**

link of levels—See **link,** definition 1.

liquid hand compass—A type of hand-held compass wherein the compass card is damped through the action of a liquid.

list—See **x-tilt.**

list of directions—A listing of objects observed at a triangulation station, together with the horizontal directions in terms of arc of the circle, referred to one of the objects observed as a zero initial.

litho copy—See **lithographic copy.**

lithographic copy—A graphic reproduced by the lithographic process. Also called **litho copy.**

lithographic drafting—See **tusching.**

lithography—A planographic method of printing based on the chemical repulsion between grease and water to separate the printing from non-printing areas. See also **offset lithography; photolithography.**

lithosphere—The solid part of the Earth or other spatial body. Distinguished from the atmosphere and the hydrosphere.

local adjustment—See **station adjustment.**

local apparent time—The apparent solar time for the meridian of the observer.

local astronomic time—Mean time reckoned from the upper branch of the local meridian.

local attraction—See **local magnetic anomaly.**

local chart—A large-scale aeronautical chart designed for contact flight in a congested area.

local civil time—See **local mean time.**

local coordinate system—A right-handed rectangular coordinate system of which the z-axis coincides with the plumb line through the origin.

local datum—The point of reference of the geodetic control used exclusively in a small area. Usually identified by a proper name.

local horizon—See **apparent horizon.**

local hour angle—Angular distance west of the local celestial meridian; the arc of the celestial equator, or the angle at the celestial pole, between the upper branch of the local celestial meridian and the hour circle of a point on the celestial sphere, measured westward from the local celestial meridian through 360°.

local lunar time—The arc of the celestial equator, or the angle at the celestial pole, between the lower branch of the local celestial meridian and the hour circle of the Moon, measured westward from the lower branch of the local celestial meridian through 24 hours; local hour angle of the Moon, expressed in time units, plus 12 hours. See also **Greenwich lunar time.**

local magnetic anomaly—Abnormal or irregular variation of the Earth's magnetic field extending over a relatively small area, due to local magnetic influences. Also called **anomalous magnetic variation; local attraction; local magnetic disturbance; magnetic anomaly.**

local magnetic disturbance—See **local magnetic anomaly.**

local mean time—**1.** (JCS) The time interval elapsed since the mean Sun's transit of the observers' antimeridian. **2.** The arc of the celestial equator, or the angle at the celestial pole, between the lower branch of the local celestial meridian and the hour circle of the mean Sun, measured westward from the lower branch of the local celestial meridian through 24 hours; local hour angle of the mean sun, expressed in time units, plus 12 hours. Called **local civil time** in United States terminology from 1925 through 1952. See also **Greenwich mean time; local astronomic time.**

local meridian—The meridian through any particular place or observer, serving as the reference for local time. Also called **reference meridian.**

local sidereal time—The local hour angle of the vernal equinox, expressed in time units. Local sidereal time at the Greenwich meridian is called Greenwich sidereal time.

local time—**1.** Time based upon the local meridian as reference, as contrasted with that based upon a time zone meridian, or the meridian of Greenwich. **2.** Any time kept locally.

local vertical—The direction of the acceleration of gravity, as opposed to the normal to a reference surface.

location survey—The establishment on the ground of points and lines in positions which have been previously determined by computation or by graphical methods, or by description obtained from data supplied by documents of record, such as deeds, maps, or other sources.

locator's hand level—A hand-held type of level used to measure approximate differences in elevation.

locking angle—In tilt analysis of oblique photographs, the complement of the interlocking angle. The depression angle of the oblique when the tilt of the vertical photograph is zero.

logical contouring—A procedure, based on the facts that contours are spaced equally along a uniform slope, which permits the sketching of contours from field notes with considerable accuracy and without the need of running a level line for every contour. Contour lines are interpolated by spacing them proportionately between spot elevations established at every point where there is a change in slope.

long chord—(route surveying) On a simple curve, the chord, or straight line, that extends from the point of curvature to the point of tangency; on a compound curve, the chord that extends from the point of compound curvature to the point of curvature or to the point of tangency. In a description of a circular land boundary, the length and bearing of the long chord is an important factor.

long line azimuth (LOLA) surveys—A measurement by use of photorecording theodolites and airborne strobe lights of long azimuth lines not visible between ground stations.

long-period constituent—A tide or tidal current constituent with a period that is independent of the rotation of the Earth but which depends upon the orbital movement of the Moon or of the Earth. The period is usually longer than a day and in general a half-month or larger.

long period perturbations—Periodic perturbations in the orbit of a planet or satellite which require more than one orbital period to execute one complete periodic variation.

long-range chart—See **long-range navigation chart.**

long-range navigation chart—Any one of a series of small scale, 1:3,000,000 or smaller, aeronautical charts designed for long flights using dead reckoning and celestial navigation as the principal means of navigation. Also called **long-range chart.**

long rod—A level rod, usually a Philadelphia rod, permitting readings of 13 feet when fully extended. Also called **high rod.** See also **short rod.**

longitude—A linear or angular distance measured east or west from a reference meridian (usually Greenwich) on a sphere or spheroid. See also **assumed longitude; astronomic longitude; celestial longitude; circle of longitude; difference of longitude; fictitious longitude; galactic longitude; geodetic longitude; geographic longitude; grid longitude; meridian; oblique longitude; terrestrial longitude; transverse longitude.**

longitude difference—See **departure,** definition 1.

longitude equation—A condition equation which expresses the relationship between the fixed longitudes of two points which are connected by triangulation or traverse.

longitude factor—The change in longitude along a celestial line of position per one minute change in latitude.

longitude of the Moon's nodes—The angular distances along the ecliptic of the Moon's nodes from the vernal equinox; the nodes have a retrograde motion, and complete a cycle of 360° in approximately 19 years.

longitude signal—A sign indicating a time event, observable at different stations, and used in comparing local times of those stations, and determining the difference of their longitudes.

longitude term gravity formula—An additional term in the formula for theoretical gravity which expresses the variation with longitude due to a triaxial ellipsoid of reference. See also **Helmert's gravity formula of 1915.**

longitudinal chromatic aberration—An aberration which affects the sharpness of all parts of an image because different colors come to a focus at different distances from the lens.

longitudinal magnification—The ratio of a length in the image, parallel to the axis, to a corresponding length in the object.

longitudinal separation—Time separation.

longitudinal tilt—See **pitch,** definition 2.

look angles—The elevation and azimuth at which a particular satellite is predicted to be found at a specified time. See also **alerts.**

loop closure—(leveling) The difference between the rod sum on the run out and the rod sum on the run back.

loop error of closure—See **error of closure,** definition 5.

loop traverse—A closed traverse that starts and ends at the same station. The traverse provides neither inherent validation of starting position and azimuth, nor validation against systematic distance error. See also **connecting traverse.**

Lorac—A trade name for a hyperbolic radio location system. This term is derived from the words "long-range accuracy."

loran—(JCS) A long-range radio navigation position fixing system using the time difference of reception of pulse type transmissions from two or more fixed stations. [This term is derived from the words "long-range navigation."]

loran C—A long-range radio navigation position fixing system using a combination of time difference of reception and phase difference of signals from two stations to provide a line of position.

loran chart—A plotting chart on which loran ground wave line of positions and sky-wave correction values have been printed, for use in loran navigation.

loran lines—Lines of constant time difference between signals from a master and a slave loran station.

loran tables—Publications containing tabular data for constructing loran hyperbolic lines of position.

lorop photography—A general term referring to any photographs taken with a long-focal-length (in excess of 100 inches) camera with a narrow-angle lens. The term is derived from the words "long-range oblique photography."

lorhumb line—A line along which the rates of change of the values of two families of hyperbolae are constants.

lost corner—(USPLS) A corner whose position cannot be determined, beyond reasonable doubt, either from traces of the original marks or from acceptable evidence or testimony that bears on the original position, and whose location can be restored only by reference to one or more interdependent corners.

Lovar—A steel alloy having a low coefficient of expansion used in construction of precise Lovar tapes. See also **Invar.**

Lovar tape—A newer version of the Invar tape used in surveying operations. Lovar tape possesses properties and cost factors between that of the less accurate steel tape and the more accurate Invar tape.

low-oblique photograph—See **oblique air photograph.**

low tide—See **low water.**

low water (LW)—The lowest limit of the surface water level reached by the lowering tide. Low water is caused by the astronomic tide-producing forces and/or the effects of meteorological conditions. Also called **low tide.**

low water datum—An approximation of the plane of mean low water, adopted as a standard datum plane for a limited area, and retained for an indefinite period, even though it might differ slightly from a better determination of mean low water from later observations.

low water full and change—The average interval of time between the transit (upper or lower) of the full or new Moon and the next low water.

low water inequality—See **diurnal inequality.**

low water interval—See **lunitidal interval.**

low water line—The line defined by the boundary of a body of water at its lowest stage (elevation). In tidal waters, the low water line is, strictly, the intersection of the plane of mean low water with the shore.

low water lunitidal interval—See **lunitidal interval.**

low water springs—See **mean low water springs.**

low water springs datum—An approximation of the plane of mean low water springs, used as a datum in local areas, and retained for an indefinite period, even though it might differ slightly from a better determination of mean low water springs from later observations.

lower branch—That half of a meridian or celestial meridian from pole to pole which passes through the antipode or nadir of a place.

lower culmination—See **culmination.**

lower high water (LHW)—The lower of two high waters of any tidal day where the tide exhibits mixed characteristics.

lower high water interval (LHWI)—The interval of time between the transit (upper or lower) of the Moon over the local or Greenwich meridian and the next lower high water. This expression is used when there is considerable diurnal inequality. See also **lunitidal interval.**

lower limb—That half of the outer edge of a celestial body having the least altitude, in contrast with the upper limb, that half having the greatest altitude.

lower low water (LLW)—The lower of two low waters of any tidal day where the tide exhibits mixed characteristics.

lower low water datum—An approximation of the plane of mean lower low water, adopted as a standard datum plane for a limited area, and retained for an indefinite period, even though it might differ slightly from a better determination of mean lower low water from later observations. Used in engineering design of harbor facilities and dredging, when there is a material difference between mean lower low and mean low datums.

lower low water interval (LLWI)—The interval of time between the transit (upper or lower) of the Moon over the local or Greenwich meridian and the next lower low water. This expression is used when there is considerable diurnal inequality. See also **lunitidal interval.**

lower motion—(surveying) The rotation of the lower plate of a repeating instrument.

lower transit—Transit of the lower branch of the celestial meridian. Also called **inferior transit.**

lowest low water—A plane of reference whose distance below mean sea level corresponds with the mean level of lowest low water of any normal tide. Also called **lowest normal low water.**

lowest low water springs—A plane of reference approximating the mean lowest low water during syzygy (spring tides).

lowest normal low water—See **lowest low water.**

loxodrome—See **rhumb line.**

loxodromic curve—See **rhumb line.**

lunar celestial equator—A great circle on the celestial sphere in the plane of the Moon's equator, i.e., in a plane perpendicular to the Moon's axis of rotation.

lunar chart—A chart showing the surface of the Moon.

lunar cycle—Any cycle related to the Moon, particularly the Callippic cycle or the Metonic cycle. See also **saros.**

lunar day—The interval between two successive upper transits of the Moon over a local meridian. The period of the mean lunar day, approximately 24.84 solar hours, is derived from the rotation of the Earth on its axis relative to the movement of the Moon about the Earth. Also called **tidal day.**

lunar declination—Angular distance of the Moon expressed in degrees north or south of the celestial equator; it is indicated as positive when north, and negative when south of the equator. Also called **declination of the Moon.**

lunar distance—The angle between the line of sight toward the Moon and the line of sight toward another celestial body at the point of an observer on the Earth.

lunar earthside chart—A chart showing that portion of the Moon's surface visible from the Earth.

lunar eclipse—The phenomenon observed when the Moon enters the shadow of the Earth. A lunar eclipse is partial if only part of its surface is obscured; and total if the entire surface is obscured.

lunar equation—A factor used to reduce observations of celestial bodies to the barycenter of the Earth-Moon system.

lunar farside chart—A chart showing that portion of the Moon's surface not visible from the Earth.

lunar inequality—**1.** Variation in the Moon's motion in its orbit, due to attraction by other bodies of the solar system. See also **evection; perturbation. 2.** A minute fluctuation of a magnetic needle from its mean position, caused by the Moon.

lunar interval—The difference in time between the transit of the Moon over the Greenwich meridian and a local meridian. The lunar interval equals the difference between the Greenwich and local intervals of a tide or current phase.

lunar month—See **synodical month.**

lunar node—A node of the Moon's orbit.

lunar parallax—The horizontal parallax or the geocentric parallax of the Moon.

lunar satellite—A manmade satellite that makes one or more revolutions about the Moon.

lunar tide—That part of the tide caused solely by the tide-producing forces of the Moon as distinguished from that part caused by the forces of the Sun.

lunar time—**1.** Time based upon the rotation of the Earth relative to the Moon. **2.** Time on the Moon. See also **Greenwich lunar time; local lunar time.**

lunation—See **synodical month.**

lune—That part of the surface of a sphere bounded by halves of two great circles.

lunicentric—See **selenocentric.**

lunisolar effect—Gravitational effects caused by the attraction of the Moon and of the Sun.

lunisolar perturbation—Perturbations of artificial satellite orbits due to the attractions of the Sun and the Moon. The most important effects are secular variations in the mean anomaly, in the right ascension of the ascending node, and in the argument of perigee. All other orbital elements, except the major semiaxis, undergo long periodic changes.

lunisolar precession—That component of general

precession caused by the combined effect of the Sun and Moon on the equatorial perturbance of the Earth, producing a westward motion of the equinoxes along the ecliptic.

lunitidal interval—The interval between the Moon's transit (upper or lower) over the local or Greenwich meridian and the following high or low water. The average of all high water intervals for all phases of the Moon, the mean high water lunitidal interval, is abbreviated to high water interval. Similarly, the mean low water lunitidal interval is abbreviated to low water interval. The high water or low water interval is described as local or Greenwich according to whether the reference is to the transit over the local or the Greenwich meridian. Also called **establishment.** .

M

magazine—(aerial camera) A component in the aerial camera system. It serves to hold the exposed and unexposed film and includes the film drive mechanism and film flattening device (platen).

magenta contact screen—A contact film screen composed of magenta dyed dots of variable density used for making halftone negatives.

magnetic amplitude—Amplitude relative to magnetic east or west. See also **amplitude.**

magnetic annual change—The amount of magnetic secular change undergone in 1 year. Also called **annual change; annual magnetic change; annual rate; annual rate of change.** See also **magnetic secular change.**

magnetic annual variation—The small regular fluctuation in the Earth's magnetism, having a period of 1 year. Also called **annual magnetic variation.**

magnetic anomaly—See **local magnetic anomaly.**

magnetic azimuth—At the point of observation, the angle between the vertical plane through the observed object and the vertical plane in which a freely suspended, symmetrically magnetized needle, influenced by no transient artificial magnetic disturbance, will come to rest. Magnetic azimuth is generally reckoned from magnetic north (0°) clockwise through 360°. Such an azimuth should be marked as being magnetic, and the date of its applicability should be given. Magnetic azimuths are frequently measured with two Wild T-0 theodolites in connection with compass rose surveys.

magnetic bearing—**1.** (navigation) Bearing relative to magnetic north; compass bearing corrected for deviation. **2.** (surveying) Bearing relative to magnetic north or south, and stated in an east or west direction. The compasses included with the engineer transit and the 1-minute theodolite can be used to read magnetic bearings. Also called **compass bearing.**

magnetic chart—A special-purpose map depicting the distribution of one of the magnetic elements, as by isogonic lines, or of its secular change.

magnetic compass—A compass depending upon the attraction of the magnetism of the Earth for its directive force.

magnetic daily variation—The transient change in the Earth's magnetic field associated with the apparent daily motions of the Sun and Moon. In most places the solar daily variation follows approximately a consistent pattern, although with appreciable and unpredictable changes in form and amplitude.

magnetic declination—(JCS) The angle between the magnetic and geographical meridians at any place, expressed in degrees east or west to indicate the direction of magnetic north from true north. In nautical and aeronautical navigation the term **magnetic variation** is used instead of **magnetic declination,** and the angle is termed **variation of the compass** or **magnetic variation.** Magnetic declination is not otherwise synonymous with **magnetic variation,** which refers to regular or irregular change with time of the magnetic declination, dip, or intensity. Because of local attraction, the magnetic declination of two close points may differ by several degrees.

magnetic deviation—See **deviation.**

magnetic dip—See **dip,** definition 2.

magnetic dip circle—An instrument for measuring the magnetic dip by the use of a needle and a graduated vertical circle.

magnetic dip needle—**1.** A dip circle or the needle thereof. Also called **dip needle. 2.** A needle arranged to disclose an intense local anomaly of the magnetic dip and useful in the recovery of lost iron survey monuments.

magnetic direction—Horizontal direction expressed as angular distance from magnetic north.

magnetic disturbance—**1.** Irregular, large amplitude, rapid time changes of the Earth's magnetic field which occur at approximately the same time all over the Earth. Also called **magnetic storm. 2.** Sometimes used to describe spatial changes in the Earth's magnetic field. See also **local magnetic anomaly.**

magnetic diurnal variation—**1.** The daily variation. **2.** The simple harmonic component of the daily variation having a period of 24 hours.

magnetic elements—The declination, the horizontal intensity, the vertical intensity, the total intensity, the inclination or dip, the strength of the force toward geographic north, and the strength of the force toward geographic east.

magnetic equator—(JCS) A line drawn on a map or chart connecting all points at which the magnetic inclination (dip) is zero for a specified epoch. Also called **aclinic line; dip equator.** See also **geomagnetic equator.**

magnetic field—See **magnetic field intensity.**

magnetic field intensity—The magnetic force exerted on an imaginary unit magnetic pole placed at any specified point of space. It is a vector quantity. Its direction is taken as the direction toward which a north magnetic pole would tend to move under the influence of the field. Also called **magnetic field; magnetic field strength; magnetic force; magnetic intensity.**

magnetic field strength—See **magnetic field intensity.**

magnetic force—See **magnetic field intensity.**

magnetic inclination—See **dip,** definition 2.

magnetic intensity—See **magnetic field intensity.**

magnetic isoporic line—See **isopor.**

magnetic latitude—See **dip,** definition 2.

magnetic lines of force—Imaginary lines so drawn

in a region containing a magnetic field to be everywhere tangent to the magnetic field intensity vector if in vacuum or non-magnetic material, or parallel to the magnetic induction vector if in a magnetic medium.

magnetic lunar daily variation—A periodic variation of the Earth's magnetic field that is in phase with the transit of the Moon.

magnetic meridian—At any point, the direction of the horizontal component of the Earth's magnetic field. Not to be confused with **geomagnetic meridian.**

magnetic moment—The quanitity obtained by multiplying the distance between two magnetic poles by the average strength of the poles.

magnetic north—(JCS) The direction indicated by the north seeking pole of a freely suspended magnetic needle, influenced only by the Earth's magnetic field. See also **compass north.**

magnetic observation—Measurement of any of the magnetic elements.

magnetic pole—Either of the two places on the surface of the Earth where the magnetic dip is 90°. Not to be confused with **geomagnetic pole.** Also called **dip pole.**

magnetic prime vertical—The vertical circle through the magnetic east and west points of the horizon.

magnetic secular change—Increase or decrease of intensity and/or change of direction of the Earth's magnetic field over a period of many years; usually given as average gammas per year for intensity values and minutes per year for directional values.

magnetic solar daily variation—A periodic variation of the Earth's magnetic field that is in phase with solar (local) time.

magnetic station—A monumented station at which a series of magnetic observations has been made. It usually consists of a bronze marker set in stone or concrete at which, in addition to latitude and longitude, the magnetic value is indicated.

magnetic storm—See **magnetic disturbance.**

magnetic survey—A survey conducted to measure the strength and/or direction of the Earth's magnetic field at specific points on or near the surface of the Earth.

magnetic variation—(JCS) The horizontal angle at a place between the true north and magnetic north measured in degrees and minutes east or west according to whether magnetic north lies east or west of true north. Also called **variation.** See also **magnetic declination.**

magnetism—The ability to attract magnetic material, notably iron and steel. Also called **terrestrial magnetism.** See also **blue magnetism; geomagnetism; horizontal intensity; red magnetism; vertical intensity.**

magnetometer—An instrument used in the study of geomagnetism for measuring a magnetic element. See also **flux-gate magnetometer; flux-meter; nuclear precession magnetometer; optical pumping magnetometer; theodolite-magnetometer; variometer.**

magnetometer survey—A survey wherein the Earth's magnetic field is mapped by the use of a magnetometer. See also **magnetic survey.**

magnification—(optics) The ratio of the size of an object to the size of its magnified image. Also called **power of a lens.** See also **angular magnification; diopter; lateral magnification; linear magnification; longitudinal magnification; unidimensional magnification.**

magnifying power—The ratio of the apparent length of a linear dimension as seen through an optical instrument, and by the unaided eye. Thus, an instrument with a magnifying power of three makes an object appear three times as high and three times as wide. Also called **diameter.** See also **diopter.**

magnitude—**1.** Relative brightness of a celestial body. Also called **stellar magnitude. 2.** The intensity of a short-period magnetic fluctuation, usually expressed in milligausses or gammas. **3.** Relating to amount, size, or greatness.

main-scheme station—A station through which the main computations and adjustments of the survey data are carried and serve for the continued extension of the survey. Also called **primary station; principal station.**

major axis—The longest diameter of an ellipse or ellipsoid.

major datum—See **preferred datum.**

major grid—The military grid(s) which is designated as the official grid for an area, and is depicted on a map sheet by full grid lines. See also **overlapping grid; secondary grid.**

major planets—The four largest planets: Jupiter, Saturn, Uranus, and Neptune. Also called **giant planets.** See also **inner planets.**

make line—An accurately scaled line denoting the size to which original copy is to be enlarged or reduced. Also called **make size.** See also **scale of reproduction.**

make size—See **make line.**

makeready—The adjustment of feeder, grippers, side guide, pressure between plate and blanket cylinder, impression plate, and ink fountain prior to a press run.

maneuvering board—A polar coordinate plotting sheet devised to facilitate solution of problems involving relative movement.

manmade features—See **culture.**

manuscript—The original drawing of a map as compiled or constructed from various data, such as ground surveys and photographs. See also **multiuse manuscript.**

map—**1.** (JCS) A graphic representation, usually on a plane surface and at an established scale, of natural and artificial features on the surface of a part or the whole of the Earth or other planetary body. The features are positioned relative to a coordinate reference system. **2.** To prepare a map or engage in a mapping operation. See also **administrative map; base map; battle map; boundary map; cadastral map; chorographic map; compiled map; contour map; controlled map; county map; distribution map; domestic map; engineering map; experimental map; flight map; flood control map; fluorescent map; forestry map; general map; general-purpose map; geological map; gravimetric map; gravity anomaly map; hemispherical map; hypsographic map (or chart); landscape map; large-scale map; line-route map; medium-scale map; morphographic map; native map; operation map; orthophotomap; orthopictomap; outline map; photo-contour map; photo-revised map; photocontrol index map; photogrammetric map; photomap; physiographic pictorial map; pictogram; pictomap; planetable map; planimetric-base map; planimetric map; planning map; plastic relief map; port plan; provisional map; quadrangle; radar intelligence map; radar map; reconnaissance map; red-light-readable map; relief model; road map; route map; shaded-relief map; situation map; sketch map; small-scale map; source map; special-purpose map; special job-cover map; standard-accuracy map; standard-content map; state base map; strategic map; Tactical Commanders' Terrain Analysis; tactical map; topical map; topographic map; Topographic Map of the United States; traffic-circulation map; wall map; weather map.**

map accuracy specifications—Specifications which set up standards to which the finished map must adhere. See also **U.S. National Map Accuracy Standards.**

map accuracy standards—See **U.S. National Map Accuracy Standards.**

map adjustment—An adjustment of the horizontal position of maps to control points or to a specific grid plotted on the map projection at compilation scale.

map catalog—A publication giving both graphical and word descriptions of all maps, charts, and related products issued by a producing agency. It contains information such as the title, scale, edition date, edition number, price, and classification of all publications issued.

map chart—See **combat chart.**

map controlled—Utilization of a map, rather than geodetic or photogrammetric data, for purposes of positioning map detail.

map-controlled mosaic—A technique of constructing mosaics by using topographic maps as the basis for control and orientation purposes. The method may be used in preparing both controlled and semicontrolled mosaics although its use is preferred with the latter.

map exchange agreement—An approved agreement between a United States and a foreign mapping organization to furnish each other specified mapping, charting, and geodetic data as published, or on a request basis. See also **cooperative mapping agreement.**

map index—(JCS) Graphic key primarily designed to give the relationship between sheets of a series, their coverage, availability, and further information on the series.

map matching—The simultaneous electronic or mechanical-optical scanning of an observed map image obtained from a space vehicle, and a reference map image, while the reference map image is being oriented and scaled until a close comparison between the two is found. An inspection of the scale and orientation of the reference map indicates the position of the vehicle. See also **pulse Doppler map matching.**

map-matching guidance—The guidance of a rocket or aerodynamic vehicle by means of a radarscope film previously obtained by a reconnaissance flight over the terrain of the route or from a radar simulation system, and used to direct the vehicle by aligning itself with radar echoes received during flight from the terrain below. See also **stellar map matching.**

map nadir—Point on a map or manuscript vertically beneath the perspective center of the camera lens at the instant of exposure.

map of standard format—A map with dimensions, layout, lettering, and symbolization in accordance with the specifications for the series.

map parallel—See **axis of homology.**

map point—A supplemental control point whose horizontal position can be obtained by scaling the coordinates from a map or chart on which the point can be identified.

map projection—A systematic drawing of lines on a plane surface to represent the parallels of latitude and the meridians of longitude of the Earth or a section of the Earth. A map projection may be established by analytical computation or may be constructed geometrically. A map projection is frequently referred to as a "projection" but the complete term should be used unless the context clearly indicates the meaning. See also **Aitoff equal-area map projection; Albers conical equal-map projection; aphylactic map projection; authalic map projection; azimuthal map projection; azimuthal equidistant map projection; Bonne map projection; Cassini map projection; Cassini-Soldner map projection; conformal map projection; conic map projection; conic map projection with two standard parallels; cylindrical equal-area map projection; cylindrical equal-spaced map projection; cylindrical map projection; doubly azimuthal map projection; equal-area map projection; equatorial map projection; globular map projection; gnomonic map projection; Goode's interrupted homolosine projection; Hammer projection; homolographic map projection; interrupted map projection; Laborde map projection; Lambert az-**

imuthal polar map projection; Lambert central equivalent map projection upon the plane of the meridian; Lambert conformal conic map projection; Lambert zenithal equal-area map projection; Mercator map projection; meridional orthographic map projection; modified Lambert conformal map projection; modified polyconic map projection; Mollweide homolographic map projection; nonperspective azimuthal map projection; oblique map projection; oblique Mercator map projection; orthembadic map projection; orthographic map projection; perspective map projection; perspective map projection upon a tangent cylinder; polar map projection; polar orthographic map projection; polar stereographic map projection; polyconic map projection; polyhedric projection; rectangular map projection; rectangular polyconic map projection; simple conic map projection; sinusoidal map projection; skewed map projection; stereographic horizon map projection; stereographic map projection; stereographic meridional map projection; transverse map projection; transverse Mercator map projection; transverse polyconic map projection; Werner map projection.

map reference code—(JCS) A code used primarily for encoding grid coordinates and other information pertaining to maps. This code may be used for other purposes when the encryption of numerals is required.

map revision—See **revision.**

map scale—See **scale,** definition 1.

map series—(JCS) A collection of sheets usually having the same scale and cartographic specifications, collectively identified by the producing agency. Also called **series.**

map sheet—(JCS) An individual map or chart, either complete in itself or part of a series.

map substitute—A hasty reproduction of aerial photographs, photomaps, or mosaics, or of provisional maps, or any other product used in place of a map, when the precise requirements of a map cannot be met.

map test—The accuracy of topographic mapping can be tested by running traverse and level lines across selected areas of any map sheet, and comparing geographic positions of map features with those determined by traverse and comparing interpolated elevations of points from the map with those determined by the level line.

mapping camera—A camera specially designed for the production of photographs to be used in mapping. The modifier **mapping** or **surveying** indicates that the camera is equipped with mechanisms to maintain and to indicate the interior orientation of the photographs with sufficient accuracy for mapping purposes. A mapping camera may be an aerial-mapping camera or terrestrial-mapping camera. Also called **surveying camera.**

mapping, charting, and geodesy (MC&G)—MC&G comprises the collection, transformation, generation, dissemination, and storing of geodetic, geomagnetic, gravimetric, aeronautical, topographic, hydrographic, cultural, and toponymic data. These data may be used for military planning, training, and operations including aeronautical, nautical, and land navigation, as well as for weapon orientation and target positioning. MC&G also includes the evaluation of topographic, hydrographic, or aeronautical features for their effect on military operations or intelligence. The data may be presented in the form of topographic, planimetric, relief, or thematic maps and graphics; nautical and aeronautical charts and publications; and in simulated, photographic, digital, or computerized formats.

mapping photography—Aerial photography obtained by precisely calibrated mapping cameras and conforming to mapping specifications, as distinguished from aerial photography for other purposes. Also called **aerial cartographic photography; cartographic photography; charting photography; survey photography.**

March equinox—See **vernal equinox.**

margin data—(JCS) All explanatory information given in the margin of a map or chart which clarifies, defines, illustrates, and/or supplements the graphic portion of the sheet. Also called **border data; border information; margin information.**

margin information—See **margin data.**

marine map—See **hydrographic chart.**

marine sextant—A sextant designed primarily for marine navigation.

marine survey—See **oceanographic survey.**

mark—**1.** A call used when simultaneous observations are being made, to indicate to the second person the moment a reading is to be made, as when the time of a celestial observation is to be noted; or the moment a reading is a prescribed value. **2.** (surveying) A definite object, such as an imprinted metal disk, used to designate a survey point and sometimes refers to the entire survey monument. Mark is used with a qualifying term such as station, reference, or bench. See also **bench mark; reference mark; station mark; witness mark.** **3.** (photogrammetry) See **floating mark; index mark.**

Marsden chart—A chart showing the distribution of meteorological data, especially over the oceans.

mask—**1.** In photomechanical processing, to block out an area by means of actinically opaque material, to prevent exposure in the part blocked out. Also, the covering material itself when so applied. **2.** A clear stable base plastic, coated with an opaque stratum which can be peeled off between photographically etched outline images, thus producing an open-window negative of the desired area. This process of masking is often identified by the trade name of the material used. **3.** A continuous-tone positive or negative made from an original negative or positive for the purpose of altering the image produced from the original. Used to alter contrast, correct color portrayal, or produce pictotone or pictoline images.

masking—A means of controlling plastic expansion locally during forming of plastic relief maps to obtain more accurate register of preprinted line work to the landforms of the mold. By masking, differential heating is achieved. Also called **screening.**

masking paper—See **goldenrod paper.**

mass attraction vertical—The normal to any surface of constant geopotential. On the Earth this vertical is a function only of the distribution of the mass and is unaffected by forces resulting from the motions of the Earth; e.g., the direction of a plumb bob on a nonrotating Earth.

master film positive—A positive made from an original negative for the purpose of making additional negatives.

master glass negative—See **calibration plate.**

master model—The developed original terrain model which bears, in miniature, the same spatial relationships as the actual ground it represents. Also called **original model.**

master plot—(JCS) A portion of a map or overlay on which are drawn the outlines of the areas covered by an air photographic sortie. Latitude and longitude, and map and sortie information are shown.

master print—(mosaicking) A photograph which is representative of the mosaic area used as a guide during the developing process to insure the tonal match of subsequent prints.

master projection—The originally computed and constructed map projection from which copies are made; one such projection serves as the master for copies circling the globe within the same set of standard parallels.

master station—That station in a given system of transmitting stations that controls the transmissions of the other stations (the slave stations) and maintains the time relationship between the pulses of the stations. In satellite surveys, positions can be upgraded considerably by translocation. See also **translocation.**

match lines—A series of grease-pencil lines drawn on a photograph, radiating from the torn edges of the print onto the adjacent areas to serve as a registration guide when laying the individual print in the mosaic.

match strip—See **tie strip,** definition 1.

matching—The act by which detail or information on the edge, or overlap area, of a map or chart is compared, adjusted, and corrected to agree with the existing overlapping chart.

matte print—Print made on photographic paper with a dull finish which is more suitable for pencil or ink annotations than a glossy print.

mean anomaly—See **anomaly,** definition 3.

mean center of Moon—**1.** A central point for a lunar coordinate system. **2.** The point on the lunar surface intersected by the lunar radius that is directed toward the Earth's center when the Moon is at the mean ascending node and when the node coincides with the mean perigee or mean apogee.

mean chart—Any chart on which isopleths of the mean value of a given oceanographic element are drawn. Also called **mean map.**

mean deviation—See **average deviation.**

mean distance—See **semimajor axis.**

mean diurnal high water inequality (DHQ)—Half the average difference between the heights of the two high waters of each tidal day over a 19-year period, obtained by subtracting the mean of all high waters from the mean of the higher high waters.

mean diurnal low water inequality (DLQ)—Half the average difference between the heights of the two low waters of each tidal day over a 19-year period, obtained by subtracting the mean of the lower low waters from the mean of all low waters.

mean equinox—A fictitious equinox whose position is that of the vernal equinox at a particular date with the effect of nutation removed. Also called **mean equinox of date.**

mean equinox of date—See **mean equinox.**

mean free-air anomaly—The representative free air gravity value for a given geographic area; i.e., $1' \times 1'$ mean, $10' \times 10'$ mean, etc.

mean ground elevation—Average elevation of the terrain above mean sea level of an area to be photographed.

mean high water (MHW)—The average height of all the high waters recorded over a 19-year period, or a computed equivalent period.

mean high water lunitidal interval—See **lunitidal interval.**

mean high water springs (MHWS)—The average height of all high waters recorded during syzygy over a 19-year period, or a computed equivalent period. Also called **high water springs.**

mean higher high water (MHHW)—The average height of all the daily higher high waters recorded over a 19-year period or a computed equivalent period. It is usually associated with a tide exhibiting mixed characteristics.

mean higher high water springs (MHHWS)—The average height of all higher high waters recorded during syzygy over a 19-year period, or a computed equivalent period.

mean low water (MLW)—The average height of all low waters recorded over a 19-year period, or a computed equivalent period.

mean low water lunitidal interval—See **lunitidal interval.**

mean low water springs (MLWS)—The average

height of all low waters recorded during syzygy over a 19-year period, or a computed equivalent period.

mean lower low water (MLLW)—The average height of all the lower low waters recorded over a 19-year period. It is usually associated with a tide exhibiting mixed characteristics.

mean lower low water springs (MLLWS)—The average height of all lower low waters recorded during syzygy over a 19-year period, or a computed equivalent period.

mean map—See **mean chart.**

mean of the errors—The average value of a set of errors.

mean place—See **mean position.**

mean position—The position of a star corrected for secular variations including proper motion, but uncorrected for short term variations. Also called **mean place.**

mean range (Mn)—The difference in height between mean high water and mean low water, measured in feet or meters.

mean refraction—The refraction effect on vertical angles given usually in the plane of a vertical circle for average conditions of temperature and barometric pressure.

mean river level—The average height of the surface of a river at any point for all stages of the tide over a 19-year period usually determined from hourly height readings. Unusual variations of river level due to discharge or runoff may be excluded in computation.

mean sea level (MSL)—(JCS) The average height of the surface of the sea for all stages of the tide, used as a reference for elevations. [Usually determined by averaging height readings observed hourly over a minimum period of 19 years.] Also called **sea level datum.**

mean sidereal time—Sidereal time adjusted for nutation, to eliminate slight irregularities in the rate.

mean solar day—The interval of time from a transit of the mean sun across a given meridian to its next successive transit across the same meridian.

mean solar time—Time measured by the diurnal motion of a fictitious body, termed the mean sun, which is supposed to move uniformly in the celestial equator, completing the circuit in 1 tropical year. The mean sun may be considered as moving in the celestial equator and having a right ascension equal to the mean celestial longitude of the true Sun. Also called **mean time.**

mean sphere depth—The uniform depth to which the water would cover the Earth if the solid surfaces were smoothed off and were parallel to the surface of the geoid.

mean-square error—The quantity whose square is equal to the sum of the squares of the individual errors divided by the number of those errors.

mean sun—See **fictitious sun.**

mean tide level (MTL)—The reference plane midway between mean high water and mean low water. Also called **half-tide level; ordinary tide level.**

mean time—See **mean solar time.**

meander corner—(USPLS) A corner established at the intersection of standard, township, or section lines with the meander line near banks of navigable streams or any meanderable body of water.

meander line—A traverse of the margin of a permanent natural body of water, along the locus of the bank or shoreline at the elevation of mean or ordinary high water, upon which bank or shoreline a riparian right may be predicated.

measured angle—An angle as read directly from an instrumental observation and without any application of corrections for local conditions. A measured angle which has been corrected for local conditions only at the point of observation is considered as an observed angle.

mechanical arm templet—See **spider templet.**

mechanical templet—Any templet which is manipulated and adjusted mechanically in laying out a radial triangulation.

mechanical templet plot—See **mechanical templet triangulation.**

mechanical templet triangulation—A graphical radial triangulation using slotted, spider, or any form of mechanical templet. Also called **mechanical templet plot.**

medium-scale map—(JCS) A map having a scale larger than 1:600,000 and smaller than 1:75,000.

memorial—(USPLS) A durable article deposited in the ground at the position of a corner to perpetuate that position should the monument be removed or destroyed. The memorial is usually deposited at the base of the monument and may consist of anything durable such as glass or stoneware, a marked stone, charred stake, or a quantity of charcoal.

Mendenhall pendulum—An invariable pendulum one-quarter meter in length, with a vibration period of one-half second, composed of a lenticular-shaped bob on a thin stem, swung in an airtight case from which the air has been largely exhausted.

mensuration—**1.** The act, process, or art of measuring. **2.** That branch of mathematics dealing with the determination of length, area or volume.

Mercator bearing—See **rhumb bearing.**

Mercator chart—A chart on the Mercator projection. This is the chart commonly used for marine navigation. Also called **equatorial cylindrical orthomorphic chart.**

Mercator direction—Horizontal direction of a rhumb line, expressed as angular distance from a reference direction. Also called **rhumb dire**

tion.

Mercator equal-area map projection—See **sinusoidal map projection.**

Mercator map projection—A conformal map projection of the cylindrical type. The Equator is represented by a straight line true to scale; the geographic meridians are represented by parallel straight lines perpendicular to the line representing the Equator; they are spaced according to their distance apart at the Equator. The geographic parallels are represented by a second system of straight lines perpendicular to the family of lines representing the meridians and therefore parallel with the Equator. Conformality is achieved by mathematical analysis, the spacing of the parallels being increased with increasing distance from the Equator to conform with the expanding scale along the parallels resulting from the meridians being represented by parallel lines. Also called **equatorial cylindrical orthomorphic map projection.**

Mercator track—See **rhumb line.**

mercury barometer—A barometer in which atmospheric pressure is balanced against the weight of a column of mercury. See also **aneroid barometer; cistern barometer; siphon barometer.**

Mercury datum—A worldwide geodetic system derived from an analysis of data from astrogeodetic, gravimetric, and satellite sources. Results of this analysis provided a best-fitting world ellipsoid used in tracking Project Mercury manned space missions and as the reference datum for the electronic navigation systems—Omega, loran, and loran C. See also **Fischer ellipsoid of 1960.**

meridian—A north-south reference line, particularly a great circle through the geographical poles of the Earth, from which longitudes and azimuths are determined; or a plane, normal to the geoid or spheroid, defining such a line. See also **astronomic meridian; auxiliary guide meridian; celestial meridian; central meridian; convergence of meridians; double meridian distance; ecliptic meridian; fictitious meridian; geodetic meridian; geographic meridian; geomagnetic meridian; Greenwich meridian; grid meridian; guide meridian; gyro meridian; local meridian; magnetic meridian; meridional difference; meridional part; meridional plane; oblique meridian; photograph meridian; prime fictitious meridian; prime grid meridian; prime meridian; prime oblique meridian; prime transverse meridian; principal meridian; standard meridian; table of meridional parts; time meridian; transverse meridian; true meridian.**

meridian altitude—The altitude of a celestial body when it is on the celestial meridian of the observer.

meridian angle—Angular distance east or west of the local celestial meridian; the arc of the celestial equator, or the angle at the celestial pole, between the upper branch of the local celestial meridian and the hour circle of a celestial body, measured eastward or westward from the local celestial meridian through 180°, and labeled E or W to indicate the direction of measurement.

meridian angle difference—The difference between two meridian angles, particularly between the meridian angle of a celestial body and the value used as an argument for entering a table. Also called **hour angle difference.**

meridian distance—**1.** (astronomy) The hour angle of a celestial body when close to but not exactly on the astronomic meridian. **2.** (plane surveying) The perpendicular distance in a horizontal plane of a point from a meridian of reference. The difference of the meridian distances of the ends of a line is called the departure of the line.

meridian extension—That portion of a meridian shown above the top construction line of a projection.

meridian line—(plane surveying) The line of intersection of the plane of the celestial meridian and the plane of the horizon. It is a horizontal direction used in surveying; its astronomic azimuth is 0° or 180°.

meridian observation—Measurement of the altitude of a celestial body on the celestial meridian of the observer, or the altitude so measured.

meridian passage—See **meridian transit.**

meridian telescope—A portable instrument so designed that it can be used as an astronomic transit, or quickly converted for use as a zenith telescope.

meridian transit—The passage of a celestial body across a celestial meridian. Also called **meridian passage.** See also **culmination; transit,** definition 1.

meridional difference—The difference between the meridional parts of any two given parallels. This difference is found by subtraction if the two parallels are on the same side of the Equator, and by addition if on opposite sides.

meridional distance—The distance between latitude lines as determined from the mid-latitude of a map projection.

meridional interval—The value of the distance between meridians of a projection at chart scale.

meridional offsets—Small distances applied to the lengths of meridians in order to create the curves of the top and bottom latitudes of a projection.

meridional orthographic map projection—A map projection having the plane of the projection parallel to the plane of some selected meridian; the geographic parallels and the central meridian are straight lines, the outer meridian is a full circle, and the other meridians are arcs of ellipses.

meridional part—The length of the arc of a meridian between the Equator and a given parallel on a Mercator chart, expressed in units of 1′ of longitude at the Equator.

meridional plane—Any plane containing the polar axis of the Earth. See also **astronomic meridian plane; geodetic meridian plane.**

meteorological chart—Any chart showing meteorological (weather) information.

meter rod—See **precise leveling rod.**

metes-and-bounds survey—A method of describing the boundaries of tracts of land by giving the bearing and length of each successive line. Much of the land in the non-public-land States has been surveyed and described by this method. This method is also used in the surveys of the public lands to define the boundaries of irregular tracts, such as claims, grants, and reservations, which are nonconformable to the rectangular system of subdivision.

method of repetitions—The determination of the angle between two marks by accumulating, on the horizontal circle of a repeating theodolite, the sum of a series of measurements of the horizontal angle between the two marks.

Metonic cycle—A period of approximately 19 years, during which all phase relationships between Moon, Sun, and Earth occur, and after the lapse of which the phases of the Moon return to a particular date in the calendar year. During any cycle, new and full Moon will recur on approximately the same day in the calendar year.

metric camera—A specially constructed and calibrated camera used to obtain geometrically accurate photographs for use in photogrammetric instruments.

metric photography—The recording of events by means of photographs, either singly or sequentially, together with appropriate coordinates, to form the basis for accurate measurements.

Metrogon lens—The trade name of a wide-angle lens for aerial cameras used in mapping, charting, and reconnaissance photography. See also **trimetrogon camera.**

microfeatures—Features of relief, drainage, and landforms which can be identified on photographs, but are too small to appear on maps.

micrometer—An auxiliary device to provide measurement of very small angles or dimensions by an instrument such as a telescope. See also **filar micrometer; ocular micrometer; transit micrometer.**

micrometer method—The determination of the astronomic azimuth of a line by measuring indirectly with an ocular micrometer attached to a theodolite or transit the horizontal angle between a selected star at its elongation and a suitable ground mark (light) placed close to the vertical plane which passes through the star, and applying that angle to the azimuth of the star computed for the epoch of the observation.

mid-latitude—**1.** See **middle latitude. 2.** (cartography) The one parallel that is at the same scale as indicated on a Mercator projection.

middle latitude—One-half of the arithmetical sum of the latitudes of two places on the same side of the Equator. Middle latitude is labeled N or S to indicate whether it is north or south of the Equator. Also called **mid-latitude.**

middle ordinate—The distance from the middle point of a chord to the middle point of the corresponding circular arc.

middle point (MP)—That point on a circular curve which is equidistant from the two ends of the curve.

middletone—In halftone, any neutral tone intermediate between the highlights and shadows of an original and the resulting reproduction. Also, the tones in a reproduction between the highlights and the shadows.

mileage chart—A chart showing distances between various points.

military city map—See **city products.**

military geography—(JCS) The specialized field of geography dealing with natural and man-made physical features that may affect the planning and conduct of military operations.

military grid—(JCS) Two sets of parallel lines intersecting at right angles and forming squares; the grid is superimposed on maps, charts, and other similar representations of the Earth's surface in an accurate and consistent manner to permit identification of ground locations with respect to other locations and the computation of direction and distance to other points. See also **military grid reference system.**

military grid reference system (MGRS)—(JCS) A system which uses a standard-scaled grid square, based on a point of origin on a map projection of the Earth's surface in an accurate and consistent manner to permit either position referencing or the computation of direction and distance between grid positions. See also **military grid.**

military level—A compact ruggedized version of the dumpy level developed specifically for military use. It is primarily used for third-order leveling, but has a second-order capability.

millemap—A quantitative distribution map on which there are 1,000 dots, representing the quantity depicted; each dot therefore represents 1/1,000 of the total, and is located as accurately as possible according to the available evidence.

milligal—A unit of acceleration equal to 1/1,000 of a gal, or 1/1,000 centimeter per second per second. This unit is used in gravity measurements, being approximately one-millionth of the average gravity at the Earth's surface. Such measurement includes the component of centrifugal acceleration in the direction of the gravitational acceleration.

milligauss—Unit of magnetic force equal to 0.001 gauss (oersted) or 100 gammas.

mine survey—A survey to determine the positions and dimensions of underground passages of a mine; also, of the natural and artificial features (surface and underground) relating to the mine. The data include both horizontal and vertical positions, lengths, directions, and slopes of tunnels; topographic and geologic characteristics of the particular vicinity; ownership of the land and of the mine.

mineral survey—(USPLS) A survey made to mark the legal boundaries of mineral deposits or ore-bearing formations on the public domain, where the boundaries are to be determined by lines other than the normal subdivision of the public lands.

minor axis—The shortest diameter of an ellipse or ellipsoid.

minor control—See **photogrammetric control.**

minor control plot—See **radial triangulation.**

minor planets—See **asteroid.**

minus angle—See **angle of depression.**

minus declination—See **declination,** definition 3.

minus sight—See **foresight,** definition 3.

minute of standard length—The length of 1 minute of longitude at the Equator. The length is variable, depending on the dimensions of the particular ellipsoid (spheroid) used as a reference surface.

mirror image—See **reverted image.**

miscellaneous chart—A chart other than a regular navigational chart; a special chart.

mismatch—A condition which occurs when detail is displaced and perfect matching cannot be achieved.

missile launch site data card—A standardized form containing launcher geodetic information which has been produced on the current world geodetic system. Used in conjunction with the missile target data card.

missile target data—Precise geodetic target positioning data required to support strategic and tactical weapon systems. See also **point position data.**

missing triangle—(pendulum) A triangle which represents the failure of the two sides of a knife edge to reach a perfect intersection in a geometric line.

mistake—See **blunder; error.**

mock-up—See **style sheet.**

model—See **airborne landing model; assault-landing model; assault models; flat model; gross model; half model; master model; neat model; perspective spatial model; relief model; stereoscopic image; stereoscopic model; strategic-planning model; tactical-planning models; terrain model; warped model.**

model coordinates—(photogrammetry) The space coordinates of any point imaged in a stereoscopic model which define its position with reference to the air base or to the instrument axes.

model datum—**1.** (photogrammetry) That surface in a stereoscopic model conceived as having been reconstructed as part of the model representing the sea-level datum of nature. Often modified to designate the type of photography used, such as convergent model datum, and transverse model datum. **2.** For relief maps or models the datum may or may not be sea level but is consistent within a relief map series.

model marriage—The rejoining of sections of a model, after the carving operation, to the original neatline limits of a relief model.

model scale—(photogrammetry) The relationship which exists between a distance measured in a stereoscopic model and the corresponding distance on the Earth.

modeling—The development of the model surface by the application of modeling clay between the step edges of the step cast. A preparatory step in producing relief models.

modified Julian day—An abbreviated form of the Julian day which requires fewer digits and translates the beginning of each day from Greenwich noon to Greenwich midnight; obtained by subtracting 2400000.5 from Julian days.

modified Lambert conformal chart—A chart on the modified Lambert conformal map projection. Also called **Ney's chart.**

modified Lambert conformal map projection—A modification of the Lambert conformal projection for use in polar regions, one of the standard parallels being at latitude 89°59′58″ and the other at latitude 71° or 74°, and the parallels being expanded slightly to form complete concentric circles. Also called **Ney's map projection.**

modified polyconic map projection—A map projection obtained from the regular polyconic projection by so altering the scale along the central meridian that the scale is exact along two standard meridians, one on either side of the central meridian and equidistant therefrom. Also called **rectangular polyconic map projection.**

modulation—A variation of some characteristic of a radio wave, called the "carrier wave," in accordance with instantaneous values of another wave called the "modulating wave." These variations can be amplitude, frequency, phase, or pulse.

modulation error—In electronic distance-measuring equipment, the difference in modulating frequencies obtained from crystals, between the actual frequencies of the crystals and the frequencies required for a correct distance measurement.

moire—An interference pattern resulting from the overlaying or overprinting of halftones or tints whose screen angles are not sufficiently separated to make the pattern inconspicuous or to preclude a pattern accuracy.

mold alterations—The slight modification of the landforms of the mold, often necessary in local areas in order to obtain proper register due to unequal stretch required to accommodate the landforms on a plastic relief map.

molded aerial photograph—A vertical aerial photograph, usually annotated with military symbols, which has been formed to show terrain configuration.

Molitor precise leveling rod—A speaking rod of T-shaped cross section, with graduation marks shaped as triangles and rectangles, the smallest division being two millimeters. Read by estimation to single millimeters. Equipped with thermometer and circular level.

Mollweide homalographic map projection—An equal-area map projection showing the Equator and geographic parallels as straight lines, and the geographic meridians as elliptical arcs, with the exception of the central meridian, represented by a straight line, and the meridian 90° from the center, shown as a full circle.

moment—A tendency to cause rotation about a point or axis, as of a control surface about its hinge or of an airplane about its center of gravity; the measure of this tendency is equal to the product of the force and the perpendicular distance between the point of axis of rotation and the line of action of the force.

moment of inertia—The quantity obtained by multiplying the mass of each small part of a body by the square of its distance from an axis, and adding all the results.

momentum—Quantity of motion. Linear momentum is the quantity obtained by multiplying the mass of a body by its linear speed. Angular momentum is the quantity obtained by multiplying the moment of inertia of a body by its angular speed.

monochromator—A dispersive device for isolation of narrow portions of the spectrum.

monochrome—A single hue or color.

monocomparator—A precision instrument, consisting of a measuring system, a viewing system, and a readout system designed for the measurement of image coordinates on a single photograph.

month—The period of the revolution of the Moon around the Earth. The month is designated as sidereal, tropical, anomalistic, nodical or dracontic, or synodical, according to whether the revolution is relative to the stars, the vernal equinox, the perigee, the ascending node, or the Sun. See also **anomalistic month; calendar month; nodical month; sidereal month; synodical month; tropical month.**

monument—1. A structure used or erected to mark the position of a survey station; permanence is implied. See also **artificial monument; natural monument. 2.** (USPLS) A physical structure, such as an iron post, marked stone, or tree in place, which marks the location of a corner point established by a cadastral survey. Objects, to be ranked as monuments, should have certain physical properties such as visibility, durability, and stability, and they must define location without resorting to measurements. "Monument" and "corner" are not synonymous, although the two terms are often used largely in the same sense. See also **corner.**

monumented bench mark—See **permanent bench mark.**

Moon position camera method—A means of determining geodetic position, that is unaffected by deflection of the vertical, by photographing the Moon against a star background.

morphographic map—A small scale map showing physiographic features by means of standardized pictorial symbols, based on the appearance such features would have if viewed obliquely from the air.

mosaic—1. (JCS) (photogrammetry) An assembly of overlapping aerial photographs which have been matched to form a continuous photographic representation of a portion of the Earth's surface. Also called **aerial mosaic.** See also **Air Target Mosaic, controlled mosaic; map-controlled mosaic; orthophotomosaic; scale-ratio mosaic; semicontrolled mosaic; strip mosaic; uncontrolled mosaic. 2.** (cartography) See **panel base.**

mosaicking board—A smooth-surfaced material, usually tempered Masonite, to which the mosaic is fastened with a suitable adhesive.

most probable value—That value of a quantity which is mathematically determined from a series of observations and is more nearly free from the effects of errors than any other value that might be derived from the same series of observations. Derivation of the most probable value is made after blunders and systematic errors have been removed from the data.

moving average—See **consecutive mean.**

multiband photography—A remote sensing system which produces more than one image of a single area in which each image shows a different wavelength band of the electromagnetic spectrum.

multicolor—Two or more colors. Also called **polychrome.** See also **process color printing.**

multiple-camera assembly—An assembly of two or more cameras mounted to maintain a fixed angle between their respective optical axes.

multiple-lens camera—A camera with two or more lenses, with the axes of the lenses systematically arranged at fixed angles in order to cover a wide field by simultaneous exposures in all chambers.

multiple-lens photograph—A photograph made with multiple-lens camera.

multiple level line—Two or more single lines of spirit leveling run between the same terminal points, but along different routes.

multiple-stage rectification—A technique employing standard equipment to rectify oblique photos by applying a series of projections to effect the desired projective transformation.

multiplex—A name applied to anaglyphic double-projection stereoplotters with the following characteristics: (1) The stereomodel is projected from diapositives reduced from aerial photograph negatives; (2) The projection system illuminates the entire diapositive format area; and (3) The stereomodel is measured and drawn by observation of a floating mark.

multiplex control—See **photogrammetric control.**

multiplex triangulation—See **stereotriangulation.**

multispectral—Remote sensing in two or more spectral bands, such as visible and infrared See also **infrared; remote sensing.**

multispectral imagery—Images obtained simultaneously in a number of discrete bands in the electromagnetic spectrum.

multispectral scanner (MSS)—A remote-sensing device which is capable of recording data in the ultraviolet and visible portions of the electromagnetic spectrum, as well as the infrared.

multispectral sensing—Employment of one or more sensors to obtain imagery from different portions (bands) of the electromagnetic spectrum.

multiuse manuscript (MUM)—A manuscript compilation that, as a minimum, establishes the contours, spot elevations, and includes the horizontal position of the significant planimetric features. It is suitable for use in completing a topographic map, or an aeronautical or nautical chart; and the integrity of its horizontal and vertical accuracy is retained in all end products made from it.

N

nadir—(JCS) That point on the celestial sphere directly beneath the observer and directly opposite the zenith. See also **ground nadir; map nadir; photograph nadir.**

nadir point—See **photograph nadir.**

nadir-point plot—See **nadir-point triangulation.**

nadir-point triangulation—Radial triangulation in which nadir points are utilized as radial centers. Also called **nadir-point plot.**

nadir radial—A radial from the nadir point.

nanotesla—(geomagnetism) A unit of magnetic field intensity generally used in describing the Earth's magnetic field. It is defined as 10^{-9} tesla = 1 gamma.

narrow-angle lens—A lens having an angle of coverage up to 60°. A lens whose focal length is equal approximately to twice the diagonal of the format.

National Geodetic Survey first-order leveling rod—See **precise leveling rod.**

National Geodetic Vertical datum of 1929—Known as "sea level datum of 1929" prior to September 1973, this datum was established by constraining the combined interconnected United States and Canadian networks of first-order leveling, as it existed in 1929, to conform to mean sea level of various epochs, as determined at 21 United States and 5 Canadian long-term tidal stations distributed along the Atlantic, Gulf of Mexico, and Pacific coasts.

National grid—The grid, based on the Transverse Mercator projection, used on all current British Ordnance Survey maps. The axes of the national grid are 2°W and 49°N, intersecting at the true origin, from which the false origin is transferred 400 km W and 100 km N.

national map accuracy standards—See **U.S. National Map Accuracy Standards.**

native map—A map of a foreign country produced by indigenous governmental or private agencies.

natural detail—The features on the Earth, such as streams, lakes, forests, and mountains; exclusive of the works of man. Also called **natural feature.** See also **culture; hydrographic detail; hypsographic detail.**

natural error—Errors arising from variations in temperature, humidity, wind, gravity, refraction, and magnetic declination.

natural feature—See **natural detail.**

natural monument—A natural feature, such as a stream, boulder, tree, etc., which serves to mark the location of a survey station or land corner. See also **monument.**

natural scale—See **representative fraction.**

natural year—See **tropical year.**

nautical chart—See **hydrographic chart.**

nautical mile—(JCS) A measure of distance equal to 1 minute of arc on the Earth's surface. The United States has adopted the International Nautical Mile equal to 1,852 meters or 6,076.11549 feet [6,076.1033 survey feet].

navigation chart—See **aeronautical chart; hydrographic chart.**

navigation sight—An auxiliary device used in the taking of aerial photography to show not only the vertical field of view but also the path ahead and behind the aircraft.

navigational planets—The four planets commonly used for celestial navigation: Venus, Mars, Jupiter, and Saturn.

navigational triangle—The spherical triangle solved in computing altitude and azimuth or great circle sailing problems. The celestial triangle is formed on the celestial sphere by the great circles connecting the elevated pole, zenith of the assumed position of the observer, and a celestial body. The terrestrial triangle is formed on a spherical Earth by the great circles connecting the pole and two places on Earth, either the assumed positon of the observer and geographic position of the body for celestial observations, or the points of departure and destination for great circle sailing problems. The expression **navigational triangle** applies to either the celestial or terrestrial triangle used for solving navigation problems.

NAVSTAR Global Positioning System (GPS)—A navigation and positioning system, under development, with which the three-dimensional geodetic position and the velocity of a user at a point on or near the Earth can be determined in real time. The system will consist of a constellation of 24 Earth-orbiting satellites which broadcast on a pair of ultrastable frequencies. The user's receiver will be able to track a minimum of four of the satellites from any location at any time, thus establishing position and velocity.

Navy Navigation Satellite System (NNSS)—A set of five or six satellites in polar orbit which broadcast time and position (ephemeris) at 2-minute intervals. Receivers pick up the broadcast during satellite passes and, from the Doppler effect, are able to compute their positions in terms of the NNSS. The satellites are tracked by a world network of integrated stations to produce a precise after-the-fact ephemeris from which more accurate positions can be computed. See also **broadcast ephemeris; Doppler navigation; precise ephemeris.**

near-certainty error (3 sigma, 3σ)—The 99.73 percent error interval based on the normal distribution function.

Nearshore Environmental Analog Prediction System—A technique used at the U.S. Naval Oceanographic Office to classify nearshore areas (shore to 30 fathoms) so that characteristics of unsurveyed locations can be inferred from surveyed locations in a similar class.

Formerly called the Harbor Analog System.

neat model—The portion of the gross overlap of a pair of photographs that is actually utilized in photogrammetric procedures. Generally, the neat model approximates a rectangle whose width equals the air base and whose length equals the width between flights. See also **gross model.**

neatlines—(JCS) The lines that bound the body of a map, usually parallels and meridians [but may be conventional or arbitrary grid lines]. Also called **sheet lines.**

negative—1. (JCS) In photography: (*a*) black and white—An image on film, plate, or paper in which the normal tones of the subject are reversed. (*b*) color—An image on film, plate, or paper, in which colors appear as their complements. **2.** In cartographic scribing, a scribed sheet is essentially a manually produced negative. See also **duplicate negative; original negative.**

negative altitude—Angular distance below the horizon.

negative component in color mixture—A component that is mixed with the sample light in order to desaturate it sufficiently to obtain a match with a mixture of the other two components.

negative corrections—Changes made directly on a negative or a scribed surface. See also **negative engraving.**

negative deflection angle—See **deflection angle,** definition 1.

negative engraving—The operation of making corrections and additions to negatives. This term should not be applied to the process of scribing on coated plastics.

negative forming—In relief model making, forming into a negative mold.

negative lens—A lens diverging a beam of parallel light rays, with no real focus being obtained. Also called **concave lens; diverging lens.**

negative mold—The cast resulting from casting over a master relief model.

negative scribing—See **scribing.**

negative titling—See **titling.**

net—See **survey net.**

neutral filter—A filter that reduces the intensity of light reaching the film or plate without affecting the tonal rendition of colors in the original scene.

new chart—A chart constructed to satisfy the needs of navigation in a particular area. It is laid out in conformity with a broad scheme to meet future needs in the adjacent areas.

new edition—Contains changes of such importance to map or chart users that all previous printings are made obsolete.

new survey—See **resurvey.**

newton—A force of 1 newton (N) acting on a mass of 1 kilogram imparts an acceleration of 1 meter per second per second. The newton is the metric (SI) unit of force. See also **dyne.**

Newtonian constant of gravitation—See **constant of gravitation.**

Newton's laws—1. (gravitation) Every particle of matter in the universe attracts every other particle with a force proportional to the product of their masses and inversely as the square of the distance between them. **2.** (motion) (*a*) Every body continues in its state of rest, or of uniform motion in a straight line, unless it is compelled to change that state by a force impressed upon it. (*b*) The rate of change of momentum is proportional to the force impressed, and takes the direction of the straight line in which the force acts. (*c*) To every action there is an equal and opposite reaction; or, the mutual actions of two bodies are always equal and oppositely directed.

Newton's rings—An interference effect arising from close, but not quite perfect, contact between two surfaces, manifested by irregular concentric rings of color.

New York leveling rod—A two-piece rod with movable target. For heights greater than 6½ feet, the target is clamped at 6½ feet and raised by extending the rod. Graduated to hundredths of a foot and read by vernier to thousandths.

Ney's chart—See **modified Lambert conformal chart.**

Ney's map projection—See **modified Lambert conformal map projection.**

night effect—(JCS) An effect mainly caused by variations in the state of polarization of reflected waves, which sometimes result in errors in direction finding bearings. The effect is most frequent at nightfall.

no-check position—See **intersection station.**

nocturnal arc—See **astronomic arc.**

nodal line—In a tide area, the line about which the tide oscillates and where there is little or no rise and fall of the tide.

nodal plane—A plane perpendicular to the optical axis at a nodal point.

nodal point—1. (optics) One of two points on the optical axis of a lens, or system of lenses, such that a ray emergent from the second point is parallel to the ray incident at the first. This first nodal point is also referred to as the **front nodal point, incident nodal point,** or **nodal point of incidence;** and the second point as the **rear nodal point, emergent nodal point,** or **nodal point of emergence.** Also called **node. 2.** (astronomy) See **node,** definition 1. **3.** (hydrography) See **amphidromic point.**

nodal point of emergence—See **nodal point,** definition 1.

nodal point of incidence—See **nodal point,** definition 1.

node—1. (astronomy) One of the two points of intersection of the orbit of a planet, planetoid, or comet with the ecliptic, or of the orbit of a satellite with the equatorial plane of the orbit of its primary. Also called **nodal point.** See also **ascending node; descending node; ecliptic node; equatorial node; line of nodes; longitude of the Moon's nodes; lunar node; regression of the nodes.** **2.** (optics) See **nodal point,** definition 1.

node cycle—The time required for the regression of the Moon's nodes to complete a circuit of 360° of longitude; a period of approximately 18.6 years.

nodical month—The interval of time between two successive passages of the Moon through the same node of its orbit, approximately 27¼ days. Also called **draconic month.**

nodical period—The interval between two successive passages of a satellite or planet through the ascending node of its orbit.

noise level—The magnitude of random errors in a particular type of measurement.

nominal focal length—(JCS) An approximate value of the focal length, rounded off to some standard figure, used for the classification of lenses, mirrors, or cameras.

nominal orbit—The true or ideal orbit upon which a space value is expected to travel.

nomogram—A diagram showing, to scale, the relationship between several variables in such manner that the value of one which corresponds to known values of the others can be determined graphically.

nonautomatic rectifier—Any rectifier which requires computation of the elements of rectification, each of which must be manually set on its corresponding circle or scale on the rectifier.

nongravitational perturbations—Perturbations caused by surface forces due to mechanical drag of the atmosphere (in case of low flying satellites), electromagnetism, and radiation pressure.

nonmonumented bench mark—See **temporary bench mark.**

nonperspective azimuthal map projection—A projection not based on imaginary lines of sight from a single point of view. Azimuthal equal-area and azimuthal equidistant map projections are nonperspective.

nonrecording gage—Any tide or stream gage which requires the presence of an attendant to observe and record the heights of the water at periodic intervals.

nonselective filter—A filter for which transmittance is substantially independent of wavelength.

nontilting-lens rectifier—A class of rectifier wherein the lens is constrained to move in the direction of its fixed axis.

nontilting-negative-plane rectifier—A class of rectifier which contains a nontilting-negative carrier. In this class of rectifiers, the negative-carrier plane remains horizontal.

normal—**1.** A straight line perpendicular to a surface or to another line. **2.** A condition of being perpendicular to a surface or line. **3.** In geodesy, the straight line perpendicular to the surface of the reference spheroid. **4.** The average, regular, or expected value of a quantity.

normal-angle lens—A lens having an angle of coverage from 60° to 75°. A lens whose focal length is equal approximately to the diagonal of the format.

normal contour—See **accurate contour.**

normal distribution function—A mathematical function describing the behavior of one-dimensional random errors.

normal equation—One of a set of simultaneous equations derived from observation, condition, or correlate equations, and expressing a condition for a least-squares adjustment. In a least-squares adjustment, values obtained from the solution of normal equations (either directly or through the correlate equations) are applied to the observation or condition equations to obtain the desired corrections.

normal gravity—A reference gravity field that is mathematically defined as a function of position. It is commonly taken as the field of a rotating level ellipsoid but may be arbitrarily defined.

normal gravity field—A mathematically derived gravity field used in geodesy to closely approximate the Earth's actual gravity field. Its level surfaces are usually exact ellipsoids of revolution.

normal orbit—The orbit of a spherical satellite about a spherical primary during which there are no disturbing effects present due to other celestial bodies, or to some physical phenomena. Also called **unperturbed orbit.**

normal section azimuth—The angle between the geodetic meridian of the observer and the plane containing the ellipsoidal normal of the observer and the projection of the observed point on the reference ellipsoid, measured in a plane perpendicular to the ellipsoidal normal of the observer, preferably clockwise from the north.

normal section line—A line on the surface of the reference spheroid, connecting two points on that surface, and traced by a plane containing the normal at one point and passing through the other point.

normal tension—(taping) The tension to be applied to a tape to compensate for the shortening effect of sag in order to bring the tape to standard length. That pull at which the tension correction and sag exactly balance each other.

normal water level—The most prevalent water level in a watercourse, reservoir, lake, or pond, generally defined by a shoreline of permanent land-type vegetation. Along large bodies of water, wave action may retard vegetation beyond the normal shoreline.

north—The primary reference direction relative

to the Earth; the direction indicated by 000° in any system other than relative. See also **compass north; grid north; magnetic north; true north.**

North American datum of 1927 (NAD 27)—The initial point of this datum is located at Meades Ranch, Kansas. Based on the Clarke spheroid of 1866, the geodetic position of triangulation station Meades Ranch and azimuth from that station to station Waldo are as follows:

Lat. of Meades Ranch	39°13′26″.686N
Long. of Meades Ranch	98°32′30″.506W
Azimuth to Waldo	75°28′09″.64

The geoid height at Meades Ranch is assumed to be zero. The geodetic positions of this system are derived from a readjustment of the triangulation of the entire country, in which Laplace azimuths were introduced. See also **preferred datum.**

North American datum of 1983 (NAD 83)—The projected datum resulting from the redefinition of the North American networks. The new adjustment of the North American networks will include a variety of geodetic data acquired since the 1927 North American datum was determined. This includes precise Geodimeter traverses, Doppler satellite positioning, astrogeodetic deflections, and gravity data. The datum parameters have not been defined. The reference ellipsoid will be Earth-centered.

north declination—See **declination,** definition 3.

north geographical pole—The geographical pole in the Northern Hemisphere, at latitude 90° N.

north geomagnetic pole—The geomagnetic pole in the Northern Hemisphere.

north magnetic pole—The magnetic pole in the Northern Hemisphere.

north point—See **celestial meridian.**

north polar circle—See **Arctic Circle.**

North Star—See **Polaris.**

northbound node—See **ascending node.**

northing—1. (JCS) Northward (that is, from bottom to top) reading of grid values on a map. See also **false northing. 2.** (plane surveying) See **latitude difference.**

nuclear precession magnetometer—A magnetometer that utilizes the precessional characteristics of hydrogen nuclei when in an ambient magnetic field. The data output of this instrument is in the form of a frequency measurement, which in turn is proportional to the magnetic field intensity.

nutation—1. The oscillation of the axis of any rotating body, as a gyroscope rotor. **2.** (astronomy) Irregularities in the precessional motion of the equinoxes because of varying positions of the Moon and, to a lesser extent, of other celestial bodies with respect to the ecliptic.

nutation in right ascension—See **equation of the equinox.**

O

objective lens—In telescopes and microscopes, the optical component which receives light from the object and forms the first or primary image. In a camera, the image formed by the objective lens is the final image. In a telescope or microscope used visually, the image formed by the objective lens is magnified by the eyepiece.

oblate ellipsoid of rotation—See **oblate spheroid.**

oblate spheroid—An ellipsoid of rotation, the shorter axis of which is the axis of rotation. The Earth is approximately an oblate spheroid. Also called **oblate ellipsoid of rotation.**

oblique air photograph—(JCS) An air photograph taken with the camera axis directed between the horizontal and vertical planes. Commonly referred to as an **oblique:** (1) **high oblique**—one in which the apparent horizon appears; and (2) **low oblique**—one in which the apparent horizon does not appear.

oblique ascension—The arc of the celestial equator, or the angle at the celestial pole, between the hour circle of the vernal equinox and the hour circle through the intersection of the celestial equator and the eastern horizon at the instant a point on the oblique sphere rises, measured eastward from the hour circle of the vernal equinox through 24 hours.

oblique chart—A chart on an oblique map projection.

oblique coordinates—Magnitudes defining a point relative to two intersecting nonperpendicular lines, called axes.

oblique cylindrical orthomorphic chart—See **oblique Mercator chart.**

oblique cylindrical orthomorphic map projection—See **oblique Mercator map projection.**

oblique equator—A great circle the plane of which is perpendicular to the axis of an oblique projection.

oblique graticule—A fictitious graticule based upon an oblique map projection.

oblique latitude—Angular distance from an oblique equator. See also **fictitious latitude.**

oblique longitude—Angular distance between a prime oblique meridian and any given oblique meridian. See also **fictitious longitude.**

oblique map projection—A map projection with an axis inclined at an oblique angle to the plane of the equator.

oblique Mercator chart—A chart on the oblique Mercator map projection. Also called **oblique cylindrical orthomorphic chart.**

oblique Mercator map projection—A conformal cylindrical map projection in which points on the surface of a sphere or spheroid, such as the Earth, are conceived as developed by Mercator principles on a cylindrical tangent along an oblique great circle. Also called **oblique cylindrical orthomorphic projection.**

oblique meridian—A great circle perpendicular to an oblique equator. The reference oblique meridian is called **prime oblique meridian.** See also **fictitious meridian.**

oblique parallel—A circle or line parallel to an oblique equator, connecting all points of equal oblique latitude. See also **fictitious parallel.**

oblique plotting instrument—An instrument for plotting from oblique photographs.

oblique pole—One of the two points 90° from an oblique equator.

oblique rhumb line—**1.** A line making the same oblique angle with all fictitious meridians of an oblique Mercator projection. Oblique parallels and meridians may be considered special cases of the oblique rhumb line. **2.** Any rhumb line, real or fictitious, making an oblique angle with its meridians. In this sense the expression is used to distinguish such a rhumb line from parallels and meridians, real or fictitious, which may be included in the expression **rhumb line.** See also **fictitious rhumb line.**

oblique sketchmaster—A type of sketchmaster in which oblique photographs are utilized.

oblique sphere—The celestial sphere as it appears to an observer between the Equator and the pole, where celestial bodies appear to rise obliquely to the horizon.

obliquity of the ecliptic—The acute angle between the plane of the ecliptic (the plane of the Earth's orbit) and the plane of the celestial equator.

obliterated corner—(USPLS) An obliterated corner is one at whose point there are no remaining traces of the monument or its accessories, but whose location has been perpetuated, or the point for which may be recovered beyond reasonable doubt, by the acts and testimony of the intersected landowners, competent surveyors, or other qualified local authorities, or witnesses, or by some acceptable record evidence.

observation equation—A condition equation which connects interrelated unknowns by means of an observed function, or a condition equation connecting the function observed and the unknown quantity whose value is sought.

observed altitude—Corrected sextant altitude; angular distance to the center of a celestial body above the horizon, corrected for instrumental errors, personal error, dip, refraction, and semidiameter and parallax if necessary. See also **true altitude.**

observed angle—An angle obtained by direct instrumental observation. A measured angle which has been corrected for local conditions only at the point of observation, is considered an observed angle.

observed gravity—The value of gravity at a station as determined from a gravity meter, a pendulum, or an instrument timing free falling bodies. The gravity obtained is either relative or absolute according to the apparatus used to make the measurements.

observed gravity anomaly—See **gravity anomaly.**

observed value—A value of a quantity that is obtained by instrumental measurement of the quantity. The term **observed value** is often applied to the value of a quantity derived from instrumental measurement after corrections have been applied for systematic errors, but before random errors have been taken out by some method of adjustment.

obsolete chart—A chart which is not considered safe to use for navigation because it does not contain the latest important navigational information.

occultation—1. (astronomy) The disappearance of a celestial body behind another body of larger apparent size. When the Moon passes between the observer and a star, the star is said to be occulted. **2.** (surveying) Name applied to a geodetic survey technique which employs the principle of occultation where repeated observations are made on an unknown position, accurately timed with similar observations at another unknown station, and mathematically reducing these data to determine the exact geodetic position of the unknown stations. See also **star occultation method.**

occupy—(surveying) To set a surveying instrument over a point for the purpose of making observations.

oceanographic station—A term used to designate oceanographic observations taken at a geographic location from a ship that is lying to or anchored at sea.

oceanographic station location—The accepted geographic position at which an oceanographic station was taken.

oceanographic survey—A study or examination of conditions in the ocean or any part of it, with reference to animal or plant life, chemical elements present, temperature gradients, etc. Also called **marine survey.**

oceanography—1. (JCS) The study of the sea, embracing and integrating all knowledge pertaining to the sea and its physical boundaries, the chemistry and physics of sea water, and marine biology. **2.** In strict usage oceanography is the description of the marine environment, whereas oceanology is the study of the oceans and related sciences.

oceanology—See **oceanography,** definition 2.

octant—A type of sextant having a range of 90° and an arc of 45°.

ocular micrometer—A filar micrometer so placed that its wire moves in the principal focal plane of a telescope. Also called **eyepiece micrometer.**

odograph—A mechanical instrument containing a distance-measuring element which is moved or turned by an amount proportional to the actual distance traveled; a compass element which provides a fixed-reference direction; and an integrator which provides for the resolution of the direction of motion into components and for the summation or integration of the distance components.

off soundings—Any area where the depth of water cannot be measured by a sounding lead, generally considered to be beyond the 100-fathom line. Opposite of **on soundings.**

office computations—Computations based on field measurements, including all calculations relative to the reduction of field survey notes to graphic form for any type of survey or for the continuation of field work.

offset—1. (cartography) In projection construction, that small distance added to the length of the meridians on each side of the central meridian in order to determine the top latitude of the constructed chart. **2.** (surveying) A short line perpendicular to a surveyed line measured to a line or point for which data are desired, thus locating the second line or point with reference to the first or surveyed line. An offset is also a job in a survey or other line, the line having approximately the same direction both before and after passing the jog. Offsets are measured from a surveyed line or lines to the edges of an irregular-shaped body of water or to any irregular line which it is desired to locate. **3.** (printing) See **offset lithography.**

offset line—A supplementary line close to and roughly parallel with a main line, to which it is referred by measured offsets. Where the line for which data are desired is in such position that it is difficult to measure over it, the required data are obtained by running an offset line in a convenient location and measuring offsets from it to salient points on the other line.

offset lithography—An indirect method of printing whereby the ink image is transferred from the pressplate to an intermediate surface of a rubber blanket, and from that to the paper or other stock. Also called **offset; offset printing.** See also **lithography; photolithography.**

offset press—A press which contains an extra cylinder, rubber covered, upon which the image is printed first and then reprinted or "offset," from this cylinder onto the paper.

offset printing—See **offset lithography.**

OK sheet—The first press impression from each color, or color combination, approved for accuracy of register and color.

Omega—A long-range hyperbolic navigation system designed to provide worldwide coverage for navigation.

omni-directional radar prediction—A radar prediction which is intended to be valid in most respects from any direction of approach. The Radar Significance Analysis Code on the Series 200 Air Target Chart is an example of omn directional radar prediction. Each coded area represents an analysis of relative radar intensity from all directions.

omni-gain radar prediction—A radar prediction containing some information about all radar responsive features within the predicted area. This is accomplished by predicting all significant radar returns in relative intensities based on the predicted probability of the return remaining on the radarscope at decreased gain. Generally, the more intense the return appears on the prediction, the more likely it will remain on the radarscope as the gain is decreased.

on soundings—Any area where the depth of water can be measured by a sounding lead, generally considered to be within the 100-fathom line. Opposite of **off soundings.**

one-projector method—See **one-swing method.**

one-swing method—The technique employed in relative orientation for clearing *y*-parallax by maintaining one projector of a pair in a fixed position and making all adjustments with the second projector in relationship to the first. Also called **one-projector method; y-swing method; single projector method.**

one-to-one (1:1) copy—See **contact size.**

opacity—See **density,** definition 1.

opaque—1. Not transmitting light. **2.** Not transmitting the particular wavelengths (which may or may not be visible) which affect given photosensitive materials. Thus, a substance may be opaque to some colors and not to others. It may be visually transparent, yet actinically opaque. **3.** A material applied to areas of a negative to make it opaque in those areas. **4.** To apply a material or block-out.

open-end traverse—See **open traverse.**

open traverse—A survey traverse which begins from a station of known or adopted position, but does not end upon such a station. There is no check whatsoever on the accuracy of such a traverse. Also called **open-end traverse.**

open window process—(cartography) A method of preparing color separation negatives or positives by peeling an opaque stratum from its base in the desired areas. It is normally used for preparing large areas covered by vegetation or open water. See also **mask,** definition 2.

operation map—(JCS) A map showing the location and strength of friendly forces involved in an operation. It may indicate predicted movement and location of enemy forces.

operational libraries—A DMA-approved, selective data file consisting of extra copies of originals, duplicate copies, computer printed catalogs, etc., obtained from any designated DoD library or from other nationally designated libraries within the non-DoD agencies, and maintained in a DoD MC&G agency for its direct use in accomplishing assigned production missions. General reference publications such as dictionaries, glossaries, atlases, periodicals, etc., are excluded from controls applied to operational libraries.

Operational Navigation Chart (ONC)—A chart at a scale of 1:1,000,000 which represents the combined requirements for a graphic to satisfy special military operations as well as general navigation uses. The World Aeronautical Chart (WAC) series is being replaced by the Operational Navigation Chart series.

opposition—1. The situation of two celestial bodies having either celestial longitudes or sidereal hour angles differing by 180°. The term is usually used only in relation to the position of a planet or moon from the Sun. **2.** The situation of two periodic quantities differing by half a cycle.

optical axis—(JCS) In a lens element, the straight line which passes through the centers of curvature of the lens surfaces. In an optical system, the line formed by the coinciding principal axes of the series of optical elements. Also called **axis of lens; lens axis; principal axis.**

optical base-line measuring apparatus—A base apparatus composed of bars whose lengths are defined by distances between lines at or near their ends, which are observed by suitably mounted and adjusted microscopes. In using any optical base-line measuring apparatus, the positions of the bars are controlled by microscopes on stable support, whose reticle lines may be brought into coincidence with the fiducial marks on the bars, either by adjusting a bar or a microscope.

optical center—The point of intersection of lines which represent within the lens those rays whose emergent directions are parallel to their respective incident directions. This point lies on the optical axis. An oblique ray, even if it passes through this point, undergoes a longitudinal displacement increasing with the thickness of the lens.

optical correlation—The process of electronically relating a stored photographic film chip of a geographic area with a real-time optical image acquired by photographic or television sensors. It is used to provide positioning information to correct or check air navigation and guidance systems.

optical density—A common logarithm of reciprocal of transmittance.

optical flat—A surface, usually of glass, ground and polished plane within a fraction of a wavelength of light. An optical element or glass blank with an optical flat is used to test the flatness of other surfaces. Also called **flat; optical plane.**

optical-mechanical scanner—A system utilizing a rotating mirror and a detector in conjunction with lenses and prisms to record reflected and/or emitted electromagnetic energy in a scanning mode along the flight path.

optical parallax—See **instrument parallax,** definition 1.

optical path—The path followed by a ray of light through an optical system.

optical plane—See **optical flat.**

optical plummet—See **vertical collimator.**

optical-projection instruments—A class of instruments which provide projected images of photographic prints or other opaque material superimposed on a map or map manuscript. Often used for transferring detail from near-vertical photographs or other source material.

optical pumping magnetometer—A highly sensitive instrument which employs a metastable helium or alkali metal absorption cell whose atoms absorb maximum energy from an infrared beam when an FM oscillator is tuned to its resonant frequency. The resonant frequency is proportional to the strength of the Earth's magnetic field.

optical rectification—The process of projecting the image of a tilted aerial photograph onto a horizontal reference plane to eliminate the image displacements caused by tilt of the aerial camera at the time of exposure.

optical square—A small hand instrument used in setting off a right angle. One form of optical square uses two plane mirrors placed at an angle of 45° to each other. In use, one object is sighted direct, and another object is so placed that its twice-reflected image appears directly in line with the first object. The lines to the point of observation from the two observed objects will then meet in a right angle. In another form of optical square, a single plane mirror is so placed that it makes an angle of 45° with a sighting line; one object is sighted direct, and the other so placed that its reflected image is seen also in the sighting line.

optical system—All the parts of a compound lens and accessory optical parts which are designed to contribute to the formation of an image on a photographic emulsion, or of a visual image, or of an image on a projection screen.

optical vernier—A microscope with vernier lines ruled on a glass slide placed in the focal plane common to the objective and the eyepiece, where it is compared with the image of the graduated circle.

optical wedge—See **wedge.**

optimum ground elevation—(photogrammetry) The elevation of an assumed horizontal surface in the area photographed that would be projected at the optimum distance in the plotting instrument.

orbit—The path of a body or particle under the influence of a gravitational or other force. For example, the orbit of a celestial body or satellite is its path relative to another body around which it revolves. The term **orbit** is commonly used to designate a closed path. See also **central force orbit; intermediate orbit; nominal orbit; normal orbit; osculating orbit; perturbed orbit; polar orbit; stationary orbit; two-body orbit.**

orbital altitude—The mean altitude above the surface of the parent body of the orbit of a satellite.

orbital elements—A set of six parameters defining the orbit of a body attracted by a central force.

orbital inclination—The direction that the path of an orbiting body takes. In the case of an Earth satellite, this path may be defined by the angle of inclination of the path to the Equator.

orbital mode—A method for determining the position of an unknown station position when the unknown position cannot be viewed simultaneously with known positions. The arc of the satellite orbit is extrapolated from the ephemeris of the satellite determined by the known stations which permits the determination of the position of the unknown station dependent completely on the satellite's orbital parameters.

orbital motion—Continuous motion in a closed path about and as a direct result of a source of gravitational attraction.

orbital path—One of the tracks on a primary body's surface traced by a satellite that orbits about it several times in a direction other than normal to the primary body's axis of rotation. Each track is displaced in a direction opposite and by an amount equal to the degrees of rotation between each satellite orbit.

orbital period—The interval between successive passages of a satellite through the same point in its orbit. Also called **period of satellite.**

orbital plane—The plane of the ellipse defined by a central force orbit.

orbital velocity—The velocity of an Earth satellite or other orbiting body at any given point in its orbit.

ordinary tide level—See **mean tide level.**

ordinates—In a system of rectangular or oblique coordinates, the linear distance of a point measured from the horizontal or x-axis, and parallel to the y-axis. Also called **total latitudes; y-coordinate.**

orientation—**1.** The act of establishing correct relationship in direction with reference to the points of the compass. **2.** The state of being in correct relationship in direction with reference to the points of the compass. **3.** A map is in orientation when the map symbols are parallel with their corresponding ground features. **4.** A planetable is in orientation when lines connecting positions on the planetable sheet are parallel with the lines connecting the corresponding ground objects. **5.** A surveyor's transit is in orientation if the horizontal circle reads 0° when the line of collimation is parallel to the direction it had at an earlier (initial) position of the instrument, or to a standard line of reference. If the line of reference is a meridian, the circle will show azimuths referred to that meridian. **6.** A photograph is in orientation when it correctly presents the perspective view of the ground or when images on the photograph appear in the same direction from the point of observation as do the corresponding map symbols. **7.** Photogrammetric orientation is the recreation of natural terrain features at a miniature scale by the optical projection of overlapping photographs. The model is formed when all corresponding light rays from the two projectors intersect in space. See also **absolute orientation; aeroleveling; astrogeodetic datum orientation; basal orientation; empirical orientation; exterior orientation; gravimetric datum orientation; interior orientation; preliminary orientation; relative orientation; single astronomic station datum orientation.**

orientation inset—See **inset.**

orientation point—A picture point selected in areas common to vertical photographs and their corresponding obliques which serves to establish the relationship between the vertical and the oblique. Two such points are usually selected in each vertical photograph and transferred to the matching oblique photo.

origin—The reference position from which angles or distances are reckoned. See also **coordinates; false origin; grid origin.**

origin of coordinates—A point in a system of coordinates which serves as an initial point in computing its elements or in prescribing its use. The term **origin of coordinates** has several definitions, each so well established that a single definition cannot be prescribed to the exclusion of others. However, the following are given in the order of preferred use; to avoid misunderstanding, the use should be defined by stating the position of the origin in the system and giving the numerical coordinates assigned it. (1) The origin of coordinates is the point of intersection of the coordinate axes, from which the coordinates are reckoned. In mathematical treatises this origin is usually given the coordinates (0,0); in surveying, however, it is standard practice to give this origin coordinates having large positive numerical values, thereby avoiding the use of negative coordinates. See also **state coordinate systems.** (2) The origin of coordinates is the point to which the coordinate values (0,0) are assigned, irrespective of its position with reference to the axes. (3) The origin of coordinates is the point from which the computation of the elements of the coordinate system (projection) proceeds.

original—See **original copy.**

original copy—The photographs, artwork, scribed material, typed matter, and/or other materials to be processed for reproduction. Also called **original.**

original model—See **master model.**

original negative—That negative developed from the film which was in a camera magazine at the instant of exposure. Synonymous with a first-generation photographic product.

original survey—See **survey.**

orthembadic map projection—An equal-area map projection.

orthochromatic—(photography) **1.** Of, or pertaining to, or producing tonal values (of light or shade) in a photograph, corresponding to the tones of nature. **2.** Designating an emulsion sensitive to blue and green light, but not to red.

orthodrome—See **great circle.**

orthogonal—At right angles; rectangularly; meeting, crossing, or lying at right angles.

orthogonal map projection—See **orthographic map projection.**

orthographic chart—A chart on the orthographic projection.

orthographic map projection—A perspective azimuthal projection in which the projecting lines, emanating from a point at infinity, are perpendicular to a tangent plane. This projection is used chiefly in navigational astronomy for interconverting coordinates of the celestial equator and horizon systems. Also called **orthogonal map projection.**

orthometric correction—A systematic correction which must be applied to a gravity oriented measured difference of elevation because level surfaces at different elevations are not exactly parallel.

orthometric elevation—A preliminary elevation to which the orthometric correction has been applied.

orthometric error—An error due to the spheroidal form of the Earth and the action of centrifugal force; level surfaces at different elevations are not exactly parallel.

orthomorphic chart—A chart on which very small shapes are correctly represented.

orthomorphic map projection—See **conformal map projection.**

orthophotograph—A photographic copy, prepared from a perspective photograph, in which the displacements of images due to tilt and relief have been removed.

orthophotomap—A photomap made from an assembly of orthophotographs. It may incorporate special cartographic treatment, photographic edge enhancement, color separation, or a combination of these.

orthophotomosaic—An assembly of orthophotographs forming a uniform-scale mosaic.

orthophotoscope—A photomechanical device, used for producing orthophotographs.

orthopictomap—A pictomap made from an orthophotomap base.

orthostereoscopy—A condition wherein the horizontal and vertical distances in a stereoscopic model appear to be at the same scale.

oscillation—A double motion, one in each direction, of a pendulum. An oscillation is composed of two successive vibrations.

osculating elements—The elements that define an osculating orbit. See also **osculating orbit.**

osculating ellipse—An ellipse that is tangent at a point (called the epoch of osculation) to a real orbit.

osculating orbit—The ellipse that a satellite would follow after a specific time "t" (the epoch of osculation) if all forces other than central inverse-square forces ceased to act from time "t" on. An osculating orbit is tangent to the real, perturbed, orbit and has the same velocity at the point of tangency.

outer orientation—See **exterior orientation.**

outer planets—The planets with orbits larger than that of Mars; i.e., Jupiter, Saturn, Uranus, Neptune, and Pluto.

outlier—A measurement which does not fit the remainder of measurements of the same quantity, where the reason for the discrepancy cannot be assessed.

outline map—(JCS) A map which represents just sufficient geographic information to permit the correlation of additional data placed upon it.

overcharging—Applying excessive additional information (aeronautical or navigational) to a map or chart resulting in clutter.

overhang—(aerial photography) The additional exposures beyond the boundary of an area to be photographed, usually two exposures at the ends of each strip to assure complete stereoscopic coverage.

overlap—1. (JCS) In photography, the amount by which one photograph includes the same area covered by another, customarily expressed as a percentage. The overlap between successive air photographs on a flight line is called **forward overlap** (or **forward lap**). Also called **end lap.** The overlap between photographs in adjacent parallel flight lines is called **side overlap** (or **side lap**). **2.** (JCS) In cartography, that portion of a map or chart which overlaps the area covered by another of the same series. **3.** An area included within two surveys of record, which by record are described as having one or more common boundary lines with no inclusion of identical parts.

overlapping grid—A military grid which is extended beyond its normal limits to map sheets located in areas bordering grid junctions and spheroid junctions. Normally, large scale maps that fall within approximately 25 miles of a grid, zone, or spheroid junction will bear an overlapping grid, depicted by ticks emanating from the map neatline. Overlapping grids are used primarily for extension of fire control and survey operations. See also **major grid; secondary grid.**

overlapping mean—See **consecutive mean.**

overlapping pair—(photogrammetry) Two photographs taken at different exposure stations in such a manner that a portion of one photograph shows the same terrain as shown on a portion of the other photograph. This term covers the general case and does not imply that the photographs were taken for stereoscopic examination. See also **stereoscopic pair.**

overlay—1. (JCS) A printing or drawing on a transparent or semitransparent medium at the same scale as a map, chart, etc., to show details not appearing, or requiring special emphasis, on the original. **2.** (lithography) Additional data, or a pattern, printed after the other features, so as to "overlay" them. See also **correction overlay; history overlay; radarscope overlays; selection overlay.**

overprint—1. (JCS) Information printed or stamped upon a map or chart, in addition to that originally printed, to show data of importance or special use. Also called **surprint. 2.** A feature of a composite map image incidentally printed so as to interfere with another feature.

overrun control—(JCS) Equipment enabling a camera to continue operating for a predetermined number of frames or seconds after normal cutoff.

oversheet—A transparency or a print of a map compilation used for recording supplemental information.

oversize chart—A chart whose neatlines have been extended slightly, thereby increasing the sheet size to include a small land area in order to avoid publishing a separate graphic of that area.

PC-1000 camera—A trade name for a geodetic stellar camera having a focal length of 1,000 mm.

PZS triangle—See **astronomic triangle.**

panchromatic—(photography) Sensitive to light of all colors, as a film or plate emulsion.

pancratic system—A variable-power optical system. Also called **zoom system.**

panel—**1.** (cartography) See **panel base. 2.** (photogrammetry) An element of a target used for control-station identification on aerial photography. Panels are made of cloth, plastics, plywood, or Masonite, and are positioned in a symmetrical pattern centered on the station. See also **target.**

panel base—(cartography) The completed assembly of pieces of film positives onto a grid or projection which is used as a base for compilation. Also called **film mosaic; panel.**

paneling—**1.** (cartography) Cutting a film positive of a map, in which some distortion is involved, into several pieces and cementing them in place, on a projection constructed on a stable-base medium, in such a way that the error is distributed in small amounts throughout the area rather than being localized. **2.** (surveying) The placement of panels on a control station to facilitate station identification on aerial photography.

panoramic camera—A camera which takes a partial or complete panorama of the terrain. Some designs utilize a lens which revolves about an axis perpendicular to the optical axis; in other designs, the camera itself is revolved by clockwork to obtain a panoramic field of view. See also **frame camera.**

panoramic distortion—The displacement of ground points from their expected perspective positions, caused by the cylindrical shape of the negative film surface and the scanning action of the lens in a panoramic camera system.

panoramic photograph—Photography obtained from a panoramic camera.

pantograph—An instrument for copying maps, drawings, or other graphics at a predetermined scale. Pantographs capable of adjustment for several scales are known as fixed-ratio pantographs. See also **two-dimensional pantograph.**

paper-strip method—(rectification) A graphical method of making a point-by-point rectification based on the invariance of the cross ratio. A modification of this technique permits map detail to be revised from an oblique aerial photograph based on the projectivity of straight lines.

parallactic aberration—See **differential aberration.**

parallactic angle—**1.** (astronomy) The angle between a body's hour circle and its vertical circle. Also called **position angle. 2.** (photogrammetry) See **angle of convergence.** Also called **angular parallax.**

parallactic error—An error caused by personal or instrument parallax.

parallactic grid—(photogrammetry) A uniform pattern of rectangular lines drawn or engraved on some transparent material, usually glass, and placed either over the photographs of a stereoscopic pair or in the optical system of a stereoscope, in order to provide a continuous floating-mark system.

parallactic inequality—A secondary effect in solar perturbations in the Moon's longitude due to the ellipticity of the Earth's orbit.

parallax—**1.** (JCS) In photography, the apparent displacement of the position of an object in relation to a reference point due to a change in the point of observation. **2.** The apparent displacement between objects on the Earth's surface due to their difference in elevation. Also called **angular parallax; want of correspondence.** See also **absolute stereoscopic parallax; age of parallax inequality; annual parallax; equatorial horizontal parallax; false parallax; geocentric parallax; horizontal parallax; instrument parallax; lunar parallax; residual parallax; solar parallax; y-parallax.**

parallax age—See **age of parallax inequality.**

parallax bar—See **stereometer.**

parallax difference—The difference in the absolute stereoscopic parallaxes of two points imaged on a pair of photographs. Customarily used in determination of the difference in elevation of objects.

parallax in altitude—Geocentric parallax at any altitude. The expression is used to distinguish the parallax at the given altitude from the horizontal parallax when the body is in the horizon.

parallax inequality—The variation in the range of tide or in the speed of tidal currents because of the continual change in the distance of the Moon from the Earth. The range of tide and speed of tidal currents tend to increase as the Moon approaches perigee and to decrease as it approaches apogee.

parallel—A circle on the surface of the Earth, parallel to the plane of the Equator and connecting all points of equal latitude, or a circle parallel to the primary great circle of a sphere or spheroid; also, a closed curve approximating such a circle. Also called **inverse parallel.** See also **astronomic parallel; ecliptic parallel; fictitious parallel; geodetic parallel; geographic parallel; grid parallel; ground parallel; isometric parallel; oblique parallel; photograph parallel; principal parallel; standard parallel; transverse parallel.**

parallel of altitude—A circle of the celestial sphere parallel to the horizon connecting all points of equal altitude. Also called **almucantar; altitude circle; circle of equal altitude.**

parallel of declination—A circle of the celestial sphere parallel to the celestial equator. Also called **celestial parallel; circle of equal declination.**

parallel of latitude—See **circle of longitude.**

parallel plate—An optical disk with optically flat, parallel surfaces; used especially in optical micrometers. Also called **plane-parallel plate.** See also **optical flat.**

parallel sphere—The celestial sphere as it appears to an observer at the pole, where celestial bodies appear to move parallel to the horizon.

parameter—In general, any quantity of a problem that is not an independent variable. More specifically, the term is often used to distinguish from dependent variables quantities which may be assigned more or less arbitrary values for purposes of the problem at hand.

parametric equations—A set of equations in which the independent variables or coordinates are each expressed in terms of a parameter.

parametric latitude—The angle at the center of a sphere which is tangent to the spheroid along the geodetic equator, between the plane of the equator and the radius to the point intersected on the sphere by a straight line perpendicular to the plane of the equator and passing through the point on the spheroid whose parametric latitude is defined. Parametric latitude is an auxiliary latitude used in problems of geodesy and cartography. In astronomic work, when the term **reduced latitude** is used, **geocentric latitude** is meant. Also called **geometric latitude; reduced latitude.** See also **geocentric latitude.**

paraxial ray—A ray whose path lies very near the axis of a lens and which intersects the lens surface at a point very close to its vertex and at nearly normal incidence.

partial tide—See **constituent.**

pass—**1.** A single circuit of the Earth by a satellite. See also **orbit. 2.** The period of time a satellite is within telemetry range of a data acquisition station. **3.** (mensuration) One complete set of pointings or measurements on a specific plate, reseau, or other media containing photographic imagery.

pass point—A point whose horizontal and/or vertical position is determined from photographs by photogrammetric methods and which is intended for use in the absolute orientation of a model. Also called **photogrammetric point.** See also **annex point; supplemental elevation; supplemental position.**

passive satellite—A satellite which contains no power sources to augment output power; a satellite which is a passive reflector. See also **active satellite.**

path—The projection of the orbital plane of the satellite on the Earth's surface; the locus of the satellite subpoint.

Peaucellier-Carpentier inversor—A modified Carpentier inversor coupled to the linkage system of a Peaucellier inversor to provide a mechanical means of solving the linear and angular elements of rectification.

Peaucellier inversor—A class of inversor providing a mechanical solution for the linear and angular elements of rectification. Also called **scissors inversor.**

peel—(negative engraving) A technique of removing the opaque stratum from its supporting base. Peeling between etched outline images produces a negative; peeling outside of the etched outline images produces a positive. See also **mask,** definition 2.

peepsight alidade—A type of alidade consisting of a peep sight mounted on a straightedge.

peepsight compass—The sights of a compass formed by standards with slits for a sighting medium rather than a telescope.

peg adjustment—A method of adjusting a leveling instrument of the dumpy level type, to make the line of collimation parallel with the axis of the spirit level, and employing two stable marks (pegs) the length of one instrument sight apart. Also called **11/10 peg adjustment.**

peg test—A method of testing the collimation adjustment of a leveling instrument.

Pemberton leveling rod—A speaking rod marked with alternate rows of circular and diamond-shaped dots, running diagonally across the rod. Read to hundredths of a foot.

pendulum—**1.** In general, a body so suspended as to swing freely to and fro under the influence of gravity and momentum. **2.** A vertical bar so supported from below by a stiff spring as to vibrate to and fro under the combined action of gravity and the restoring force of the spring. See also **compound pendulum; dummy pendulum; free-swinging pendulum; idle pendulum; Invar pendulum; invariable pendulum; Mendenhall pendulum; quartz pendulum; receiver; relative pendulum; reversible pendulum; simple pendulum; working pendulum.**

pendulum alidade—A telescopic alidade in which a pendulum device replaces the conventional bubble for establishing a horizontal reference line from which vertical angles may be measured.

pendulum astrolabe—An astronomic instrument using a constant altitude for position determination. Its distinctive feature is a mirror suspended on top of a pendulum to form the artificial horizon.

pendulum level—A leveling instrument in which the line of sight is automatically maintained horizontal by means of a built-in pendulum device. Also called **automatic level.**

percent of enlargement/reduction—The factor by which an original is to be enlarged or reduced in reproduction. A 50 percent linear enlarge ment of a 4- by 5-inch original would be 6 by 7½ inches, while a 50 percent reduction of the same original would be 2 by 2½ inches. See also **scale of reproduction.**

percent of slope—See **gradient.**

periapsis—See **pericenter.**

periastron—That point of the orbit of one member of a double star system at which the stars are nearest together. Opposite of **apastron.**

pericenter—In an elliptical orbit, the point in the orbit which is the nearest distance from the focus where the attracting mass is located. The pericenter is at one end of the major axis of the orbital ellipse. Opposite of **apoapsis; apocenter; apofocus.** Also called **periapsis; perifocus.**

pericynthion—See **perilune.**

perifocus—See **pericenter.**

perigee—(JCS) The point at which a satellite orbit is the least distance from the center of the gravitational field of the controlling body or bodies. Opposite of **apogee.**

perigee-to-perigee period—See **anomalistic period.**

perihelion—The point in the elliptical orbit of a planet which is the nearest to the Sun, when the Sun is the center of attraction. Opposite of **aphelion.**

perilune—The point of closest approach of an orbiting body to the Moon. Opposite of **aplune; apocynthion.** Also called **pericynthion.**

period—**1.** The interval needed to complete a cycle. **2.** The interval between passages at a fixed point of a given phase of a simple harmonic wave; the reciprocal of frequency. See also **anomalistic period; nodical period; orbital period; sidereal period; synodic period.**

period of satellite—See **orbital period.**

periodic errors—In a complete set of observations there corresponds to every individual error another error which is necessarily more or less equal and opposite. In a limited series the cancellation may not be quite exact, but the error of the mean of n observations may be expected to be $1/n$ of that of a single measure, or less.

periodic perturbations—Perturbations to the orbit of a planet or satellite which change direction in regular or periodic fashion in time, such that the average effect over a long period of time is zero.

periodic terms—In the mathematical expression of an orbit, terms which vary with time in both magnitude and direction in a periodic fashion.

permanent bench mark (PBM)—A bench mark of as nearly permanent character as it is practicable to establish. Usually designated simply as a bench mark or BM. A permanent bench mark is intended to maintain its elevation with reference to an adopted datum without change over a long period of time. Also called **monumented bench mark.**

perpendicular—A perpendicular line, plane, etc. A distinction is sometimes made between perpendicular and normal, the former applying to a line at right angles to a straight line or plane, and the latter referring to a line at right angles to a curve or curved surface.

perpendicular equation—(traverse) A condition equation to reduce to zero the algebraic sum of the projections of the separate lines of a traverse upon perpendiculars to a fixed line with which the traverse forms a closed figure.

personal equation—The time interval between the sensory perception of a phenomenon and the motor reaction thereto. A personal equation may be either positive or negative, as an observer may anticipate the occurrence of an event, or wait until he actually sees it occur before making a record. This is a systematic error, treated as the constant type.

personal error—An error caused by an individual's personal habits, his inability to perceive or measure dimensional values exactly, or by his tendency to react mentally and physically in a uniform manner under similar conditions. It may be a systematic error, if it occurs regularly or a blunder if it occurs once. A certain amount of minor personal error can be included with random errors.

personal parallax—See **instrument parallax,** definition 2.

perspective—The appearance to the eye of objects in respect to their relative distance and position.

perspective axis—See **axis of homology.**

perspective center—The point of origin or termination of bundles of perspective rays. The two such points usually associated with a survey photograph are the interior perspective center and the exterior perspective center. In a perfect lens-camera system, perspective rays from the interior perspective center to the photographic images enclose the same angles as do the corresponding rays from the exterior perspective center to the objects photographed. In a lens having distortion, this is true only for a particular zone of the photograph. In a perfectly adjusted lens-camera system, the exterior and interior perspective centers correspond, respectively, to the front and rear nodal points of the camera lens. Also called **center of projection.**

perspective chart—A chart on a perspective projection.

perspective grid—(JCS) A network of lines, drawn or superimposed on a photograph, to represent the perspective of a systematic network of lines on the ground or datum plane. Also called **Canadian grid.** See also **grid method.**

perspective map projection—A map projection produced by straight lines radiating from a selected point and passing through points on the sphere to the plane of projection. The plane of projection is usually tangent to the sphere which represents the Earth at the center of the area being mapped; the point of projection is on the diameter of the sphere which passes through the point of tangency, and at some selected point on that diameter. Also called **geometric map projection.**

perspective map projection upon a tangent cylinder—A cylindrical map projection upon a cylinder tangent to a sphere, by means of straight lines radiating from the center of the sphere. The geographic meridians are represented by a family of equally spaced parallel straight lines, perpendicular to a second family of parallel straight lines which represent the geographic parallels. The spacing, with respect to the Equator, of the lines which represent the parallels, increases as the tangent of the latitude; the line representing 90° latitude is at an infinite distance from the line which represents the Equator. Not to be confused with the **Mercator map projection** to which it bears some general resemblance.

perspective plane—Any plane containing the perspective center. The intersection of a perspective plane and the ground will always appear as a straight line on an aerial photograph.

perspective projection—The projection of points by straight lines drawn through them from some given point to an intersection with the plane of projection. Unless otherwise indicated, the point of projection is understood to be at a finite distance from the plane of projection.

perspective ray—A line joining a perspective center and a point object. See also **image ray.**

perspective spatial model—Optical reconstruction of an area of terrain showing depth by viewing a pair of aerial photographs through a stereoscope.

perspectivity—The correspondence between the points, lines, or planes of two geometric configurations in perspective. Usually referred to as linear perspectivity because the true perspective center must be recoverable before angular perspectivity can be included.

perturbation—In celestial mechanics, differences of the actual orbit from a central force orbit, arising from some external force such as a third body attracting the other two; a resisting medium (atmosphere); failure of the parent body to act as a point mass, and so forth. See also **gravitational perturbations; long period perturbations; lunisolar perturbations; nongravitational perturbations; periodic perturbations; secular perturbations; short period perturbations; terrestrial perturbations.**

perturbed orbit—The orbit of a satellite differing from its normal orbit due to various disturbing effects, such as nonsymmetrical gravitational effects, atmospheric drag, radiation pressure, and so forth. See also **perturbation.**

perturbing factors (forces)—In celestial mechanics, any force that acts on the orbiting body to change its orbit from a central force orbit.

phase—1. (general) Of a periodic quantity, for a particular value of the independent variable, the fractional part of a period through which the independent variable has advanced, measured from an arbitrary reference. **2.** (surveying) The apparent displacement of an object or signal caused by one side being more strongly illuminated than the other. The resultant error in pointing is similar to the error caused by observing an eccentric signal. **3.** (astronomy) A stage in a cycle of recurring aspects, caused by a systematic variation of the illumination of an object. The Moon passes through its phases, new Moon to full Moon and back to new Moon, as its position relative to the Sun and Earth changes.

phase age—See **age of phase inequality.**

phase angle—1. The phase difference of two periodically recurring phenomena of the same frequency, expressed in angular measure. **2.** The angle at a celestial body between the Sun and Earth.

phase inequality—Variations in the tide or tidal currents associated with changes in the phase of the Moon. At new and full Moon (springs) the tide-producing forces of the Sun and Moon act in conjunction, resulting in greater than average tide and tidal currents. At first and last quarters of the Moon (neaps) the tide-producing forces oppose each other, resulting in smaller than average tide and tidal currents.

Philadelphia leveling rod—A two-piece target rod, with graduation marks so styled that it may also be used as a speaking rod. For heights greater than 7 feet the target is clamped at 7 feet, and raised by extending the rod. As a target rod, it is read by vernier to thousandths of a foot; as a speaking rod, to half-hundredths of a foot.

photo altitude—Height of an aircraft above the mean elevation of the terrain to be photographed.

photo-contour map—Essentially, a topographic map upon which the planimetric detail is depicted photographically in its correct position. It is usually prepared from convergent photography although conventional vertical photography can be used.

photo-contour process—A process developed to combine, in a photo-contour map, that information normally portrayed on a topographic drawing and an aerial photograph. The system usually is composed of three elements: a conventional stereoplotter for contouring, a rectifier for tilt rectification of the aerial photographs, and a zone printer to eliminate relief displacement. It is designed to utilize convergent photographs although normal vertical photographs can be utilized as well.

photo index—1. An index map made by assembling the individual aerial photographs into their proper relative positions and copying the assembly photographically at a reduced scale. Also called **index to photography; photo plot; plot map.** **2.** See **sortie plot.**

photo plot—See **photo index,** definition 1.

photo pyramid—A component of an analytical method of precise determination of photographic tilt which represents a specific spatial configuration formed by three control points of known position on the photograph (forming a triangle) and the exposure station. When used with the ground pyramid, it permits the exact position of the exposure station to be determined and, by analytical techniques, the exact tilt of the photograph. See also **ground pyramid.**

photo-revised map—A topographic or planimetric map which has been revised by photoplanimetric methods.

photo revision—The process of making changes on a map based upon information obtained from a study of aerial photographs.

photo scale—See **scale,** definition 1.

photoalidade—A photogrammetric instrument having a telescopic alidade, a plateholder, and a hinged ruling arm mounted on a tripod frame. It is used for plotting lines of direction and measuring vertical angles to selected features appearing on oblique and terrestrial photographs.

photoangulator—See **angulator.**

photobase—The distance between the principal points of two adjacent prints of a series of vertical aerial photographs. It is usually measured on one print after transferring the principal point of the other print. See also **base line,** definition 2.

photocompose—To mechanically impose one or more images by step-and-repeat exposures in predetermined positions on a pressplate or negative by means of a photocomposing machine.

photocontrol base—See **control base.**

photocontrol diagram—Any selected base map or photo index on which proposed ground control networks, to include proposed positions for pass points, are delineated. See also **photocontrol index map.**

photocontrol index map—Any selected base map or photo index on which ground control and photo identified ground points are depicted and identified. See also **photocontrol diagram.**

photocontrol point—See **picture control point.**

photogoniometer—An instrument for measuring angles from the true perspective center to points on a photograph.

photogrammetric camera—A general term applicable to any camera used in any of the several branches of photogrammetry.

photogrammetric compilation—See **compilation,** definition 2.

photogrammetric control—(JCS) Control established by photogrammetric methods as distinguished from control established by ground methods. Also called **minor control; multiplex control.**

photogrammetric control point—A horizontal control point which has been established by photogrammetric triangulation.

photogrammetric map—A topographic map produced from aerial photographs and geodetic control data by means of photogrammetric instruments. Also called **stereometric map; stereotopographic map.**

photogrammetric point—See **pass point.**

photogrammetric pyramid—An analytical method for the precise determination of photographic tilt, consisting of a ground pyramid and a photo pyramid, which represent a spatial configuration formed by three control points of known position on the photograph (forming a triangle) and the exposure station. See also **ground pyramid; photo pyramid.**

photogrammetric rectification—See **rectification.**

photogrammetric survey—A survey utilizing either terrestrial or aerial photographs.

photogrammetric triangulation—See **phototriangulation.**

photogrammetry—**1.** (JCS) The science or art of obtaining reliable measurements from photographic images. **2.** The preparation of charts and maps from aerial photographs using stereoscopic equipment and methods. See also **aerial photogrammetry; analytical photogrammetry; stereophotogrammetry; terrestrial photogrammetry.**

photograph—A general term for a positive or negative picture made with a camera on sensitized material, or prints from such a camera original. See also **aerial photograph; annotated photograph; equivalent vertical photograph; homologous photographs; horizon photograph; horizontal photograph; molded aerial photograph; multiple-lens photograph; oblique air photograph; orthophotograph; panoramic photograph; pinpoint photograph; terrestrial photograph; vertical photograph; wing photograph.**

photograph center—The center of a photograph as indicated by the images of the fiducial mark or marks of the camera. In a perfectly adjusted camera, the photograph center and the principal point are identical.

photograph coordinates—A system of coordinates, either rectangular or polar, describing the position of a point on a photograph.

photograph meridian—The image on a photograph of any horizontal line in the object space which is parallel to the principal plane. Since all such lines meet at infinity, the image of the meeting point is at the intersection of the principal line and the horizon trace and all photograph meridians pass through that point.

photograph nadir—The point at which a vertical line through the perspective center of the camera lens pierces the plane of the photograph. Also called **nadir point; plumb point.**

photograph parallel—The image on a photograph of any horizontal line in the object space which is perpendicular to the principal plane. All photograph parallels are perpendicular to the principal line.

photograph perpendicular—The perpendicular from the interior perspective center to the plane of the photograph.

photograph plane—The plane in the camera in which the plate or film is held. It is not exactly the primary focal plane of the lens, but is a plane placed so as to secure the best balance of sharp focus on all parts of the plate or film. Also called **image plane.**

photograph plumb point—See **photograph nadir.**

photographic coverage—(JCS) The extent to which an area is covered by photography from one mission or a series of missions or in a period of time. Coverage in this sense conveys the ideal of availability of photography and is not a synonym for the word **photography.**

photographic datum—The effective datum for each photograph. It is a horizontal plane at the average elevation of the terrain, on which distances measured will be at the average scale of the photograph.

photographic exposure—The time of exposure multiplied by irradiance or illuminance.

photographic interpretation—The examination of photographic images for the purpose of identifying objects and deducing their significance. Also called **photointerpretation.** See also **imagery interpretation.**

photographic reading—(JCS) The simple recognition of natural or cultural features from photographs not involving imagery interpretation techniques.

photographic reduction—The production of a negative, diapositive, or print at a scale smaller than the original.

photographic survey—A survey accomplished from either aerial photographs or terrestrial photographs, or from a combination of both.

photographic zenith tube (PZT)—The most precise instrument for meridian observations. No corrections are required for level, azimuth, collimation, or flexure. Each observation gives a measure of both the time and the latitude.

photography—The art or process of producing images on sensitized material through the action of light. The term **photography** is sometimes incorrectly used in place of the term **photographs.** See also **analytical photography; composite air photography; continuous-strip photography; control-point photography; convergent photography; cross-flight photography; direct photography; fan-camera photography; horizontally controlled photography; indirect photography; inertial reference photography; lorop photography; mapping photography; metric photography; multiband photography; positional camera photography; process photography; radar photography; radarscope photography; reconnaissance photography; shoran-controlled photography; split-vertical photography; supplemental photography; terrain profile photography; tricamera photography.**

photoidentification—(surveying) The detection, identification, and marking of ground survey stations on aerial photographs. Positive identification and location is required if survey data are to be used to control photogrammetric compilation. Also called **control-station identification.**

photointerpretation—See **photographic interpretation.**

photointerpretation key—Reference materials designed to facilitate rapid and accurate identification and the determination of the significance of objects or conditions from an analysis of their photo images.

photointerpretometer—A device, used in conjunction with a pocket stereoscope, for making vertical and horizontal measurements.

photolithography—A lithographic process in which photographic products are used to produce an image on the printing surface. See also **lithography; offset lithography.**

photomap—(JCS) A reproduction of a photograph or photomosaic upon which the grid lines, marginal data, contours, place names, boundaries, and other data may be added.

photomap backup—A photomap printed on the back of a line map of the same area and at the same scale.

photomapping—The process of making maps or charts from various types of photographs, with reference to other source maps, charts, or surveys.

photomechanical—Pertaining to or designating any reproduction process by a combination of photographic and mechanical operations.

photosphere—The intensely bright portion of the Sun visible to the unaided eye.

phototheodolite—A ground-surveying instrument combining a survey camera and a transit; used for measuring the angular orientation of the camera at the moment of exposure. Also called **camera transit.**

phototopography—The science of surveying in which the detail is plotted entirely from photographs taken at suitable ground stations. See also **terrestrial photogrammetry.**

phototriangulation—The process for the extension of horizontal and/or vertical control whereby the measurements of angles and/or distances on overlapping photographs are related into a spatial solution using the perspective principles of the photographs. Generally, this process involves using aerial photographs and is called **aerotriangulation, aerial triangulation,** or **photogrammetric triangulation.** See also **analytical nadir-point triangulation; analytical phototriangulation; analytical radial triangulation; Arundel method; bridging; cantilever extension; direct radial triangulation; extension of control; graphical radial triangulation; hand-templet triangulation; isocenter triangulation; mechanical templet triangulation; nadir-point triangulation; radial triangulation; slotted-templet triangulation; spider-templet triangulation; stereotemplet triangulation; stereotriangulation; strip radial triangulation; templet method.**

phototrig traverse—A vertical-angle traverse employing phototrig methods; a procedure for determining elevations trigonometrically, wherein horizontal distances are determined photogrammetrically and vertical angles are either measured instrumentally in the field, or are obtained from measurements on terrestrial photographs.

phototypesetter—A type setting unit comprising two separate and independent units: the keyboard unit and the photographic unit. Composition is accomplished at the keyboard unit, essentially an electric typewriter, which produces a typewritten proof copy and a perforated tape. The tape is then fed at any convenient time thereafter to the photographic unit which produces a right-reading film positive suitable for stick-up work.

physical characteristic—(target) The visible material aspects of a target or installation, including, but not limited to, dimensions, structural materials, predominant height, configuration, and orientation of its various components such as buildings, structures, runways, and associated facilities and services.

physical geodesy—See **gravimetric geodesy.**

physiographic pictorial map—A map with relief depicted by the systematic application of a standardized set of conventional pictorial symbols, based on the simplified appearance of the physical features they represent, as viewed obliquely from the air at an angle of about 45°.

piano-wire tape—Piano wire used instead of a metallic ribbon tape when it is advisable to control hydrography by precise traverse rather than by a weak extension of triangulation.

pictochrome process—The process employed to produce pictomaps. Consists of three tonal separations photographically extracted from a photomosaic, block-out masks, drafted symbols, and names data.

pictogram—A map of distributions, especially commodities, in which small pictorial representative symbols (e.g., sacks, bricks, barrels) are located over the area of production.

pictoline process—A photographic masking process utilizing a rotating vacuum frame to produce an edge-enhanced line image from a continuous tone image.

pictomap—(JCS) A topographic map in which the photographic imagery of a standard mosaic has been converted into interpretable colors and symbols by means of a pictomap process. See also **pictochrome process.**

pictorial symbolization—(JCS) The use of symbols which convey the visual character of the features they represent.

pictotone process—A photolithographic method from which film for reproduction and transfer to printing plates is derived for the printing of monochrome photomaps and pictomaps. The process provides a random granular-like effect which visibly sharpens the definition of features and separation of tones, and, in many instances, is superior to halftone printing.

picture control points—Supplementary horizontal and vertical control points that are required for the immediate control of mapping operations in a given area. These points are established by field survey parties in specific locations and are precisely identified on the aerial photographs for the project. Also called **photocontrol point; picture point.**

picture plane—A plane upon which can be projected a system of lines or rays from an object to form an image or picture. In perspective drawing, the system of rays is understood to converge to a single point. In photogrammetry, the photograph is the picture plane.

picture point—See **control point; picture control points.**

pilot—See **sailing directions.**

pilot chart—Special charts, covering the oceans of the world for each month of the year, issued on one sheet for 3 months at a time on a quarterly basis. They show meteorologic, oceanographic, and hydrographic data for use in conjunction with conventional charts. Timely articles of professional interest to the seafarer are published on the backs.

pilot sheet—A sample of a new series, made as a trial in anticipation of a map series, to disclose the problems which occur in the various stages of compilation, drafting, and reproduction. It is later used as a guide in developing the series. Also called **prototype.** See also **experimental map.**

pilotage chart (PC)—A chart at the scale of 1:500,000 used for preflight planning and inflight navigation and for short-range flights using dead reckoning and visual pilotage.

pilot's trace—(JCS) A rough overlay to a map made by the pilot of a photographic reconnaissance aircraft during or immediately after a sortie. It shows the location, direction, number, and order of photographic runs made, together with the camera(s) used on each run.

pin—(surveying) A metal pin used for marking taped measurements on the ground. A set consists of 11 pins. Also called **chaining pin;**

surveyor's arrow; taping arrow; taping pin. See also **turning-point pin.**

pinholes–Tiny clear spots on negative images caused by dust, air bubbles, or undissolved chemicals.

pinpoint photograph–(JCS) A single photograph or one or more stereopairs of a specific object or target.

pinpoint target–(JCS) In artillery and naval gunfire support, a target less than 50 meters in diameter. See also **area target; precise installation position.**

pitch–1. (JCS) The rotation of an aircraft or ship about its lateral axis. **2.** (JCS) In air photography, the camera rotation about the transverse axis of the aircraft. **3.** (photogrammetry) A rotation of the camera, or of the photograph-coordinate system, about either the photograph *y*-axis or the exterior *y*-axis. In some photogrammetric instruments and in analytical applications, the symbol phi (ϕ) may be used. Also called **longitudinal tilt; tip** (which is an obsolete term); **y-tilt.**

place–See **position,** definition 2.

place name–See **toponym.**

plan position indicator (PPI)–1. (JCS) A cathode-ray tube on which radar returns are so displayed as to bear the same relationship to the transmitter as the objects giving rise to them. **2.** A cathode-ray indicator in which a signal appears on a radial line. Distance is indicated radially and bearing as an angle.

plan range–(JCS) In air photographic reconnaissance, the horizontal distance from the point below the aircraft to an object on the ground.

plane–See **astronomic meridian plane; basal plane; collimation plane; epipolar plane; focal plane; geodetic meridian plane; ground plane; hill plane; horizontal plane; meridional plane; nodal plane; orbital plane; perspective plane; photograph plane; picture plane; principal plane; tangent plane; vertical plane.**

plane coordinates–See **plane rectangular coordinates.**

plane elliptical arc–Any part of the line formed by the intersection of a plane and an ellipsoid. Also called **plane curve.**

plane parallel plate–See **parallel plate.**

plane polar coordinates–A system of polar coordinates in which the points all lie in one plane. In the terminology of analytical geometry, the distance from the origin to the point is the magnitude of the radius vector and the polar distance is the vectorial angle.

plane rectangular coordinates–A system of coordinates in a horizontal plane, used to describe the positions of points with respect to an arbitrary origin. The origin is established by a pair of axes which intersect at right angles. The position of a point is determined by the perpendicular distances to these axes. Also called **plane coordinates.**

plane survey–A survey in which the surface of the Earth is considered a plane. For small areas, precise results may be obtained with plane-surveying methods, but the accuracy and precision of such results will decrease as the area surveyed increases in size.

planet–A celestial body of the solar system revolving around the Sun in a nearly circular orbit, or a similar body revolving around a star. See also **asteroid; inferior planets; inner planets, major planets; navigational planets; outer planets; principal planets; superior planets; terrestrial planet.**

planetable–A field device for plotting the lines of a survey directly from observations. It consists essentially of a drawing board mounted on a tripod, with a leveling device designed as part of the board and tripod. See also **alidade; Philadelphia leveling rod; stadia.**

planetable map–A map compiled by planetable methods. The term includes maps made by complete field mapping on a base projection and field contouring on a planimetric-base map.

planetable traverse–A graphical traverse accomplished by planetable methods.

planetary aberration–The angular displacement of the geometric direction between the object and the observer at the instant of light-emission, from the geometric direction at the instant of observation.

planetary configurations–Apparent positions of the planets relative to each other and to other bodies of the solar system, as seen from the Earth.

planetary geometry–1. Mathematical treatment of the shape and figure of a planet. **2.** Mathematical treatment of relationships between two or more planets and/or their orbits.

planetary precession–That component of general precession caused by the effect of other planets on the equatorial protuberance of the Earth, producing an eastward motion of the equinoxes along the ecliptic.

planetoid–See **asteroid.**

planimeter–A mechanical integrator for measuring the area of a plane surface. See also **polar planimeter.**

planimetric-base map–A map prepared from aerial photographs by photogrammetric methods, as a guide or base for contouring.

planimetric map–(JCS) A map representing only the horizontal position of features. Also called **line map.** See also **topographic map.**

planimetry–1. The science of measuring plane surfaces; horizontal measurements. **2.** Parts of a map which represent everything except relief that is, works of man, and natural features such as woods and water.

planisphere–A representation, on a plane, of the celestial sphere, especially one on a polar projection, with means provided for making certain measurements such as altitude and azimuth. Also, a map representation, on a plane, of the Earth's sphere.

planispheric astrolabe—An astrolabe consisting of a full graduated circle with a centrally mounted alidade and accessory adjustable plates on which are engraved stereographic projections of the heavens and of the sphere for local latitudes.

planning chart—A chart designed specifically for planning flight operations.

planning map—Small-scale military map used for general planning purposes.

plastic block—The block of bonded cellulose-acetate sheets, each sheet equal in thickness to the contour interval at the scale of the relief model, from which the terrain base is cut. Also called **laminate.**

plastic relief map—A topographic map printed on plastic and molded into a three-dimensional form. The plastic medium is generally formed by heat and vacuum over a terrain model to achieve the three-dimensional representation.

plat—A diagram drawn to scale showing land boundaries and subdivisions, together with all data essential to the description and identification of the several units shown thereon, and including one or more certificates indicating due approval. A plat differs from a map in that it does not necessarily show additional cultural, drainage, and relief features. See also **cadastral map.**

plate—**1.** (lithography) A thin metal, plastic, or paper sheet, that carries the printing image and whose surface is treated to make only the image areas ink-receptive. Also called **pressplate.** See also **color plate; combination plate.** **2.** (photography) A transparent medium, usually glass, coated with a photographic emulsion. See also **diapositive; stellar plate.**

plate coordinates—The *x*- and *y*-coordinates of control points appearing on a photographic plate.

plate level—A spirit level attached to the plate of a surveying instrument for leveling the graduated circle or, indirectly, making the vertical axis truly vertical.

plate reduction—Scaling of control point images on a stellar plate.

platform—The vehicle that holds a sensor. It is usually a satellite, but may be a plane or a helicopter. Sensors can be mounted on tripods for certain uses, such as examining electromagnetic radiation from various types of vegetation.

Platonic year—See **great year.**

plot—(JCS) **1.** A map, chart, or graph representing data of any sort. **2.** To represent on a diagram or chart the position or course of a target in terms of angles and distances from known positions; locate a position on a map or chart. **3.** The visual display of a single geographical location of an airborne object at a particular instant of time. **4.** A portion of a map or overlay on which are drawn the outlines of the areas covered by one or more photographs.

plot map—See **photo index.**

plotting chart—**1.** (JCS) A chart designed for the graphical processes of navigation. **2.** A chart designed primarily for plotting and dead reckoning or lines of position from celestial observations or radio aids. Relief, culture, and drainage are shown as necessary.

plotting scale—The relationship of the size of the compilation to the size of the ground area it represents.

plumb bob—A conical device, usually of brass and suspended by a cord, by means of which a point can be projected vertically into space over relatively short distances.

plumb line (plumbline)—**1.** The line of force in the geopotential field. The continuous curve to which the direction of gravity is everywhere tangential. **2.** A cord with a plumb bob at one end for determining the direction of gravity.

plumb point—See **photograph nadir.**

plunge—See **transit,** definition 3.

plus angle—See **angle of elevation.**

plus declination—See **declination,** definition 3.

plus distance—Fractional part of 100 feet used in designating the location of a point on a survey line as "4 + 47.2," meaning 47.2 feet beyond Station No. 4 or 447.2 feet from the initial point, measured along a specified line. See also **plus station.**

plus point—An intermediate point on a traverse course located by a plus distance from the beginning of the course.

plus sight—See **backsight.**

plus station—An intermediate point on a traverse, not at an even tape length distance from the initial point. See also **plus distance; taping station.**

point—A position on a reference system determined by a survey. See also **amphidromic point; angle point; annex point; antisolar point; astrogravimetric points; cardinal points; check point; control point; datum point; detail points; distant points; fix; image point; initial point; intercardinal point; map point; middle point; nodal point; orientation point; pass point; plus point; principal point; sublunar point; subsatellite point; subsolar point; substellar point; tie point; turning point; wing point; witness point.**

point anomaly—The value of the gravity anomaly at a specific location as observed or predicted.

point base—A manuscript which contains radial centers, picture points, pass points, control points, and tie points from the photographs used in the radial triangulation method.

point-designation grid—(JCS) A system of lines having no relation to the actual scale or orientation, drawn on a map, chart, or air [aerial] photograph, dividing it into squares so that points can be more readily located.

point marker—A device used for identifying points on diapositives by either marking a small hole in the emulsion or marking a small ring around the detail point itself. Also called **snap marker.** See also **point-transfer device.**

point-matching method—(rectification) The technique of utilizing an autofocus rectifier for tilt removal by the manual matching of projected image points to those plotted in their correct horizontal position on a film templet.

point of certainty—In a simple two-point intersection problem, that point where the two intersecting rays cross and the point is confirmed by the intersection of a third or check ray passing through the same point.

point of compound curvature (PCC)—The point on a line survey where a circular curve of one radius is tangent to a circular curve of a different radius, both curves lying on the same side of their common tangent.

point of contact—Any level surface along a terrain profile recorder (TPR) flight line that can be flown over both before and after the changing or adjustment of a TPR positional camera magazine, a chart roll, or a recording pen.

point of curvature (PC)—The point in a line survey where a tangent ends and a circular curve begins. See also **point of tangency.**

point of cusp—The point of tangency of two curves, the direction of the extension of said curves being of opposite sign; such as the vertex of a Y of a railroad track or a point on the edge of a convex-concave lens. Can also be applied to the point of tangency of a straight line and a curve where the direction of extension of the line and curve are of opposite sign.

point of inflection—The point at which a reversal of direction of curvature takes place.

point of intersection (PI)—The point where the two tangents of a circular curve meet. Also called **vertex of curve.**

point of origin—See **initial point.**

point of reverse curvature (PRC)—The point of tangency common to two curves, the curves lying on the opposite side of the common tangent.

point of symmetry—The point in the focal plane of a camera about which all lens distortions are symmetrical. If the lens were perfectly mounted, the point of symmetry would coincide with the principal point.

point of tangency (PT)—The point in a line survey where a circular curve ends and a tangent begins. The point of tangency and point of curve are both points of tangency, their different designations being determined by the direction of progress along the line; the point of curvature is reached first.

point of vertical curve (PVC)—The point of change from a line of uniform slope to a vertical curve.

point of vertical intersection (PVI)—The point of intersection of two lines, each having different uniform slopes.

point of vertical tangent (PVT)—The point of change from a vertical curve to a line of uniform slope.

point position—(Doppler) The geocentric or geodetic position of a point determined from satellite tracking data by a Doppler receiver and the satellite(s) ephemerides.

point position data (PPD)—The collective result of an analytical triangulation effort that provides evaluated geodetic positions of photo-identifiable ground points or reseau intersections. These positions are the result of an evaluated adjustment of the points to a specific mathematical surface and are expressed in terms of latitude, longitude, elevation, and positional accuracy for each point.

point positioning—(surveying) The process of observing Navy navigation satellites with a single Doppler survey receiver to produce the position (latitude, longitude, and height) of the receiver's antenna. See also **short arc; short arc geodetic adjustment; translocation.**

Point Positioning Data Base (PPDB)—Sets of geodetically controlled photographs and accompanying data that enable trained personnel using appropriate hardware and software to derive accurate coordinates for any feature identifiable within the PPDB.

point the instrument—Turning the survey instrument to where the crosshairs (vertical, horizontal, or both) are accurately aligned with the target.

point-transfer device—A stereoscopic instrument used to make corresponding image points on overlapping photographs. Also called **transcriber.** See also **point marker.**

pointing—**1.** (mensuration) Placing the reticle or index mark of a precision measuring instrument, such as a comparator, within the symmetrical center or center of gravity of a point being measured to determine its position relative to the position of other points in some system of coordinates. **2.** (stereocompilation) A general term applied to the movement of the tracing table of a stereoplotting instrument to specific control and/or picture points on the datum during orientation of a stereomodel. **3.** See **line of sight,** definition 2.

pointing accuracy—The exactness, in surveying or photogrammetry, with which the line of sight or floating mark can be directed toward a target or image point.

pointing errors—Errors which reflect the accuracy with which the floating mark of a stereoplotting system can be located on a sharp model point. These errors generally follow a more or less random distribution but show a systematic trend with progressive working time on the instrument due to eye fatigue and its effect on stereoscopic perception.

pointing line—See **line of collimation.**

polar axis—The primary axis of direction in a system of polar or spherical coordinates.

polar bearing—In a system of polar or spherical coordinates, the angle formed by the intersection of the reference meridional plane and the meridional plane containing the point.

polar chart—**1.** A chart of polar areas. **2.** A chart on a polar projection. The projections most used for polar charts are the gnomonic, stereographic, azimuthal equidistant, transverse Mercator, and modified Lambert conformal.

polar circle—Either the Arctic Circle (north polar circle) or the Antarctic Circle (south polar circle).

polar coordinates—(JCS) In artillery and naval gunfire support, the direction, distance, and vertical correction (shift) from the observer's or spotter's position to the target.

polar diameter—The diameter of the Earth between the poles.

polar distance—Angular distance from a celestial pole; the arc of an hour circle between a celestial pole, usually the elevated pole, and a point on the celestial sphere, measured from the celestial pole through 180°. See also **codeclination.**

polar grid—A grid system utilized for aerial navigation in the polar regions. It consists of a rectangular grid with x- and y-axes aligned with the 0°–180° and the 90°E–90°W meridians respectively. When plotted on a transverse Mercator map projection of the polar regions, it represents a system of transverse meridians and parallels whose poles are at the intersections of the Equator and the 0°–180° meridian.

polar map projection—A map projection centered on a pole.

polar motion—See **variation of the pole.**

polar orbit—An Earth satellite orbit that has an inclination of about 90° and, hence, passes over the Earth's poles.

polar orthographic map projection—A map projection having the plane of the projection perpendicular to the axis of rotation of the Earth (parallel with the plane of the Equator); in this projection, the geographic parallels are full circles, true to scale, and the geographic meridians are straight lines.

polar planimeter—An instrument used in measuring areas from a drawing. The instrument rotates about a pole, hence its name.

polar radius—The radius of the Earth measured along its axis of rotation.

polar satellite—Any satellite that passes over the north and south poles of the Earth; i.e., one that has an inclination of about 90° with respect to the Earth's Equator.

polar stereographic map projection—A stereographic projection having the center of the projection located at a pole of the sphere.

Polaris—The second-magnitude star, Alpha, in the constellation Ursa Minor (Little Dipper). Also called **North Star; polestar.**

Polaris correction—A correction to be applied to the observed altitude of Polaris to obtain the latitude.

polarization—(optics) The act or process of modifying light in such a way that the vibrations are restricted to a single plane. According to the wave theory, ordinary (unpolarized) light vibrates in all planes perpendicular to the direction of propagation. On passing through or contacting a polarizing medium (such as Polaroid or a Kerr cell) ordinary light becomes plane-polarized, that is, its vibrations are limited to a single plane.

polarization filter—Any of the manufactured plastic fibers which plane-polarize ordinary light when it passes through the filter. Usually identified by a trade name.

polastrodial—A mechanical counter for determining the azimuth and altitude of Polaris at any time.

pole—**1.** Either of the two points of intersection of the surface of a sphere or spheroid and its axis. **2.** The origin of a system of polar coordinates. See also **average terrestrial pole; celestial pole; depressed pole; ecliptic pole; elevated pole; fictitious pole; galactic pole; geographic pole; geomagnetic pole; magnetic pole; north geographical pole; north geomagnetic pole; north magnetic pole; oblique pole; south geographical pole; south geomagnetic pole; south magnetic pole; terrestrial pole; transverse pole.**

pole of the Milky Way—The pole in the galactic system of coordinates.

polestar—See **Polaris.**

polhody—A chart depicting the motion of the terrestrial pole as a function of time. See also **variation of the poles.**

polychrome—See **multicolor.**

polyconic chart—A chart on the polyconic map projection.

polyconic map projection—A map projection having the central geographic meridian represented by a straight line, along which the spacing for lines representing the geographic parallels is proportional to the distances between the parallels; the parallels are represented by arcs of circles which are not concentric, but whose centers lie on the line representing the central meridian, and whose radii are determined by the lengths of the elements of cones which are tangent along the parallels. All meridians except the central ones are curved. The projection is neither conformal nor equal area, but it has been widely used for maps of small areas because of the ease with which it can be constructed.

polyhedric projection—A projection used for a large scale topographic map whereby a small quadrangle on the spheroid is projected onto a plane trapezoid. Scale is made true either on the central meridian or along the sides.

Porro-Koppe principle—The principle applied in some photogrammetric instruments to eliminate the effect of camera-lens distortion. The photographic positive or negative is observed through a lens or optical system identical in distortion characteristics to the camera objective which made the original exposure. In effect, this method of observation is a reverse use of the camera, with the focal plane becoming the object which is imaged at infinity by parallel bundles of rays emerging from the lens. The chief ray of each bundle assumes its correct direction, and the cone of rays is identical to that whose vertex was the incident node of the camera lens at the instant of exposure. The parallel bundles may be observed by means of a telescopic system focused at infinity and made rotatable about the incident node of the lens. This method of eliminating lens distortion is utilized in photogrammetric instruments of both the monoscopic type, such as the photogoniometer, and the stereoscopic type used for stereoplotting.

Porro prism—A prism that deviates the axis 180° and inverts the image in the plane in which the reflection takes place. It may be described as two right-angle prisms cemented together.

port plan—A special-purpose large-scale map of a port area showing piers, railroad extensions, repair facilities, pilot office, customhouse, and other applicable non-navigational features.

portable automatic tide gage—A small automatic tide gage, designed for use where a short series of observations is necessary for the reduction of soundings to a common datum.

position—**1.** Data which define the location of a point with respect to a reference system. **2.** The place occupied by a point on the surface of the Earth or in space. Also called **place**. **3.** The coordinates which define the location of a point on the geoid or spheroid. **4.** A prescribed setting (reading) of the horizontal circle of a direction theodolite which is to be used for the observation on the initial station of a series of stations which are to be observed. Also called **circle position.** See also **adjusted position; apparent position; astrometric position; astronomic position; celestial fix; celestial line of position; circle of position; convergent position; electronic line of position; field position; fix; geocentric station position; geodetic position; geographic position; line of position; mean position; point position; point positioning; precise installation position; preliminary position; relative position; supplemental position; transverse position; true position.**

position angle—See **parallactic angle,** definition 1.

position plotting sheet—A blank chart, usually on the Mercator projection, showing only the graticule and a compass rose, so that the chart can be used for any longitude. See also **universal plotting sheet.**

positional accuracy—(cartography) A term used in evaluating the overall reliability of the positions of cartographic features on a map or chart relative to their true position, or to an established standard.

positional camera photography—Photography obtained with a camera aligned with the TPR radar beam, used for correlation and transfer of recorded vertical data to the cartographic photography.

positional error—(cartography) The amount by which a cartographic feature fails to agree with its true position.

positioning camera—A camera used for correlation purposes in the airborne profile recorder system. It is mounted on the radar antenna and records the area illuminated by the radar beam.

positive—(JCS) In photography, an image on film, plate or paper having approximately the same total rendition of light and shade as the original subject.

positive altitude—Angular distance above the horizon.

positive deflection angle—See **deflection angle.**

positive forming—In relief model making, forming over a positive mold.

positive lens—A lens that converges a beam of parallel light rays to a point focus. Also called **converging lens; convex lens.**

positive mold—The cast pulled from a negative mold when making a relief model.

potential—A scalar function, the gradient of which results in a vector field. Use of the scalar function simplifies investigation and description of the phenomenon considered. Used extensively for magnetic, gravitational, and gravity field investigations. In celestial mechanics and geodesy, the negative of the potential, sometimes called the **force function,** is usually employed. See also **disturbing potential; gravitational potential.**

potential disturbance—See **disturbing potential,** definition 1.

potential of disturbing masses—See **disturbing potential,** definition 1.

potential of random masses—See **disturbing potential,** definition 1.

power of a lens—See **diopter; magnification.**

power of telescope—(surveying) The magnification of a telescope when focused at infinity.

Pratt-Hayford theory of isostasy—A theory of isostatic compensation which assumes that every topographic excess or defect of mass is compensated by an equal and opposite defect or excess, evenly distributed immediately below it between ground level or sea-bottom level and a fixed depth, called the depth of compensation, commonly 113.7 km. Also called **fermenting dough theory.** See also **Airy theory of isostasy.**

precession—Change in the direction of the axis of rotation of a spinning body, as a gyroscope, when acted upon by a torque. The direction of motion of the axis is such that it causes the direction of spin of the gyroscope to tend to coincide with that of the impressed torque. See also **drift; general precession; planetary precession; topple; topple axis; total drift.**

precession in declination—The component of general precession along a celestial meridian, amounting to about 20.″0 of arc per year.

precession in right ascension—The component of general precession along the celestial equator, amounting to about 46.″1 of arc per year.

precession of the equinoxes—The conical motion of the Earth's axis about the vertical to the plane of the ecliptic, caused by the attractive force of the Sun, Moon, and other planets on the equatorial protuberance of the Earth. See also **general precession.**

precise ephemeris—Coordinates and velocity of an artificial satellite, computed for uniform time intervals from data acquired from a worldwide tracking network. The ephemeris is computed from observations taken from many stations spaced worldwide and adjusted together by least-squares methods for maximum accuracy. See **broadcast ephemeris; ephemeris; Navy Navigation Satellite System.**

precise installation position (PIP)—Geodetic coordinates of installation reference points reflecting their maximum possible refinement by utilizing all intelligence sources and optimum computer techniques of analytical area adjustment.

precise level—An instrument designed specifically for obtaining precise results by direct leveling techniques. It is essentially the same as an engineer's level except that it contains micrometer screws for more precise leveling of the instrument and contains a prism arrangement whereby the level bubble can be observed simultaneously with the rod reading.

precise leveling rod—A rod used for precise leveling. The graduations are on an Invar ribbon which is maintained under constant tension and which, for all practical purposes, eliminates the need for correcting for changes in length. These rods are usually graduated in whole and fractional meters. The back side of the rod is graduated in feet and tenths of feet. Also called **Invar leveling rod; meter rod; Molitar precise leveling rod; National Geodetic Survey first-order leveling rod.**

precise radar significant location (PRSL)—The horizontal and vertical values derived for a selected ground feature that is radar significant. The return may be either positive or negative.

precision—The degree of refinement in the performance of an operation, or the degree of perfection in the instruments and methods used when making the measurements. **Precision** relates to the quality of the operation by which a result is obtained, and is distinguished from **accuracy** which relates to the quality of the result.

precision altimeter—A sensitive aneroid barometer. In surveying, it will produce results accurate to within a meter only when the two-base method is carefully applied. See also **two-base method.**

Precision Automatic Photogrammetric Intervalometer System (PAPI)—Automatic intervalometer utilizing radar in determining interval for desired aerial photography forward lap.

precision camera—A relative term used to designate any camera capable of giving high resolution and dimensional results of a high order of accuracy.

precision depth recorder—A device which records sounding information on electrosensitive paper for depths up to 6,000 fathoms. It provides a trigger to the sonar and performs a time-measuring function.

predominant height—(JCS) In air reconnaissance, the height of 51 percent or more of the structures within an area of similar surface material.

preferred datum—A geodetic datum selected as a base for consolidation of local independent datums within a geographical area. Also called **major datum.**

preliminary—Not of the desired accuracy and precision, and adopted for temporary use with the proviso of later being superseded.

preliminary edition—See **provisional edition.**

preliminary elevation—An elevation arrived at in the office after the index, level, rod, and temperature corrections have been applied to the observed differences of elevation and new elevations have been computed.

preliminary orientation—An initial, rough orientation of projectors prior to accomplishing relative orientation of a stereomodel. It is the approximate leveling and scaling of the instrument frame and projectors, based on the best estimate of what their ultimate orientations are assumed to be.

preliminary position—In the adjustment of triangulation, the term **preliminary** is applied to geographic positions derived from selected observations for use in forming latitude and longitude condition equations.

preliminary survey—The collection of survey data on which to base studies for a proposed project. See also **reconnaissance survey.**

preliminary triangle—In the adjustment of triangulation, the term **preliminary** is applied to triangles derived from selected observations for use in forming latitude and longitude condition equations.

prepunch register system—A method in which a system of precisely located holes are punched in the margins of map or chart materials (such as films, vinyls, etc.) prior to their actual use. Exact register of materials can be accomplished by placing register studs (small plastic or metal pins) through the holes, thereby assuring exact register of detail. See also **register marks.**

press proof—A lithographed impression taken from among the first copies run on the press and used for checking purposes. Also called **press pull.**

press pull—See **press proof.**

pressplate—See **plate.**

pressure altimeter—See **barometric altimeter.**

pressure altitude—(JCS) An atmospheric pressure, expressed in terms of altitude which corresponds to that pressure in the standard atmosphere. See also **altitude.**

pressure gage—A tide gage that is operated at the bottom of the body of water being gaged and which records tidal height changes by the difference in pressure due to the rise and fall of the tide.

primary—See **primary body.**

primary bench mark—A bench mark close to a tide station to which the tide staff and tidal datum originally were referenced.

primary body—The celestial body or central force field about which a satellite or other body orbits, or from which it is escaping, or towards which it is falling. Also called **primary.**

primary circle—See **primary great circle.**

primary compilation—A specially prepared matte plastic material used to depict sounding data corrected to true depths in bathymetric compilations.

primary great circle—A great circle used as the origin of measurement of a coordinate; particularly such a circle 90° from the poles of a system of spherical coordinates, as the Equator. Also called **fundamental circle; primary circle.**

primary station—See **main-scheme station.**

primary tide station—A place at which continuous tide observations are made over a number of years to obtain basic tidal data for the locality.

prime fictitious meridian—The reference meridian (real or fictitious) used as the origin for measurement of fictitious longitude.

prime grid meridian—The reference meridian of a grid. In polar regions it is usually the 180°–0° geographic meridian, used as the origin for measuring grid longitude.

prime inverse meridian—See **prime transverse meridian.**

prime meridian—The meridian of longitude 0°, used as the origin for measurement of longitude. The meridian of Greenwich, England, is almost universally used for this purpose.

prime oblique meridian—The reference fictitious meridian of an oblique graticule.

prime transverse meridian—The reference meridian of a transverse graticule. Also called **prime inverse meridian.**

prime vertical—See **prime vertical circle.**

prime vertical circle—The vertical circle through the east and west points of the horizon. It may be true, magnetic, compass, or grid, depending upon which east or west points are involved. Also called **prime vertical.**

prime vertical plane—The plane perpendicular to the meridian plane (astronomic or geodetic) containing the normal. The intersections of the astronomic prime vertical plane with the horizon are the east and west points.

principal axis—See **optical axis.**

principal distance—1. The perpendicular distance from the internal perspective center to the plane of a particular finished negative or print. This distance is equal to the calibrated focal length corrected for both the enlargement or reduction ratio and the film or paper shrinkage or expansion. It maintains the same perspective angles at the internal perspective center to points on the finished negative or print, as existed in the taking camera at the moment of exposure. This is a geometrical property of each particular finished negative or print. Also called **effective focal length. 2.** (multiplex) The perpendicular distance from the internal perspective center of the projector lens to the plane of the emulsion side of the diapositive.

principal-distance error—In a stereoplotting system, an instrument error resulting from improper calibration of the aerial camera, diapositive printer, or projector. The error is of little importance in a flat surface model but the effects are increased in proportion to the relief in the model.

principal focus—See **focus.**

principal line—The trace of the principal plane upon a photograph (e.g., the line through the principal point and the nadir point).

principal meridian—1. (USPLS) The meridian extended from an initial point, upon which regular quarter-quarter section, section, and township corners have been or are to be established. See also **auxiliary guide meridian; guide meridian. 2.** (photogrammetry) See **principal point.**

principal parallel—(JCS) On an oblique photograph, a line parallel to the true horizon and passing through the principal point.

principal plane—1. (JCS) (photogrammetry) A vertical plane which contains the principal point of an oblique photograph, the perspective center of the lens, and the ground nadir. **2.** (optics) A plane through a principal point and perpendicular to the optical axis. See also **axis of homology; axis of tilt; ground parallel; ground trace; horizon trace; isometric parallel; line of constant scale; line of equal scale;**

photograph meridian; photograph parallel; principal line; principal meridian; principal parallel; vanishing line; vanishing point.

principal planets—The larger bodies revolving about the Sun in nearly circular orbits. The known principal planets, in order of their distance from the Sun, are: Mercury, Venus, Earth, Mars, Jupiter, Saturn, Uranus, Neptune, and Pluto.

principal point—(JCS) (photogrammetry) The foot of the perpendicular to the photo plane through the perspective center. Generally determined by intersection of the lines joining opposite collimating or fiducial marks. [If the fiducial marks are not visible on the photograph, the principal point may be found by drawing diagonals between opposite corners or by measuring one-half the distance along each side of the photograph and connecting these marks in the same manner as fiducial marks, or by reseau marks.] Also called **indicated principal point.** See also **photograph center.**

principal-point assumption—The assumption with respect to approximately vertical photographs that radial directions are correct if measured from the principal point.

principal point (calibrated)—The center of radial lens distortion, usually given as x and y distances from indicated principal point.

principal-point error—A personal error in which the principal points in a stereoplotting system are displaced in such a manner that they have unequal x-components with a resultant error in vertical scale. Such errors are usually introduced into the system by either improper orientation of the diapositive plate in the printer, in the projector, or both.

principal-point radial—A radial from the principal point of a photograph.

principal-point triangulation—See **radial triangulation.**

principal station—See **main-scheme station.**

principal vertical—(JCS) On an oblique photograph, a line perpendicular to the true horizon and passing through the principal point.

principal vertical circle—The vertical circle through the north and south points of the horizon, coinciding with the celestial meridian.

print—A photographic copy made by projection or contact printing from a negative or transparency. See also **contact print; diapositive; enlargement; matte print; photographic reduction; projection print; ratio print; rectified print; transformed print.**

prism—A transparent body bounded in part by two plane faces that are not parallel; used to deviate or disperse a beam of light. See also **dove prism; horizon prism; index prism; Porro prism; reflecting prism; refracting prism; retrodirective prism; rhomboidal prism; right-angle prism; roof prism; wedge.**

prismatic astrolabe—An astrolabe consisting of a telescope in a horizontal position, with a prism and artificial horizon attached at its objective end, used for determining astronomic positions.

prismatic compass—A small magnetic compass held in the hand when in use and equipped with peep sights and glass prism so arranged that the magnetic bearing or azimuth of a line can be read at the same time that the line is sighted over.

prismatic error—That error due to lack of parallelism of the two faces of an optical element, such as a mirror or a shade glass.

probability interval—See **confidence interval.**

probable error—**1.** In a measured value, it is the most probable value of the resultant error in the measurement. This is a plus-or-minus quantity that may be larger than the resultant error or smaller than the resultant error, and its probability of being larger is equal to its probability of being smaller. **2.** The 50 percent error interval based on the normal distribution function. See also **standard error.**

process camera—See **copy camera.**

process color printing—(lithography) A technique for the reproduction of a subject or chart in full color rendition, by combining tones of the primary colors, (yellow, magenta, cyan) and black. See also **process plates.**

process lens—A lens for photochemical copying, enlarging, or projection purposes, free from aberrations, usually of low aperture and of symmetrical construction.

process photography—Line and halftone photography in which the resulting negatives and positives are subsequently used in the preparation of pressplates.

process plates—Two or more color pressplates combined to produce other colors and shades. See also **color plate; combination plate; process color printing.**

production priority list (PROP)—A publication incorporating all ATMP graphic requirements of all Unified and Specified Commands and Military Departments. In addition to the requirements, the PROP includes information relative to availability of the required coverage, in-work status, and evaluations of improvability through the application of newly acquired source material. The priority sequence of the list, used as a guide in determining the order for production assignment, is computer developed through the application of point values to the various data fields of the PROP.

profile—A vertical section of the surface of the ground, or of underlying strata, or both, along any fixed line.

profile leveling—The determination of elevations of points at short measured intervals along a definitely located line, such as the center line of a highway.

progress sketch—A map or sketch showing work accomplished. In triangulation and traverse surveys, each point established is shown on the progress sketch, and also lines observed over and base lines measured. In a leveling survey, the progress sketch shows the route followed and the towns passed through, but not necessarily the locations of the bench marks.

progressive motion—Motion in an orbit in the usual orbital direction of celestial bodies within a given system. Specifically, motion of a satellite in the same direction to the direction of the primary. Opposite of **retrograde motion.**

progressive proofs—A series of color prints that show the individually separated color printings of a job and their progressive combinations as each color is overprinted.

projected map display (PMD)—An inflight navigation aid which uses a continuous color or black-and-white sprocketed filmstrip containing chart imagery, projected on a display device and driven by the aircraft's computer. The aircraft's position is displayed with correlated chart image to show location and direction actually being flown.

projection—1. (geometry) The extension of lines or planes to intersect a given surface; the transfer of a point from one surface to a corresponding position on another surface by graphical or analytical methods. See also **map projection. 2.** (photography) The process of placing a negative or positive photograph in a projecting camera and reproducing the image on a screen or on a sensitized photographic medium. **3.** (surveying) The extension of a line beyond the points which determine its character and position. The transfer of a series of survey lines to a single theoretical line by a series of lines perpendicular to the theoretical line. In surveying a traverse, a series of measured short lines may be projected onto a single long line, connecting two main survey stations, and the long line is then treated as a measured line of the traverse. See also **prolongation.**

projection computation—The determination, from a set of tables derived from formulas, of the true shape and dimensions of a map projection, for the purpose of constructing such a projection. See also **grid computation.**

projection distance—The distance from the external node of a projection lens to the plane onto which the image is projected.

projection print—(JCS) An enlarged or reduced photographic print made by projection of the image of a negative or a transparency onto a sensitized surface.

projection printer—An optical device for enlarging or reducing the image of a negative or positive transparency by projecting it onto a sensitized surface.

projection tables—Data made available in tabular form for determining a definite relationship which exists between any grid intersection and any adjacent intersection of latitude and longitude lines on the map projection.

projection ticks and crosses—Ticks perpendicular to and inside the neatline of a map placed to indicate points through which parallels and meridians would pass if they had been extended. Small cross marks indicate where the lines intersect within the map.

projector—An optical instrument which throws the image of a negative or print upon a screen or other viewing surface, usually at a larger scale. See also **reflecting projector.**

projector station—The position of a projector unit of a stereoplotter when absolute orientation has been accomplished. This position recreates the conditions existing at the corresponding camera station at the instant of exposure.

prolate ellipsoid of rotation—See **prolate spheroid.**

prolate spheroid—An ellipsoid of rotation, the longer axis of which is the axis of rotation. Also called **prolate ellipsoid of rotation.**

prolongation—In surveying, a line is prolonged when the last segment of the surveyed line is extended in the same direction as the segment itself. A prolongation of a curve under such a definition of extension would be a line tangent to the curve at the point of extension, although the term frequently is used to mean a continuation along the curvature of the curve.

proof—A trial print, produced by any method, for examination or editing, to be marked for necessary corrections or approval. See also **color composite; color proof; color-proof process; composite; final composite; galley proof; hand proof; OK sheet; press proof; progressive proofs; proofing.**

proofing—The operation of pulling proofs of plates for proofreading, revising, approval, and other purposes prior to production printing.

proper motion—That component of the space motion of a celestial body perpendicular to the line of sight, resulting in the change of a star's apparent position relative to a coordinate system such as right ascension and declination. This change is expressed as a velocity, such as seconds of arc per century.

property map—See **cadastral map.**

property survey—See **land survey.**

proportionate measurement—A measurement that applies an even distribution of a determined excess or deficiency of measurement, ascertained by retracement of an established line, to provide concordant relations between all parts. See also **double proportionate measurement; single proportionate measurement.**

prototype—See **experimental map; pilot sheet.**

provisional edition—A map or chart printed and distributed for temporary use with the proviso that it will later be superseded. Also called **preliminary edition.**

provisional map—Any nonofficial map, photo, or

other material which is used as a map. It may vary from a highly accurate captured enemy map, which has not been sanctioned for use, to a hastily made drawing or sketch. It is usually a hastily made line map based on aerial photographs, used as a map substitute.

pseudoscopic stereo—A three-dimensional impression of relief which is the reverse of that actually existing when the positions of a stereo pair of photographs are interchanged. Also called **inverted stereo; reverse stereo.** See also **false stereo.**

publication scale—See **reproduction scale.**

Pulkovo 1932 datum—The geodetic datum defined by the geographic position of the center point of the round hall at Pulkovo Observatory, U.S.S.R. and the azimuth from that point to Signal "A" of the Sablino Base, computed on the Bessel spheroid. The same origin corrected for the deflection resulting from an astrogeodetic adjustment and computed on the Krasovsky ellipsoid is known as Pulkovo 1942 datum.

pull up—See **selection overlay.**

pulse Doppler map matching (PDMM)—An image-matching concept employing a pulse Doppler mapping technique to locate three preselected unique edges within the terrain scene viewed prior to reentry of the vehicle. Edges are defined by differences in radar reflectance. Range and range-rate data are obtained from a small elemental area of the terrain illuminated by the antenna beam spot during a fix action. A range/range-rate map of the spot area is then correlated against a reference map of the particular edge to provide a range and velocity update to the onboard vehicle inertial measurement unit (IMU). Three such successive fix correlations are executed to enable a trilateration solution of the vehicle's position relative to the target scene by the sensor's computer. Therefore, the IMU can calculate a corrective maneuver prior to impact.

pygmy meter—A small cup type current meter for use in low-velocity measurements in shallow streams. This meter is used in conjunction with wading rods only.

Pythagorean right-angle inversor—A simple device which provides a mechanical solution for linear and angular elements of rectification, thus permitting any enlarger to be made autofocusing provided negative, lens, and easel planes are parallel.

Q

quad—Abbreviated form of quadrangle or quadrilateral.

quadrangle—A rectangular, or nearly so, area covered by a map or plat, usually bounded by given meridians of longitude and parallels of latitude. Also called **quad; quadrangle map.** See also **standard quadrangle.**

quadrangle map—See **quadrangle.**

quadrangle report—A brief history of the mapping of a specific quadrangle. It accompanies the mapping material through each phase of production, and is filed with the map material. The narrative summary for each operational phase stresses conditions that may affect later phases.

quadrant—**1.** (mathematics) A sector having an arc of 90°. **2.** (surveying) A surveying or astronomic instrument composed of a graduated arc about 90° in length (180° in range), equipped with a sighting device. The quadrant may be considered a form of sector. Some survey quadrants combine both surveying and astronomic functions.

quadrature—**1.** The position in the phase cycle when the two principal tide-producing bodies (Moon and Sun) are nearly at a right angle to the Earth; the Moon is then in quadrature in its first quarter or last quarter. **2.** The situation of two periodic quantities differing by a quarter of a cycle.

quarter-quarter section corner—(USPLS) A corner at an extremity of a boundary of a quarter-quarter section; midpoint between or 20 chains from the controlling corners on the section or township boundaries. Written as 1/16 section corner. Also called **sixteenth-section corner.**

quarter section—(USPLS) One-fourth of a section, containing 160 acres more or less.

quarter-section corner—(USPLS) A corner at an extremity of a boundary of a quarter section. Written as ¼ section corner, not as one-fourth section corner.

quartz pendulum—A pendulum of fused quartz used for determining the acceleration of gravity. Quartz is employed in the construction because its thermal expansion coefficient is only one-fourth that of Invar.

quintant—A sextant having a range of 144°, or an arc of 72°.

R

radar—(JCS) Radio detection and ranging equipment that determines the distance and usually the direction of objects by transmission and return of electromagnetic energy.

radar altimeter—An instrument used for determining aircraft flying height above terrain by measurement of time intervals between emission and return of electromagnetic pulses.

radar altitude—The altitude of an aircraft or spacecraft as determined by a radar altimeter; thus, the actual distance from the nearest terrain feature.

radar chart—A chart intended primarily for use with radar, or one suitable for this purpose.

radar clutter—(JCS) Unwanted signals, echoes, or images on the face of the display tube which interfere with observation of desired signals.

radar correlation—The process of electronically relating real-time radar images with stored digital data on the radar reflectance and position of terrain and features on the Earth's surface. It is used to provide positioning information to correct or check air navigation and guidance systems.

radar coverage—(JCS) The limits within which objects can be detected by one or more radar stations.

radar fix point (RFP)—The most significant radarscope ground feature for a given geographic area. The feature may be radar reflective or completely void of reflectivity to show contrast with the surrounding area; e.g., land/water—show/no show. Positioning data established for RFPs are used in offset aiming procedures or for enroute and final update of navigation systems.

radar horizon—(JCS) The line at which direct radar rays are tangential to the Earth's surface.

radar intelligence item—(CENTO) A feature which is radar significant but which cannot be identified exactly at the moment of its appearance as homogeneous.

radar intelligence map (RIM)—An intermediate element in the process of light-optical radar simulation and in the production of analytical predictions.

radar map—A map produced through the application of radar techniques.

radar photography—A combination of the photographic process and radar techniques. Electrical impulses are sent out in predetermined directions and the reflected or returned rays are utilized to present images on cathode-ray tubes. Photographs are then taken of the information displayed on the tubes.

radar prediction categories—In the broadest sense, radar prediction is separated into two major categories: experience prediction and analytical prediction.

radar prediction formats—Radar predictions appear in a wide variety of formats, generally indicative of intended application. The four most common formats are: spot predictions; strip predictions; radar intelligence maps (RIM); and Series 200 Air Target Charts.

radar prediction types—Each major radar prediction category is divided into three types of predictions: single heading predictions; omnidirectional predictions; and omni-gain predictions.

radar reconnaissance—(JCS) Reconnaissance by means of radar to obtain information, on enemy activity and to determine the nature of terrain.

radar reflectivity plate—A scaled, three-dimensional model of a target area constructed of radar reflective materials on a transparent plastic plate, used in a radar trainer to simulate the radar returns of that area. Also called **radar simulation plate; radar trainer plate.**

radar reflector—A device capable of or intended for reflecting radar signals.

radar return analysis—Those items of the radar significance analysis code (RSAC), the special area (SA) information, and the radar significant powerline (RSPL) information which have been developed from an analysis of cartographic, photographic, and intelligence sources.

radar return code (RRC)—An omni-directional radar prediction based on the decibel radar prediction system and depicted in a color code on certain air target charts.

radar shadow—A condition in which radar signals do not reach a region because of an intervening obstruction.

radar significance analysis code (RSAC)—The unique radar intensity categories of built-up areas and other radar reflective objects and structures based on surface material/height factors, and depicted by a system of color coding.

radar significant power line (RSPL)—A power transmission line which, because of its unique physical characteristics and/or voltage capacity, is known to possess radar reflective qualities, and is therefore distinctively displayed on a target graphic.

radar simulation plate—See **radar reflectivity plate.**

radar target—An object which reflects a sufficient amount of a radar signal to produce an echo signal on the radar screen.

radar trainer plate—See **radar reflectivity plate.**

radargrammetry—That branch of photogrammetry which utilizes radar images.

radarscope overlays—(JCS) Transparent overlays for placing on the radarscope for comparison and identification of radar returns.

radarscope photography—(JCS) A film record of the returns shown by a radar screen.

radial—(photogrammetry) A line or direction from the radial center to any point on a photograph. The radial center is assumed to be the principal point, unless otherwise designated (e.g., nadir radial). See also **isoradial; nadir radial; principal-point radial.**

radial assumption—In an aerial photograph containing both tilt displacement and relief displacement, neither the nadir point nor the isocenter is the theoretically correct radial center. The photographic nadir point should be used as the radial center if relief is the major consideration and the isocenter should be used if tilt is the major consideration.

radial center—The selected point on a photograph from which radials (directions) to various image points are drawn or measured; that is, the origin of radials. The radial center is either the principal point, the nadir point, the isocenter, or a substitute center. Also called **center of radiation; center point.**

radial distortion—Linear displacement of image points radially to or from the center of the image field, caused by the fact that objects at different angular distances from the lens axis undergo different magnifications.

radial line—(surveying) A radius line of a circular curve to a designated point in the curve; if the line is extended beyond the convex side of the curve, it is a prolongation of the radial line.

radial-line intersection—That point at which two or more radial lines cross or intersect.

radial-line plotter—See **radial plotter.**

radial plot—See **radial triangulation.**

radial plotter—A device whereby two overlapping photographs are viewed stereoscopically, and the planimetric details in their common area can then be transferred to a map or base sheet through a mechanical linkage utilizing the radial line principle. Also called **radial-line plotter.**

radial secator—See **templet cutter.**

radial triangulation—The aerotriangulation procedure, either graphical or analytical, in which directions from the radial center, or approximate radial center, of each overlapping photograph are used for horizontal-control extension by the successive intersection and resection of these direction lines. A radial triangulation also is correctly called a radial plot or a minor-control plot. If made by analytical methods, it is called an analytical radial triangulation. A radial triangulation is assumed to be graphical unless prefixed by the word analytical. It is also assumed to be based on the principal point unless prefixed by definitive terms such as **isocenter** or **nadir point.**

radiant energy—The energy of any type of electromagnetic radiation. See also **radiation,** definition 2.

radiation—**1.** (surveying) The process of locating points by a knowledge of their direction and distance from a known point. The directions may be azimuths or bearings read from a theodolite or graphical directions determined by alidade and planetable. The distances may be taped or measured by stadia. **2.** The process by which electromagnetic energy is propagated through free space by virtue of joint undulatory variations in the electric and magnetic fields in space.

radio acoustic ranging—A means of determining distance by a combination of radio and sound; the radio being used to determine the instant of transmission or reception of the sound, and the distance being determined by the time of transit of sound, usually in water.

radio beacon—(JCS) A radio transmitter which emits a distinctive, or characteristic, signal used for the determination of bearings, courses, or location.

radio direction finding—(JCS) Radio location in which only the direction of a station is determined by means of its emissions.

radio-facility chart—See **enroute chart.**

radio fix—(JCS) The location of a ship or aircraft by determining the direction of radio signals coming to the ship or aircraft from two or more sending stations, the locations of which are known.

radio interferometer—An interferometer operating at radio frequencies; used in radio astronomy and in satellite tracking.

radio navigation—(JCS) Radio location intended for the determination of position or direction or for obstruction warning in navigation.

radio range findings—(JCS) Radio location in which the distance of an object is determined by means of its radio emissions, whether independent, reflected or retransmitted on the same or other wavelength.

radio range station—(JCS) A radio navigation land station in the aeronautical radio navigation service providing radio equi-signal zones. In certain instances a radio range station may be placed on board a ship.

radiometric camera calibration—The calibration of a camera for spectral recording characteristics.

radius vector—The line (distance) and direction connecting the origin with the point whose position is being defined. See also **polar coordinates.**

random error—Random errors are those not classified as blunders, systematic errors, or periodic errors. They are numerous, individually small, and each is as likely to be positive as negative. Also called **accidental error; casual error.**

random line—A trial line, directed as closely as possible toward a fixed terminal point which is invisible from the initial point. The error of closure permits the computation of a correction to the initial azimuth of the random line;

it also permits the computation of offsets from the random line to establish points on the true line.

random traverse—A survey traverse run from one survey station to another station which cannot be seen from the first station in order to determine their relative positions.

range—1. (JCS) The distance between any given point and an object or target. **2.** The difference between the maximum and minimum of a given set of quantities. See also **distance. 3.** Two or more objects in line. Such objects are said to be in range. **4.** (USPLS) Any series of contiguous townships situated north and south of each other; also sections similarly situated within a township. **5.** The well-defined lines or courses whose positions are known and are used in determining soundings in a hydrographic survey.

range finder—An instrument, using the parallax principle, for finding the distance from a place of observation to points at which no instruments are placed.

range line—(USPLS) A boundary of a township surveyed in a north-south direction. See also **township lines.**

range marker—(JCS) A single calibration blip fed on to the time base of a radial display. The rotation of the time base shows the single blips as a circle on the plan position indicator scope. It may be used to measure range.

Range Only Correlation System (ROCS)—An all-weather terminal guidance system that uses ranging information to determine its position by comparing two radar images. The images are taken 90° apart and compared with a reference (or prediction) of contrast edge information (bright spots) in the range return signal. The checkpoint is a geographic reference location, selected prior to the mission, by which the vehicle can determine course correction.

range pile—Any pile serving as a guide for marine surveying.

range pole—See **range rod.**

range-rate data—Information gathered by an instrument that measures the rate of change in the distance (range) to a moving object.

range resolution—(JCS) The ability of the radar equipment to separate two reflecting objects on a similar bearing, but at different ranges from the antenna. The ability is determined primarily by the pulse length in use.

range rod—A slender wood or metal rod, 6 to 8 feet long, with a pointed metal shoe, usually painted in contrasting colors (red and white), alternately, at 1-foot intervals. It is frequently used as a sighting signal at the ends of traverse courses. Also called **line rod; lining pole; range pole; ranging pole; sight rod.**

range signal—A buoy, rod, flag, or other similar object used to mark and identify range points when taking soundings during a hydrographic survey.

ranging data—Information gathered by an instrument that measures the distance (range) to the object in question.

ranging-in—See **wiggling-in.**

ranging pole—See **range rod.**

rate station—See **drift station.**

ratio print—A print in which the scale has been changed from that of the negative by photographic enlargement or reduction.

ratiograph—See **ratiometer.**

ratiometer—An instrument used to help solve the mathematical relationship of a photograph to a mosaic. It determines scale ratios from which, through mathematical formulas, a rectified print can be made on a properly calibrated rectifying printer.

rational horizon—See **celestial horizon.**

rationalization method—A technique of relative orientation which takes into consideration the limiting factors of the equipment being used, the nature and variations of tilt and crab angles at successive camera stations, and providing approximate projector adjustments based on these data.

ray of light—The geometrical concept of a single element of light propagated in a straight line and of infinitesimal cross section; used in analytically tracing the path of light through an optical system. See also **beam of light.**

ray tracing—(optics) A trigonometric calculation of the path of a light ray through an optical system.

Raydist—The trade name of an electronic distance-measuring system. A non-line-of-sight system capable of simultaneous multiparty, range-range operation; it gives continuous range information from two base stations operating simultaneously with one or more aircraft and surface vessels.

real image—An image actually produced and capable of being shown on a surface, as in a camera.

rear element—See **lens element.**

rear nodal point—See **nodal point.**

recast—To change a map from one horizontal datum to another by appropriately changing the geographic values of the map graticule.

receiver—1. (pendulum) A heavy cast-metal box within which the pendulum is suspended and some auxiliary equipment placed when making observations for the intensity of gravity. **2.** (satellite surveying) The equipment necessary to receive signals broadcast from the Navy Navigation Satellite System, including an antenna, preamplifier, processor, oscillator, output device, and power system.

reciprocal bearing—See **back bearing,** definition 1.

reciprocal leveling—Trigonometric leveling wherein vertical angles have been observed at both ends of the line to eliminate errors.

reciprocal observations—Observations taken backward and forward such as vertical angles at both termini of a line for trigonometric leveling.

reciprocal vertical angle—A vertical angle measured over a line at both ends in trigonometric leveling to eliminate (at least partly) the effects of curvature and refraction. Reciprocal observations must be made as simultaneously as practicable to obviate error caused by changing refractive conditions.

recognition—In photointerpretation, the act of discovering the true identity of an object.

recompilation—The process of producing a map or chart that is essentially a new item and which replaces a previously published item. Normally, recompilation of a map or chart involves significant change to the horizontal position of features, revision of vertical values, improvement in planimetric or navigational data, or any combination of these factors.

reconnaissance—(JCS) A mission undertaken to obtain, by visual observation or other detection methods, information about the activities and resources of an enemy or potential enemy; or to secure data concerning the meteorological, hydrographic, or geographic characteristics of a particular area. See also **aerial reconnaissance; hydrographic reconnaissance; radar reconnaissance; triangulation reconnaissance.**

reconnaissance map—The plotted results of a reconnaissance survey and data obtained from other sources.

reconnaissance photography—(JCS) Photography taken primarily for purposes other than making maps, charts, or mosaics. It is used to obtain information on the results of bombing, or on enemy movements, concentrations, activities, and forces.

reconnaissance sketch—A drawing which resembles a reconnaissance map but is lacking in some map element.

reconnaissance survey—A preliminary survey, usually executed rapidly and at relatively low cost. The information obtained is recorded, to some extent, in the form of a reconnaissance map or sketch.

recording statoscope—A statoscope equipped with a recording camera whose shutter is synchronized with that of the aerial camera and the image of the statoscope is recorded on each individual frame.

recover—(surveying) To visit a survey station, identify its mark as authentic and in its original location, and verify or revise its description. The term is usually modified to indicate the type or nature of the recovery, such as recovered bench mark, or a recovered triangulation station.

recovered control—See **recover.**

recovery of station—See **recover.**

rectagraver—A scribing instrument which rests on the scribing surface during the operation and only the cutter arm moves to scribe each symbol.

rectangular chart—A chart on the rectangular projection.

rectangular coordinates—Coordinates on any system in which the axes of reference intersect at right angles. See also **Cartesian coordinates.**

rectangular coordinate plotter—See **coordinatograph.**

rectangular map projection—A cylindrical map projection with uniform spacing of the parallels.

rectangular polyconic map projection—A modified polyconic map projection having a line representing a standard parallel divided to exact scale, through whose division points pass the lines representing the geographic meridians, intersecting the lines which represent the geographic parallels in right angles.

rectangular space coordinates—The perpendicular distances of a point from places defined by each pair of a set of three axes which are mutually perpendicular to each other at a common point of origin. In photogrammetry, space coordinates are also called **survey coordinates,** and are the x-coordinates and y-coordinates which define the horizontal position of a point on a ground system, and the z-coordinate, which is the elevation of the point with reference to the ground system. Also called **air coordinates.**

rectangular surveys—A system of surveys in which an area is divided by a base line intersected at right angles by a principal meridian, the intersection termed the initial point from which the partitions are subdivided into equal size townships containing 36 sections of land each.

rectification—(JCS) In photogrammetry, the process of projecting a tilted or oblique photograph onto a horizontal reference plane. [Although the process is applied principally to aerial photographs, it may also be applied to the correction of map deformation.] See also **analytical orientation; empirical orientation; graphical rectification; multiple-stage rectification; optical rectification; paper-strip method; point-matching method; transformation.**

rectified altitude—See **apparent altitude.**

rectified print—A photograph in which tilt displacement has been removed from the original negative, and which has been brought to a desired scale.

rectifier—A specially designed projection printer whose geometry is variable in order to eliminate tilt from an aerial negative. There are two basic types: those in which the optical axis of the rectifier lens is the common reference or base direction of the instrument, and those in which the line between the principle point of the negative and the rectifier lens is the common reference. Also called **automatic rectifier; nonautomatic rectifier; nontilting-lens rectifier; nontilting-negative-plane rectifier; rectifying camera; tilting-lens rectifier.** See also **autofocus rectifier; transforming printer.**

rectifying camera—See **rectifier.**

rectifying latitude—The latitude on a sphere such that a great circle on it has the same length as a meridian on the spheroid and such that all lengths along a meridian from the Equator are exactly equal to the corresponding lengths on the spheroid. Rectifying latitude is an auxiliary latitude used in problems of geodesy and cartography.

rectifying printer—See **rectifier.**

rectilinear coordinates—See **rectangular coordinates.**

rectoblique plotter—See **angulator.**

rectoplanigraph—An instrument utilizing a vertical photograph mounted in a vertical position, and used in the preparation of planimetric maps.

red-light-readable map—A map printed with special inks which can be read under conditions requiring special lighting; e.g., in a tank or aircraft during nighttime operations.

red magnetism—The magnetism of the north-seeking end of a freely suspended magnet. This is the magnetism of the Earth's south magnetic pole. See also **blue magnetism.**

reduced gravity—Observed gravity that has been reduced to the geoid or to some other reference surface by one of the gravity reductions.

reduced latitude—See **parametric latitude.**

reduction factor—See **scale of reproduction.**

reduction printer—See **diapositive printer.**

reduction-to-center—**1.** The amount which must be applied to a direction observed at an eccentric station or to an eccentric signal, to reduce such direction to what it would be if there were no such eccentricity. **2.** (astronomy) One of the values used in finding the equation of time.

reduction to ellipsoid—The correction subtracted from or added to the measured horizontal length of a line at average topographic elevation to reduce it to the corresponding length on the pertinent ellipsoid. The height used in this reduction will differ by the amount of the geoid height from the height used in reduction to sea level.

reduction to sea level—A reduction applied to a measured horizontal length on the Earth's surface to reduce it to the surface of the local sea level datum.

reduction to the meridian—The process of applying a correction to an altitude observed when a body is near the celestial meridian of the observer, to find the altitude at meridian transit. The altitude at the time of such an observation is called an ex-meridian altitude.

reference datum—A general term applied to any datum, plane, or surface used as a reference or base from which other quantities can be measured.

reference direction—A direction used as a basis for comparison of other directions.

reference ellipsoid—See **reference spheroid.**

reference frame—See **coordinates.**

reference grid—See **grid.**

reference level—See **datum level.**

reference line—Any line which can serve as a reference or base for the measurement of other quantities. Also called **datum line.**

reference mark—A permanent supplementary mark close to a survey station to which it is related by an accurately measured distance and direction, and/or a difference in elevation.

reference meridian—See **local meridian.**

reference monument—(USPLS) An iron post or rock cap accessory used where the point for a corner monument is such that, for practical purposes, a permanent corner monument cannot be established, or if monumented, a full complement of bearing trees or bearing objects are not obtainable.

reference plane—See **datum level.**

reference point—See **datum point.**

reference signal—In telemetry, the signal against which data-carrying signals are compared to measure differences in time, phase, frequency, or other values or quantities.

reference spheroid—A theoretical figure whose dimensions closely approach the dimensions of the geoid; the exact dimensions are determined by various considerations of the section of the Earth's surface concerned. Also called **reference ellipsoid; spheroid of reference.** See also **World Geodetic System.**

reference station—A place where tide or tidal current constants have been determined from observations, and which is used as a standard for the comparison of simultaneous observations at a subordinate station. Also, a place for which independent daily predictions are obtained for other locations by means of differences or factors. Also called **standard port; standard station.**

referencing—The process of measuring the horizontal distances and directions from a survey station to nearby landmarks, reference marks, and other objects which can be used in the recovery of the station.

reflected ray—A ray extending outward from a point of reflection.

reflecting prism—A prism that deviates a light beam by internal reflection. Practically all prisms used in photogrammetric instruments are of this type.

reflecting projector—An instrument which is used to project the image of photographs, maps, or other graphics onto a copying table. The scale of the projected image can be varied by raising or lowering the projector or, in some models, the copy board. These latter models also allow the tilting of the copy board in x- and y-directions in order to compensate for tip and tilt distortion in aerial photographs.

reflection—The return or change in the direction of travel of particles or radiant energy which impinges on a surface but does not enter the substance providing the reflecting surface. See also **diffraction; diffuse reflection**; **refraction; specular reflection.**

reflector constant—The amount that a distance measurement must be reduced when using glass reflectors because the velocity of light is slower in glass than it is in air. The constant will also include the distance difference between the reflector housing plumbing point and the effective reflecting plane of the prism.

reflight—Another flight over the same course to secure photographs to fill in for those missing or defective.

refracted ray—A ray extending onward from the point of refraction.

refracting prism—A prism that deviates a beam of light by refraction. The angular deviation is a function of the wavelength of light; therefore, if the beam is composed of white light, the prism will spread the beam into a spectrum. Refracting prisms can be used in optical instruments only for small deviations. See also **wedge.**

refraction—The change in direction of motion of a ray of radiant energy as it passes obliquely from one medium into another in which the speed of propagation is different. See also **angle of incidence; angle of refraction; astronomic refraction; atmospheric refraction; coastal refraction; coefficient of refraction; electronic refraction; horizontal refraction; lateral refraction; mean refraction; refracted ray; refraction angle; refraction line; Snell's law of refraction; terrestrial refraction.**

refraction angle—That portion of an observed zenith distance, which is due to the effect of atmospheric refraction.

refraction displacement—Displacement of images radially outward from the photograph nadir because of atmospheric refraction. It is assumed that the refraction is symmetrical about the nadir direction.

refraction line—A line of sight to a survey signal which becomes visible only by the effect of atmospheric refraction.

regional gradient—See **regional gravity.**

regional gravity—In gravity prospecting, contributions to the observed anomalies due to density irregularities at much greater depths than those of the possible structures, the location of which was the purpose of the survey. Also called **regional gradient.**

register—(JCS) The correct position of one component of a composite map image in relation to the other components, at each stage of production.

register hole punch--See **prepunch register system.**

register marks—(JCS) Designated marks, such as small crosses, circles, or other patterns applied to original copy prior to reproduction to facilitate registration of plates and to indicate the relative positions of successive impressions. Also called **corner marks; corner ticks; register ticks; registration ticks; ticks.**

register studs—See **prepunch register system.**

register ticks—See **register marks.**

register trials—The test runs necessary to obtain the proper combination of the degree of partial vacuum and the length of the heating cycle required for individual models in forming a plastic relief map.

registration ticks—See **register marks.**

regression of the nodes—Precessional motion in a direction opposite to the direction of revolution of a set of nodes. See also **precession.**

regular error—See **systematic error.**

relative accuracy—**1.** (general) An evaluation of the random errors in determining the positional orientation (e.g., distance, azimuth) of one point or feature with respect to another. **2.** (chart, feature to graticule) An evaluation of the random errors in chart features with respect to the graticule excluding any error in the graticule or the datum defined by the graticule. **3.** (chart, feature to feature) An evaluation of the random errors in determining the positional orientation of one chart feature to another feature on the same chart.

relative aperture—(JCS) The ratio of the equivalent focal length to the diameter of the entrance pupil of photographic lens, expressed as f:8, etc. Also called **aperture ratio; f-number; lens speed; speed of lens; stop numbers.**

relative coordinate system—Any coordinate system which is moving with respect to an inertial coordinate system.

relative deflection—See **astrogeodetic deflection.**

relative direction—Horizontal direction expressed as angular distance from a heading.

relative distance—Distance relative to a specified reference point, usually one in motion.

relative error of closure—The value obtained by dividing the total error of closure by the total length of the traverse, commonly expressed by a fraction having a numerator equal to unity, e.g., 1/1,540. It is used for determining the degree of accuracy of a survey.

relative gravity—Gravity determined from gravity difference measurements (e.g., gravimeter, relative pendulum) between the observer and a reference station. The value obtained is relative with respect to the reference station.

relative motion—See **apparent motion.**

relative movement—Motion of one object or body relative to another. The expression is usually used when describing relative move ments other than that of a celestial body. See also **apparent motion; direction of relative movement.**

relative orientation—The reconstruction (analytically or in a photogrammetric instrument) of the same geometric conditions between a pair of photographs that existed when the photographs were taken. In the instrument, this is achieved by a systematic procedure of rotational and translational movements of the projectors. Also called **clearing y-parallax.** See also **absolute orientation; vertical deformation.**

relative pendulum—A device for measuring relative gravity through the difference in the period of a pendulum at two stations.

relative position—The location of a point or feature with respect to other points or features, either fixed or moving.

relative relief—The relation of the altitudes of the highest and lowest points of land in any area. The difference between the highest and lowest points is the amplitude of relative relief. Various types of maps have been devised to show this, usually depending on gridding the area on a map, finding a value for the amplitude in each grid square, and producing an isopleth or dot map to depict the distribution of these values.

relative setting—In tilt analysis of oblique photography, the dihedral angle between the two planes passing through the principal point of the opposite obliques, the principal point of the vertical photograph, and the common exposure station. This angle is measured on the vertical photograph as the angle between the two isolines, or as the deflection angle between the perpendiculars from the principal point of the vertical photograph to the two isolines.

relative swing—In the tilt analysis of oblique photographs, the angle of rotation of the oblique camera about its own axis with respect to the plane of the vertical photograph, measured on the oblique photograph by the angle between the isoline and a line joining the fore and aft fiducial marks.

relative tilt—The angular relationship between two overlapping vertical photographs with no reference to an established datum.

releasability code letter—A code letter prefix to the chart identification number which limits the releasability of a particular chart to specific users.

reliability diagram—A diagram included on some MC&G products depicting horizontal and vertical accuracies and date(s) of information.

relief—(JCS) Inequalities of elevation and the configuration of land features on the surface of the Earth which may be represented on maps or charts by contours, hypsometric tints, shading, spot elevations, or hachures.

relief displacement—Displacement radial from the nadir point of a photograph caused by differences in elevation of the corresponding ground objects. Also called **height displacement; relief distortion.**

relief distortion—See **relief displacement.**

relief model—A general category which denotes any three-dimensional representation of an object or geographic area, modeled in any size or medium. See also **plastic relief map; terrain model.**

relief stretching—See **hyperstereoscopy.**

remote sensing—The measurement or acquisition of information of some property of an object or phenomenon by a recording device that is not in physical or intimate contact with the object or phenomenon under study. Sometimes restricted to the practice of data collection in the wavelengths from ultraviolet to radio regions.

remote station—See **slave station.**

repeatability—A measure of the variation in the accuracy of an instrument when tests are made over the same line(s) at different times of the year, with different operators, and with different but equivalent instruments, all using the same procedures. See also **external error.**

repeating instrument—See **repeating theodolite.**

repeating theodolite—A theodolite so designed that successive measures of an angle may be accumulated on the graduated circle, and a final reading of the circle made which represents the sum of the repetitions. Also called **double-center theodolite; repeating instrument.**

repetition of angles—The accumulation of a series of measures of the same angle on the horizontal circle of a repeating theodolite or surveyor's transit.

representative fraction (RF)—(JCS) The scale of a map or chart expressed as a fraction or ratio. [Relates unit distance on the map to distance measured in the same unit on the ground.] Also called **fractional scale; natural scale.**

representative pattern—(cartography) **1.** An accurate portrayal of the surface of the Earth in the area being compiled. **2.** The selection and portrayal of the most prominent of a dense group of similar features.

reprint—The process of using existing reproducibles without change to print additional quantities of a product.

reproducible—Any copy capable of being used as a master-to-be. May be either a negative or positive transparency.

reproduction—**1.** The summation of all the processes involved in printing copies from an original drawing. **2.** A printed copy of an original drawing made by any of the processes of reproduction.

reproduction material—(JCS) Material, generally in the form of positive or negative copies on film or glass for each color plate, from which a map or chart may be reproduced without redrafting. Also called **repromat.**

reproduction positive mold—The positive mold

which has been drilled through with vacuum holes, and over which the plastic relief map is formed.

reproduction ratio—See **scale of reproduction.**

reproduction scale—The scale at which a map or chart is published or is to be published. Also called **publication scale.**

repromat—See **reproduction material.**

Repsold base-line measuring apparatus—An optical base-line measuring apparatus, composed of a steel bar approximately 4 meters long, whose exact length at any temperature is known, and whose temperature is determined by means of a metallic thermometer composed of the steel measuring bar and a similar bar of zinc, the two being fastened together at their middle points.

reseau—**1.** A glass plate on which is etched an accurately ruled grid. Sometimes used as a focal-plane plate to provide a means of calibrating film distortion; used also for calibrating plotting instruments. Also called **grid plate. 2.** Intersecting orthogonal lines superimposed over photo imagery.

resection—**1.** The graphical or analytical determination of a position, as the intersection of at least three lines of known direction to corresponding points of known position. **2.** (surveying) The determination of the horizontal position of a survey station by observed directions from the station to points of known positions. Also, the line drawn through the plotted location of a station to the occupied station. **3.** (photogrammetry) The determination of the position and/or attitude of a camera, or the photograph taken with that camera, with respect to the exterior coordinate system. Also called **three-point method.** See also **intersection,** definition 2.

resection station—A station located by resection methods.

residual—A general term denoting a quantity remaining after some other quantity has been subtracted. It occurs in a variety of particular contexts. For example, if the true value of a variable is subtracted from an observed value then the difference may be called a residual; it is also frequently called an error. Similarly, if a mathematical model is fitted to data, the values by which the observations differ from the model values are called residuals.

residual deviation—Deviation of a magnetic compass after adjustment or compensation.

residual error—The difference between any value of a quantity in a series of observations, corrected for known systematic errors, and the value of the quantity obtained from the combination or adjustment of that series. Frequently used as the difference between an observed value and the mean of all observed values of a statistically valid set. See also **error; residual.** The latter term is generally used in referring to actual values in a specific computation.

residual gravity—In gravity prospecting, the portion of a gravity effect remaining after removal of some type of regional gravity, usually the relatively small or local anomaly components of the total or observed gravity field.

residual parallax—Small amounts of *y*-parallax which may remain in a model after relative orientation is accomplished.

resolution—**1.** (JCS) The measure of the ability of a lens, a photographic material, or a photographic system to distinguish detail under certain specific conditions. The measure of this ability is normally expressed in lines per millimeter or angular resolution. **2.** The minimum distance between two adjacent features, or the minimum size of a feature, which can be detected by a remote sensory system. **3.** In gravity or magnetic prospecting, the indication in some measured quantity, such as the vertical component of gravity, of the presence of two or more close but separate disturbing bodies.

resolution in bearing—The minimum detectable separation of objects at the same range and the same elevation, expressed in terms of the horizontal angular distance between such objects.

resolution in elevation—The minimum detectable separation of objects at the same range and same bearing, expressed in terms of the vertical angular distance.

resolution in range—The minimum detectable separation of objects in the same line of sight, expressed in terms of the distance between them.

resolution limit—In gravity and magnetic prospecting, the separation of two disturbing bodies at which some obvious indication, in a measured quantity, of the presence of two separate bodies, ceases to be visible.

resolving power—A mathematical expression of definition in an imaging system, usually stated as the maximum number of lines per millimeter that can be seen as separate lines in the image.

resolving power target—A test chart used for the evaluation of photographic, optical, and electro-optical systems. The design usually consists of ruled lines, squares, or circles varying in size according to a specified geometric progression.

responder—In general, an instrument that indicates reception of an electric or electromagnetic signal. See also **transponder.**

responsor—A radio receiver which receives the reply from a transponder and produces an output suitable for feeding to a display system.

restitution—(JCS) The process of determining the true planimetric position of objects whose images appear on photographs. [Restitution corrects for distortion resulting from both tilt and relief displacement.]

restoration—The recovery of one or more lines or corner positions, or both, of a prior survey; the replacement of one or more lost corners or obliterated monuments by approved methods, including the substantial renewal of one or more monuments, as required for the purpose of a survey.

resultant error—The error in any measurement that is the difference between the measured value and the true value for a quantity. Also called **true error.**

resurvey—A retracing on the ground of the lines of an earlier survey, in which all points of the earlier survey that are recovered are held fixed and used as a control. If too few points of the earlier survey are recovered to satisfy the control requirements of the resurvey, a **new survey** may be made. A resurvey is related directly to an original survey, though several resurveys may interpose between them. See also **dependent resurvey; independent resurvey; retracement.**

reticle (reticule)—1. (surveying) A system of wires, hairs, threads, etched lines, or the like, placed normal to the axis of a telescope at its principal focus, by means of which the telescope is sighted on a star, or target, or by means of which appropriate readings are made on some scale, such as a leveling or stadia rod. **2.** (optics) A mark, such as a cross or system of lines, lying in the image plane of a viewing apparatus and used singly as a reference mark in certain types of monocular instruments or as one of a pair to form a floating mark, as in certain types of stereoscopic instruments. See also **floating mark; graticule; index mark; parallactic grid.**

retouching—Corrective treatment of a plate, negative, positive, or copy by means of brush, pencil, pen, airbrush, or other method.

retracement—A term applied to a survey that is made for the purpose of verifying the direction and length of lines, and identifying the monuments and other marks of an established prior survey. See also **resurvey.**

retrodirective prism—A prism consisting of a solid glass element having three mutually perpendicular reflecting surfaces and a fourth surface oblique to the three reflecting surfaces. A light beam entering through the oblique surface is reflected on each of the three other surfaces and turned 180° to be returned along the same airpath which it traveled to the retrodirective prism.

retrograde motion—1. Motion in an orbit opposite to the usual orbital direction of celestial bodies within a given system. Specifically, of a satellite, motion in a direction opposite to the direction of rotation of the primary. **2.** The apparent motion of a planet westward among the stars. Also called **retrogression.** Opposite of **progressive motion.**

retrograde vernier—A vernier scale which has spaces or divisions slightly longer than those of the primary scale. The numbers on the vernier scale run in the opposite direction from those on the primary scale.

retrogression—See **retrograde motion.**

reverse stereo—See **pseudoscopic stereo.**

reversible level—A spirit level having a bubble tube with the inner surface ground barrel-shaped so that the tangent lines to curves on the upper and lower sides are parallel when exactly opposite, permitting the level to be used in either the erect or inverted positions.

reversible pendulum—A pendulum so designed and equipped with means of support that it may be used with either end up or down.

reversing in azimuth and altitude—See **double centering.**

reverted image—(optics) An image in which detail is in reverse order, from left to right, compared to the corresponding detail of the object. The order of detail from top to bottom remains unchanged. A mirror image.

revision—The process of updating a product to reflect current information. Typically, revision of a map or chart does not require significant changes to the horizontal position of features or vertical data values; rather, improvement in planimetric data is provided. Normally, publications are revised, not recompiled.

revolution—The turning of a body about an exterior point or axis. The correct distinction between **revolution** and **rotation** is given in the statement "the Earth revolves around the Sun, and rotates on its axis."

rhomboidal prism—A prism that displaces the axis of the beam of light only laterally.

rhumb bearing—The direction of a rhumb line through two terrestrial points, expressed as angular distance from a reference direction. It is usually measured from 000° at the reference direction clockwise to 360°. Also called **Mercator bearing.**

rhumb direction—See **Mercator direction.**

rhumb line—(JCS) A line on the surface of the Earth cutting all meridians at the same angle. [A loxodrome or loxodromic curve spiraling toward the poles in a constant true direction. Parallels and meridians, which also maintain constant true directions, may be considered special cases of the rhumb line. A rhumb line is a straight line on a Mercator projection.] Also called **equiangular spiral; loxodrome; loxodromic curve; Mercator track.**

rhumb-line distance—Distance along a rhumb line, usually expressed in nautical miles.

ridge line—A graphic representation of major ridges used to give more definition to the topographic character of an area for the determination of low altitude radar predictions.

They are shown only in areas of rise gradient to depict those places in which the elevated terrain forms a sufficient background to partially screen vision at low altitude.

right-angle prism—A prism that turns a beam of light through a right angle. It inverts (turns upside down) or reverts (turns right for left) according to the orientation of the prism.

right ascension—The angular distance measured eastward on the Equator from the vernal equinox to the hour circle through the celestial body, from 0 to 24 hours.

right ascension system—An equatorial system of curvilinear celestial coordinates which has the Equator as the primary reference plane and the perpendicular hour circle through the vernal equinox as the secondary reference plane. The direction to a body is given by its right ascension and declination.

right bank—That bank of a stream or river on the right of the observer when he is facing in the direction of flow, or downstream.

right-reading—A descriptive term for an image which, when viewed through the base, reads the same as the original. Other terms sometimes used to identify image direction, such as **normal reading, natural readings,** etc., are not recommended because of possible confusion in negative-positive relationship.

right sphere—The celestial sphere as it appears to an observer at the Equator, where celestial bodies appear to rise vertically above the horizon.

rigid tripod engraver—A scribing instrument with three points of contact surface, to absorb the normal imbalance of the operator's hand pressure.

rise—**1.** To cross the visible horizon while ascending. **2.** (satellite surveying) To cross the observer's horizon while ascending; detectable by broadcast data received.

rise gradient—A color-coded omni-directional graphic representation of those terrain slopes which are predicted to be low altitude radar significant.

rise time—The time at which a satellite's broadcast can be picked up by a suitably equipped observer, as taken from an alert. Set time and time of closest approach are also given. See also **alerts; rise.**

rising tide—See **flood tide.**

river crossing—(leveling) Carrying a line of levels across a stream or other body of water, when no suitable bridge is available and the width of the body of water is greater than the maximum allowable length of sight for the leveling, requires a special series of observations which taken collectively is known as a river crossing.

riverine area—(JCS) An inland or coastal area comprising both land and water, characterized by limited land lines of communication, with extensive water surface and/or inland waterways that provide natural routes for surface transportation and communications.

road map—A medium- or small-scale special-purpose map, generally showing only planimetric detail, with emphasis upon the road network and related data. Its main purpose is to furnish pertinent road information for tactical and administrative troop movement.

road net—(JCS) The system of roads available within a particular locality or area.

roamer—(JCS) Grids constructed to common map scales used for determination of map coordinates.

rod—See **leveling rod.**

rod correction—(leveling) That correction which is applied to an observed difference of elevation to correct for the error introduced when the leveling rods are not actually of the length indicated by the graduations.

rod float—A small cylindrical tube of any material, closed at the bottom and weighted with shot until it floats in an upright position with about 2 to 6 inches projecting above the water surface. Current velocities are determined by direct observations.

rod level—An accessory for use with a leveling rod or a stadia rod to assure a vertical position of the rod prior to instrument reading.

rod sum—(leveling) The algebraic total of plus and minus sights in a given level line.

roll—**1.** (JCS) The rotation of an aircraft or ship about its longitudinal axis. **2.** (photogrammetry) A rotation of a camera or a photograph-coordinate system about either the photograph x-axis or the exterior x-axis. In some photogrammetric instruments and in analytical applications, the symbol omega (ω) may be used. See also **tilt.**

romanization—**1.** The process of recording in roman script either the sounds of a language or the graphic symbols of a non-roman writing system. **2.** An item of a language which has undergone this process. See also **transcription; transliteration.**

roof prism—A type of prism in which the image is reverted by a roof, that is, two surfaces inclined at 90° to each other.

root-mean-square error—See **standard error.**

roots of mountain theory—See **Airy theory of isotasy.**

rotating prism—See **dove prism.**

rotating prism camera—A class of panoramic camera in which a double dove prism is rotated while the lens system remains fixed. This configuration can achieve a scan of 180° or more.

rotation—**1.** (astronomy) A turning of a body about a self-contained axis, as the daily rotation of the Earth. See also **revolution. 2.** (surveying) A turning of an instrument or part of an instrument.

rotational movement—(photogrammetry) The

systematic rotation of projectors or projector assemblies. When applied to the projector body within the gimbal inner ring, the movement is about the *z*-axis and is called **swing.** Rotation of the inner ring is about an *x*-axis (secondary axis) and is called **x-tilt.** Rotation of the outer ring is about a *y*-axis (primary axis) and is called **y-tilt.**

route chart—1. A chart showing routes between various places, usually with distances indicated. **2.** An aeronautical chart covering the route between specific terminals, and usually of such scale as to include the entire route on a single chart. Also called **flight chart.**

route map—A map showing roads to be followed and nearby points of military significance.

route survey—Surveys for linear construction such as railroads, highways, and transmission lines which include the layout of lines and grades for these projects.

rubber blanket—See **offset lithography; offset press.**

run—1. (lithography) The number of impressions made on a press for a given sheet. **2.** (micrometer) See **error of run. 3.** (JCS) (aerial photography) That part of a flight of one photographic reconnaissance aircraft during which photographs are taken.

run of micrometer—See **error of run.**

running—(leveling) A continuous series of measured differences of elevation, made set-up by a set-up in one direction along a section of a line of levels, which results in a measurement of the difference of elevation between the bench marks or other points, either temporary or permanent, at the ends of the section.

running fix—(JCS) The intersection of two or more position lines, not obtained simultaneously, adjusted to a common time.

running mean—See **consecutive mean.**

S

sag correction—(taping) The difference between the effective length of a tape, or part of a tape, when supported continuously throughout its length and when supported at a limited number of independent points. Base tapes usually are used with three or five points of support, and hang in curves (catenaries) between adjacent supports. Correction for sag is not required when the method of support in use is the same as was used in the standardization of the tape; only the standardization correction is applied. A base tape may also be used supported throughout or with four points of support, as on a railway rail. Also called **catenary correction.**

sailing chart—A small-scale chart used for offshore sailing between distant coastal ports and for plotting the navigator's position out of sight of land and as he approaches the coast from the open ocean. They show offshore soundings and the most important lights, outer buoys, and natural landmarks which are visible at considerable distances.

sailing directions—A descriptive book for the use of mariners, containing detailed information of coastal waters, harbor facilities, etc. Also called **coast pilot; pilot.**

Sanson-Flamsteed map projection—See **sinusoidal map projection.**

saros—The eclipse cycle of about 18 years, almost the same length as 223 synodical months. At the end of each saros, the Sun, Moon, and line of nodes return to approximately the same relative positions and another series of eclipses begins, closely resembling the series just completed. See also **lunar cycle.**

satellite—An attendant body, natural or manmade, that revolves about another body, the primary. See also **active satellite; communications satellite; Earth satellite; equatorial satellite; geodetic satellite; lunar satellite; NAVSTAR Global Positioning System; Navy Navigation Satellite System; passive satellite; polar satellite; synchronous satellite.**

satellite geodesy—The discipline which employs observations of an Earth satellite to extract geodetic information.

satellite surveying—1. (Doppler) The process of positioning one or more points on the Earth's surface by collecting Doppler shift data from passes of Navy navigation satellites. See also **point positioning; short arc; translocation. 2.** (NAVSTAR) The position (by resection) of a point receiving signals from four satellites of the Global Positioning System.

satellite trail—A streak-like image of a satellite recorded on a stellar plate by a photographic time exposure.

satellite triangulation—The determination of the angular relationships between two or more stations by the simultaneous observation of an Earth satellite from these stations.

satellite triangulation stations—Triangulation stations whose angular positions relative to one another are determined by the simultaneous observation of an Earth satellite from two or more of them.

saturable reactor—See **flux-gate magnetometer.**

scalar—Any physical quantity whose field can be described by a single numerical value at each point in space. A scalar quantity is distinguished from a vector quantity by the fact that a scalar quantity possesses only magnitude, whereas a vector quantity possesses both magnitude and direction.

scale—1. (JCS) The ratio or fraction between the distance on a map, chart, or photograph and the corresponding distance on the surface of the Earth. **2.** A series of marks or graduations at definite intervals on a measuring device or instrument. **3.** Measurement by means of a scale. See also **compilation scale; equivalent scale; graphic scale; Invar scale; model scale; plotting scale; representative fraction; reproduction scale; scaling; x-scale; y-scale; z-scale.**

scale checking—1. The process of determining the scale of an aerial photograph, or more correctly, the altitude above sea level which best fits the print. **2.** (stereocompilation) The process of determining the scale of a vertical photograph for points at a specific elevation and the subsequent measurement of direction and distance therefrom.

scale error—A systematic error in the lengths of survey lines usually proportional to the lengths of the lines. See also **instrument error.**

scale factor—A multiplier for reducing a distance obtained from a map by computation or scaling to the actual distance on the datum of the map. Also, in the state coordinate systems, scale factors are applied to geodetic lengths to obtain grid lengths, or to grid lengths to obtain geodetic lengths. Both are lengths on a sea level datum, but the grid lengths are affected by the scale change of the map projection.

scale indicator—A logarithmic scale devised as a rapid and convenient method of determining the natural scale of a map from the divisions marked on the graphic scales, or from the intervals of latitude on a map.

scale of reproduction—The enlargement or reduction ratio of an original to the final copy. This ratio is expressed as a diameter, percent, times (X), or a fraction. Also called **enlargement factor; reduction factor; reproduction ratio.** See also **contact size; diameter enlargement; make line; percent of enlargement/reduction; times (X) enlargement.**

scale-ratio mosaic—An assembly of photographic prints brought to a common scale by projection printing to scale factors obtained from map distances to allow the best possible fit of contiguous photo detail.

scaling—**1.** Alteration of the scale in photogrammetric triangulation to bring the model into agreement with a plot of horizontal control. **2.** Fitting a stereoscopic model to a horizontal control plot. A step in absolute orientation. Also called **scaling the model.** **3.** Determining the scale of a photograph or graphic. **4.** (cartography) See **cartometric scaling.**

scaling the model—See **scaling,** definition 2.

scan positional distortion—In a panoramic camera system, the displacement of images of ground points from their expected cylindrical positions caused by the forward motion of the vehicle as the lens scans.

scatterometry—A method of using radar to measure the variation of radar scattering coefficients. These variations may be used by geoscientists to discriminate between surfaces with different roughness and materials. The scatterometer is distinguished from other radars by its ability to measure amplitude.

scene matching area correlator (SMAC)—An electro-optical correlation system which uses photographic reference material matched with onboard real-time scenes to achieve correlation. The reference scene is prepared from reconnaissance photographs and is stored in the form of a photographic transparency. The reference is placed around the inner circumference of a drum that rotates at a high angular velocity. The reference then forms a continuous strip which modulates or chops the projected image being focused on it by the optical system. The real-time energy from the ground within the field-of-view is spatially modulated by the transmissivity of the reference transparency.

Scheimpflug condition—The requirement that object, lens, and image planes intersect in a common line for sharp focus in any direct projection system.

Schott base-line measuring apparatus—A contact, compensating base-line measuring apparatus composed of three parallel bars; the middle bar of zinc, the outer bars of steel. One end of each steel bar is free; the other end is fastened to an end of the zinc bar, a different end for each steel bar. The lengths of the bars are so proportioned with respect to their coefficients of thermal expansion that a constant distance is maintained between the free ends of the steel bars.

scintillation—See **shimmer.**

scissors inversor—See **Peaucellier inversor.**

screen—See **area pattern screen; biangle screen; dot screen; halftone screen; magenta contact screen.**

screen angle—(photography) The angle which the rows of halftone dots make with the vertical when right-reading. The angle is measured clockwise with 0° at 12 o'clock.

screening—See **masking.**

scribed plate—See **scribed sheet.**

scribed sheet—A scribing surface on which the reproduction scribing has been completed. Also called **scribed plate.**

scriber—An instrument holding a scribing point; used for scribing on coated plastics. Also called **engraver; graver; scribing instrument.** See also **engraver subdivider; rectagraver; rigid tripod engraver; straight line graver; swivel graver; turret graver.**

scribing—(JCS) A method of preparing a map or chart by cutting the lines into a prepared coating. [The process of preparing a negative which can be reproduced by contact exposure. Portions of a photographically opaque coating are removed from a transparent base with specially designed tools.] Also called **negative scribing.**

scribing guide—See **guide.**

scribing instrument—See **scriber.**

scribing points—Needles or blades in various diameters or cross-section shapes ground and sharpened to prescribed dimensions. Used in scribers for scribing on coated plastics.

sea level—The height of the surface of the sea at any time. See also **ideal sea level; mean sea level.**

sea level contour—A contour line delineating points at sea level.

sea level datum—See **mean sea level.**

sea level datum of 1929—See **National Geodetic Vertical datum of 1929.**

sea level variation—Sea level varies from day to day, from month to month, and from year to year. This variation is attributed to meteorological conditions and should not be confused with the lunar tides.

search-and-rescue chart—A chart designed primarily for directing and conducting search and rescue operations.

secant—(USPLS) **1.** A line that cuts a geometric curve or surface at two or more points. **2.** A trigonometric function of an angle. See also **secant method.**

secant conic chart—See **conic chart with two standard parallels.**

secant conic map projection—See **conic map projection with two standard parallels.**

secant method—(USPLS) A method of determining the parallel of latitude for the survey of a base line or standard parallel by offsets from a great circle line which cuts the parallel at the first and fifth mile corners of the township boundary. See also **secant.**

secator—See **templet cutter.**

second-order bench mark—A bench mark connected to the datum (usually mean sea level) by continuous second-order leveling or by a continuous combination of first-order leveling and second-order leveling.

second-order leveling—Spirit leveling which does not attain the quality of first-order leveling but does conform to the current specification for second-order (class I or class II) leveling per *Classification Standards of Accuracy and General Specifications of Geodetic Control Surveys.* Recommended for densification of the National Network and for localized crustal movement and engineering projects.

second-order traverse—A survey traverse which extends between adjusted positions of the first-order or second-order control surveys and which conforms to the current specifications for second-order (class I or class II) traverse per "Classification Standards of Accuracy and General Specifications of Geodetic Control Surveys." Recommended for densification of the National Network and metropolitan area surveys.

second-order triangulation—Second-order triangulation was at one time known as secondary triangulation; changed in 1921 to primary triangulation, and in 1925 to second-order triangulation. These surveys conform to the current specifications for second-order (class I or class II) triangulation per "Classification Standards of Accuracy and General Specifications of Geodetic Control Surveys." Recommended for densification of the National Network and metropolitan area surveys.

second-order work—The designation given any survey work of next-to-the-highest order of accuracy and precision.

secondary—**1.** See **secondary great circle. 2.** A celestial body revolving around another body, its primary.

secondary circle—See **secondary great circle.**

secondary compilation—A specially prepared matte plastic material used to depict uncorrected or discrete soundings in bathymetric compilation.

secondary control point (SCP)—A point photographically identifiable, positioned to a high degree of accuracy using an average of several shiran horizontally controlled photographs. Seven to thirteen controlled photographs, each containing the point, are taken over each of four quadri-directional passes intersecting perpendicularly over the point. Photogrammetric measurements are used to determine the position of the point by relating it to the shiran positional nadir point on each photograph. These positions are mathematically averaged to obtain the most probable coordinates for the point.

secondary great circle—A great circle perpendicular to a primary great circle, such as a meridian other than the prime meridian. Also called **secondary; secondary circle.**

secondary grid—An obsolete grid, occasionally shown on maps in conjunction with the major grid(s).

secondary station—An additional triangulation station, usually marked and identified, established to strengthen horizontal map control. Secondary stations are connected to the main-scheme stations but are not considered as being part of the main-scheme net. They are often used for providing means for checks and for photogrammetric purposes.

secondary tide station—A tide station which is operated for a short period of time to obtain data for a specific purpose.

SECOR—A phase-comparison electronic long-range distance-measuring system used to determine positions and orbits of satellites or flight vehicles that contain the necessary transponders. This term is an acronym for "sequential collation of range" (now obsolete).

section—**1.** (USPLS) The unit of subdivision of a township with boundaries conforming to the rectangular system of surveys, nominally 1 mile square, containing 640 acres. See also **fractional section. 2.** (leveling) That portion of a line of levels which is recorded and abstracted as a unit. See also **fractional section; half section; quarter section.**

section corner—(USPLS) A corner at the extremity of a section boundary.

sectional chart—A series of aeronautical charts at a 1:500,000 scale covering the entire United States, suitable for contact or visual flying.

sectorial harmonics—The set of spherical harmonics which change from positive to negative as a function of longitude only. See also **tesseral harmonics; zonal harmonics.**

secular aberration—The aberration due to the motion of the center of mass of the solar system in space. Also called **aberration of fixed stars.**

secular perturbations—Perturbations to the orbit of a planet or satellite that continue to act in one direction without limit, in contrast to periodic perturbations which change direction in a regular manner.

secular terms—In the mathematical expression of an orbit, terms which are proportional to time, resulting in secular perturbations.

selection overlay—A tracing of selected map source detail compiled on transparent material; usually described by the name of the features or details depicted, such as contour overlay, vegetation overlay. Also called **lift; pull up; trace.**

selenocentric—Relating to the center of the Moon; referring to the Moon as a center. Also called **lunicentric.**

selenocentric coordinates—Quantities which express the position of a point with respect to the center of the Moon.

selenodesy—(JCS) That branch of applied mathematics which determines, by observation and measurement, the exact positions of points and the figures and areas of large portions of the

Moon's surface, or the shape and size of the Moon.

selenodetic—(JCS) Of or pertaining to, or determined by, selenodesy.

selenographic—**1.** Relating to the physical geography of the Moon. **2.** Specifically, referring to positions on the Moon measured in latitude from the Moon's equator and in longitude from a reference meridian.

selenology—That branch of astronomy that deals with the Moon, its magnitude, motion, constitution, and the like.

selenotrope—A device used in geodetic surveying for reflecting the Moon's rays to a distant point, to aid in long-distance observations. See also **heliotrope.**

self-leveling level—A level utilizing the action of gravity in its operation. A prismatic device, called a compensator, is an integral part of the instrument which, once the instrument has been roughly leveled, causes the optical system to swing into proper horizontal line of sight and to maintain that position during readings at a given station.

self-reading leveling rod—A rod with graduation marks designed to be read by the observer at the leveling instrument. Also called **speaking rod.**

self-registering gage—Any tide or stream gage which provides a continuous record of the variation of tide or stream level with the passage of time and which will operate, unattended, for a number of days. Also called **automatic gage.**

semianalytical triangulation—The measurement of *x*-, *y*-, and *z*-model coordinates on an analog instrument and the transformation from model coordinates to grid coordinates by a computational procedure.

semicontrolled mosaic—(JCS) A mosaic composed of corrected or uncorrected prints laid so that major ground features match their geographical coordinates. See also **controlled mosaic; mosaic; uncontrolled mosaic.**

semidiameter—**1.** The radius of a closed figure. **2.** Half the angle at the observer subtended by the visible disk of a celestial body.

semidiameter correction—A correction due to semidiameter, particularly that sextant altitude correction resulting from observation of the upper or lower limb of a celestial body, rather than the center of that body.

semidiurnal—Having a period of, occurring in, or related to approximately half a day.

semidiurnal constituent—A tidal constituent that has two maximums and two minimums each constituent day.

semimajor axis—**1.** One-half the longest diameter of an ellipse. Also called **mean distance. 2.** (geodesy) Equatorial axis of a spheroid or ellipsoid.

semiminor axis—One-half the shortest diameter of an ellipse.

sensibility—(spirit level) See **sensitivity.**

sensible horizon—That circle of the celestial sphere formed by the intersection of the celestial sphere and a plane through any point, such as the eye of an observer, and perpendicular to the zenith-nadir line.

sensitive altimeter—See **barometric altimeter.**

sensitivity—(spirit level) The accuracy and precision which a spirit level is capable of producing. Sensitivity depends on the radius of curvature of its longitudinal section; the longer the radius, the more sensitive the level. Sensitivity is rated by equating the linear length of a division between graduation marks on the level tube and its angular value at the center of curvature of the tube. Also called **sensibility.**

sensitometric curve—See **characteristic curve.**

sensitometry—The measurement of the response of a photosensitive material to the action of light.

sensor—A technical means to extend man's natural senses. Also a sensing device or equipment which detects and records in the form of imagery, the energy reflected or emitted by environmental areas, features, objects, and events, including natural and cultural features and physical phenomena, as well as manmade features, objects and activities. The energy may be nuclear, electromagnetic, including the visible and invisible portions of the spectrum, chemical, biological, thermal, or mechanical, including sound, blast, and Earth vibration.

sensor simulation system—A device, such as navigation simulator or trainer, in which specific types of sensor simulator materials are utilized for training purposes. See also **sensor simulator materials.**

sensor simulator materials—Those terrain models or maps, factored transparencies or radar reflectivity plates developed or produced from mapping, charting, geodetic, and/or intelligence data or compilations for use in weapon system or navigation simulators or trainers.

September equinox—See **autumnal equinox.**

series—See **coordinated series; map series.**

series designation—A descriptive title, a number, or a combination of a letter and number, used individually or collectively to identify a group or family of maps, charts, or related publications.

series specifications—See **specifications.**

set—**1.** The direction toward which the current flows. Usually indicated in degrees true or points of the compass. **2.** A finite or infinite number of objects of any kind, of entities, or of concepts, that have a given property or properties in common. **3.** (surveying) A specified number of observations, as of astronomic azimuth, astronomic longitude.

set-back—The horizontal distance from the fiducial mark on the front end of a tape or part of tape, which is in use at the time, back to the point on the ground mark or monument to which the particular measure is being made.

set forward—See **set-up.**

set-up—1. The instrument (transit or level) placed in position and leveled, ready for taking measurements; or a point where an instrument is to be or has been placed. Also called **instrument station. 2.** In base-line measurements, the horizontal distance from the fiducial mark on the front end of a tape or part of tape which is in use at the time, measured in a forward direction to the point on the ground mark or monument to which the particular measure is being made. Also called **set forward.**

sexagesimal system—A system of notation by increments of 60; as the division of the circle into 360 degrees, each degree into 60 minutes, and each minute into 60 seconds.

sextant—A double-reflecting instrument for measuring angles, primarily altitudes of celestial bodies. As originally used, the term applied only to instruments having an arc of 60° (and a range of 120°), from which the instrument derived its name. In modern practice the term applies to similar instruments, regardless of range. Also called **hydrographic sextant.** See also **bubble sextant; marine sextant; octant; quadrant; quintant; surveying sextant.**

sextant altitude—The altitude of a celestial body as indicated by a sextant or a similar instrument before corrections are applied.

sextant chart—A chart with curves enabling a graphical solution of the three-point problem rather than using a three-arm protractor.

shade error—That error of an optical instrument due to refraction in the shade glasses.

shaded relief—(JCS) A cartographic technique that provides an apparent three-dimensional configuration of the terrain on maps and charts by the use of graded shadows that would be cast by high ground if light were shining from the northwest. Shaded relief is usually used in combination with contours. See also **hill shading.**

shaded-relief map—A map on which hypsography is made to appear three-dimensional by the use of graded shadow effects. Generally, the features are shaded as though illuminated from the northwest. A shaded-relief map may also contain contours or hachures in combination with the shading.

shadow factor—(JCS) A multiplication factor derived from the Sun's declination, the latitude of the target, and the time of photography, used in determining the heights of objects from shadow length. Also called **tan alt.**

shadow projector—An optical device developed for checking dimensional accuracy of the various casts of relief models.

sheer—Transformation of a rectangle into a parallelogram.

sheet—A single map, either a complete map in one sheet, or belonging to a series.

sheet lines—See **neatlines.**

shift—(radar) (JCS) The ability to move the origin of a radial display away from the center of the cathode-ray tube.

shimmer—An atmospheric effect due to atmospheric turbulences. It may be more critical in photographic observations of celestial objects than refraction. The shimmer makes the image fluctuate rapidly. It averages out in the case of long exposures but is serious in case of flashes. Shimmer affects both right ascension and declination in a random manner and, unlike regular refraction, is not zero at the zenith. Also called **scintillation.**

ship-to-shore triangulation—A method of triangulation involving simultaneous observations from three shore stations on a target carried by a ship offshore of the middle station. The middle station must be visible from each of the two end stations.

Ships Inertial Navigation System (SINS)—A precise dead-reckoning system which maintains ships' position and heading through measurements made with gyroscopes and accelerometers.

shiran—An electronic distance-measuring system for measuring distances with geodetic accuracy from an airborne station to each of four ground stations. This term is an acronym for "S-band high precision short-range electronic navigation."

shoot—1. (astronomy; surveying) To make an observation with an instrument. **2.** (photography) A slang term used to denote photographing copy, such as a map manuscript, with a copy camera.

shop calibration—Adjustments to precision instruments made in an instrument maintenance shop having a limited amount of specialized testing equipment.

shoran—(JCS) A precise short-range electronic navigation system which uses the time of travel of pulse-type transmissions from two or more fixed stations to measure slant-range distance from the stations. The term is an acronym for the phrase "short-range navigation."

shoran-controlled photography—A method by which the positions of aircraft are determined by distance measurements to two shoran ground stations simultaneously with photographic exposures.

shoran-line crossing—A method of determining distance between two points by flying across the adjoining line.

shoran range—The maximum possible operating distance between shoran aircraft and ground stations as limited by flying height, ground-station elevation, terrain, and Earth curvature.

shoran reduction—The computation process of converting from a shoran-distance reading to an equivalent geodetic distance.

shoran trilateration—A method of extending horizontal control in which the sides of appropriate figures are measured by the shoran line-crossing method.

shoran-wave path—The path taken by the shoran wave as it travels from the mobile (airborne or shipborne) station to the ground station.

shoreline effect—See **coastal refraction.**

short arc—A small portion (usually less than half) of the orbital arc traversed by a satellite in making 1 revolution about the Earth.

short arc geodetic adjustment (SAGA)—The least-squares adjustment for position, elevation, azimuth, and distance of a number of stations using Doppler satellite observations of the same passes. The satellite positions are permitted to vary. Only portions of satellite arcs are observed. Points along these short arcs are computed for the times of the observations as an intermediate step towards deriving the station positions. See also **point positioning; short arc; short arc network.**

short arc network—A network of positions established by adjustment of simultaneous satellite observations.

short arc reduction method—A computational procedure in which only short arcs of the satellite orbit are employed in order to minimize the effects of secular and long period perturbations.

short distance navigational aids—(JCS) An equipment or system which provides navigational assistance to a range not exceeding 200 statute miles/320 kilometers.

short period perturbations—Periodic perturbations in the orbit of a planet or satellite which execute one complete periodic variation in the time of one orbital period or less.

short rod—A level rod, usually a Philadelphia rod, permitting readings of 7 feet or less. See also **long rod.**

side equation—A condition equation which expresses the relationship between the various sides in a triangulation figure as they can be derived by computation from one another.

side-equation tests—Side-equation tests are a modification of side equations which are helpful to geodetic field parties in checking the accuracy of observations and in locating the points where horizontal-direction observations may be in error.

side lap—See **overlap,** definition 1.

side line—Applied to a strip of land such as a street or right-of-way, it defines the boundaries of that strip; not applied to the ends of a strip.

side-looking airborne radar (SLAR)—(JCS) An airborne radar, viewing at right angles to the axis of the vehicle, which produces a presentation of terrain or moving targets.

side shot—A reading or measurement from a survey station to locate a point which is not intended to be used as a base for the extension of the survey. A side shot is usually made for the purpose of determining the position of some object which is to be shown on the map.

side-sight—A sight made with the transit to a point not on the line of traverse. It may be a side shot or an observation to locate an in-and-out station, an azimuth mark, or an intersected point.

side test—In triangulation of a quadrilateral or similar figure, where distances can be computed two different ways, the ratio of the difference between the two computed results to the length of the line.

sidereal—Of or pertaining to the stars. Although sidereal generally refers to the stars and tropical to the vernal equinox, sidereal time and the sidereal day are based upon the position of the vernal equinox relative to the meridian.

sidereal day—The interval of time from a transit of the (true) vernal equinox across the upper branch of a given meridian to its next successive transit across the upper branch of the same meridian. Also called **equinoctial day.**

sidereal focus—The position of the principal focal plane of a lens system. A camera or telescope is in sidereal focus when incident rays from a great distance come to a focus in the plane of the photographic plate or of the reticle. Also called **solar focus.**

sidereal hour angle (SHA)—Angular distance west of the vernal equinox; the arc of the celestial equator, or the angle at the celestial pole, between the hour circle of the vernal equinox and the hour circle of a point on the celestial sphere, measured westward from the hour circle of the vernal equinox through 360°.

sidereal month—The interval of time between two successive passages of the Moon past a fixed star.

sidereal period—**1.** The time taken by a planet or satellite to complete one revolution about its primary and as referred to a fixed star. **2.** Specifically, the interval between two successive returns of an Earth satellite in orbit to the same geocentric right ascension.

sidereal time—Time based upon the rotation of the Earth relative to the vernal equinox.

sidereal year—The period of one apparent revolution of the Earth around the Sun, with respect to the fixed stars, with an accepted value of 365 days, 6 hours, 9 minutes, 9.5 seconds in 1900; and increasing at the rate of 0.0001 second annually.

sight—Observation of the altitude, and sometimes also the azimuth, of a celestial body for a line of position; or the data obtained by such observation.

sight line—See **line of collimation.**

sight reduction—The process of deriving from observation of a celestial body the information needed for establishing a line of position.

sight reduction tables—Tables for performing sight reductions, particularly those for determining computed altitude for comparison with the observed altitude of a celestial body to determine the altitude difference for establishing a line of position.

sight rod—See **range rod.**

sight tree—See **line tree.**

signal lamp—A compact, portable, battery-operated electric lamp used as a target for observations on surveys of high precision (usually on first- and second-order geodetic triangulation). The parabolic reflector is mounted in a special case to facilitate pointing and adjusting.

signature—The characteristics or patterns of physical features that permit objects to be recognized on aerial imagery. A category is said to have a signature only if the characteristic pattern is highly representative of all units of that category.

significant date—The date that represents the best approximation of the date of the product's information; e.g., the date of compilation or the date of the source materials used to revise the product.

simple conic chart—A chart on a simple conic projection.

simple conic map projection—A conic map projection in which the surface of a sphere or spheroid, such as the Earth, is conceived as developed on a tangent cone, which is then spread out to form a plane.

simple harmonic motion—The projection of uniform circular motion on a diameter of the circle of such motion.

simple pendulum—A theoretical concept: a heavy particle suspended from a fixed point by a fine thread which is inextensible and without weight. A simple pendulum cannot be realized in actual work. A simple pendulum is, however, the basis of reductions of observations made with an actual pendulum. Those observations have corrections applied to them to obtain results which would have been produced by an equivalent simple pendulum.

Simpson's 1/3 rule—A mathematical expression for determining areas between an irregular boundary and a traverse line where equally spaced offset measurements have been taken.

simultaneous altitudes—Altitudes of two or more celestial bodies observed at the same time.

simultaneous double line—See **simultaneous level line.**

simultaneous level line—A line of spirit leveling composed of two single lines run over the same route, both in the same direction, but using different turning points. Also called **simultaneous double line.**

simultaneous mode—A satellite method for determining the position of an unknown station by the simultaneous ranging from three stations of known position and the unknown station, or simultaneously observing direction from two stations of known position and the unknown station, and mathematically reducing the data to solve for a line or surface of position of the unknown. This technique permits position determination independent of a satellite's orbital parameters.

simultaneous observations—(satellite) Observations of a satellite that are made from two or more distinct points or tracking stations at exactly the same time.

single astronomic station datum orientation—The orientation of a geodetic datum by accepting the astronomically determined coordinates of the origin and the azimuth to one other station without any correction.

single-base method—A technique of barometric leveling utilizing two barometers. One barometer is designated as a base and a second, or roving, barometer is used to determine pressures at specific points. Time and pressure are recorded at each position occupied by the roving barometer and time and pressure are recorded every 5 minutes by the base barometer. Data are reduced to elevations by office computations. See also **barometric leveling.**

single-heading radar prediction—A radar prediction made for a single aircraft position or from one specific point in relation to the target. It may be either an experience or an analytical prediction.

single-model instrument—A general class of stereoscopic plotting instruments with a capability for projecting a single stereomodel per setup. This class of plotter is designed for compilation only and is dependent upon supplementary photogrammetric techniques to accomplish necessary stereotriangulation.

single-point transfer instrument—Any instrument used for the transfer of planimetric detail from a single photograph. These instruments are of two general types, **reflecting projector** and **camera lucida.**

single-projector method—See **one-swing method.**

single proportionate measurement—A method of proportioning measurements in the restoration of a lost corner whose position is determined with reference to alignment in one direction. Examples of such corners are—quarter-section corners on the line between two section corners, all corners on standard parallels, and all intermediate positions on any township boundary line. The ordinary field problem consists of distributing the excess of deficiency between two existent corners in such a way that the amount given to each interval shall bear the same proportion to the whole difference as the record length of the interval bears to the whole record distance. After having applied the proportionate difference to the record length of each interval, the sum of the several parts will equal the new measurement of the whole distance.

single-target leveling rod—Any target rod having graduations on one face only.

sinusoidal map projection—A particular type of the Bonne map projection, employing the Equator as the standard parallel, and showing all geographic parallels as truly spaced parallel straight lines, along which exact scale is preserved. This is an equal-area map projection. Also called **Mercator equal-area map projection.** See also **Sanson-Flamsteed map projection.**

siphon barometer—A mercury barometer consisting of a column of mercury in a glass tube which is bent so as to have two vertical branches, one about one-fourth the length of the other. The end of the longer branch is closed, and the air in it is displaced by the mercury, but the shorter branch is left open, and the mercury is thereby subjected to atmospheric pressure. The difference of the height of the mercury in the two branches is a measure of the atmospheric pressure.

situation map—(JCS) A map showing the tactical or administrative situation at a particular time.

sixteenth-section corner—(USPLS) A corner at an extremity of a boundary of a quarter-quarter section; midpoint between or 20 chains from the controlling corners on the section or township boundaries. Written as 1/16 section corner. Also called **quarter-quarter section corner.**

size—**1.** To coat with any of the various glutinous materials used for filling the pores in the surface of paper, fiber, or of a mosaicking board. **2.** To calculate the measurements required in photographing a map to a desired scale.

sizing the litho—An operation performed in order to determine the actual measurement of the original lithographic maps to be used as source for a map revision in order to determine what distortion and changes of dimensions are necessary to fit the old map inside the new projection.

sketch map—A map made from loose, uncontrolled surveys. The information thereon is generally sparse.

sketchmaster—A form of camera lucida that permits superimposition of a rectified virtual image of a photograph over a map manuscript. See also **oblique sketchmaster; universal analog photographic rectification system; universal sketchmaster; vertical sketchmaster.**

skewed map projection—Any standard projection used in map or chart construction which does not conform to a general north-south format with relation to the neatlines of the map or chart.

slant range—(JCS) The line-of-sight distance between two points not at the same elevation.

slave station—That station in a given system of stations that is controlled by the master station. Also called **remote station; slave.**

slope—See **gradient.**

slope angle—The angle between a slope and the horizontal.

slope chaining—See **slope taping.**

slope correction—**1.** (hydrographic surveying) The correction applied to soundings erroneously positioned as a result of an echo sounder receiving its initial return from a point upslope from its recorded position. **2.** (land surveying) See **grade correction.**

slope correction of tape—See **grade correction.**

slope taping—Taping wherein the tape (or chain) is held as required by the slope of the ground, the slope of the tape measured, and the horizontal distance computed. Also called **slope chaining.**

slot cutter—See **templet cutter.**

slotted templet—A templet on which the radials are represented as a slot cut in a sheet of cardboard, metal, or other material.

slotted-templet plot—See **slotted-templet triangulation.**

slotted-templet triangulation—A graphical radial triangulation made by the use of slotted templets. Also called **slotted-templet plot.**

small circle—A circle on the surface of the Earth, the plane of which does not pass through the Earth's center.

small-scale map—A map having a scale of 1:600,000 or smaller.

smooth sheet—A final plot of field control and hydrographic development such as: soundings, fathom curves, wire drag areas, etc., to be used in chart construction.

snakeslip—See **etch slip.**

snap marker—See **point marker.**

Snell's law of refraction—This law states that the sine of the angle of incidence divided by the sine of the angle of refraction equals a constant termed the index of refraction when one of the media is air. The index of refraction can also be explained as the ratio of the velocity of light in one medium to that in another.

solar altitude—Angular distance of the Sun above the horizon.

solar attachment—An auxiliary instrument which may be attached to an engineer's transit, permitting its use as a solar compass.

solar day—**1.** The duration of one rotation of the Earth on its axis, with respect to the Sun. This may be either a mean solar day, or an apparent solar day, as the reference is the mean or apparent Sun, respectively. See also **apparent solar day; mean solar day. 2.** The duration of one rotation of the Sun.

solar declination—Angular distance of the Sun expressed in degrees north or south of the celestial equator; it is indicated as (+) when north and (−) when south of the Equator. Also called **declination of the Sun.**

solar eclipse—The obscuration of the light of the Sun by the Moon. A solar eclipse is partial if the Sun is partly obscured; total if the entire surface is obscured; or annular if a thin ring of the Sun's surface appears around the obscuring body.

solar eclipse method—A means of determining the angular distance between two observers along the center line of the path of a solar eclipse.

solar ephemeris—A daily tabulation of astronomic positions of the Sun.

solar focus—See **sidereal focus.**

solar occultation—An occultation of the Sun by the Moon.

solar parallax—The angle subtended by the equatorial radius of the Earth at a distance of one astronomic unit (i.e., the equatorial horizontal parallax of the Sun).

solar-radiation pressure—A perturbation of high-flying artificial satellites of large diameter. The greater part is directly from the Sun, a minor part is from the Earth, which is usually divided into direct (reflected) and indirect terrestrial (radiated) radiation pressures.

solar time—**1.** Time based upon the rotation of the Earth relative to the Sun. **2.** Time on the Sun.

solar transit—A regular transit to which has been added a solar attachment, which effects the instantaneous mechanical solution of the astronomic triangle (Sun-zenith-pole) and permits the establishment and surveying of the astronomic meridian or astronomic parallel directly by observation.

solar year—See **tropical year.**

solid angle—The integrated angular spread at the vertex of a cone, pyramid, or other solid figure.

solstice—One of two points of the ecliptic farthest from the celestial equator; one of the two points on the celestial sphere occupied by the Sun at maximum declination. Also called **solstitial point.** See also **summer solstice; winter solstice.**

solstitial colure—The hour circle through the solstices.

solstitial point—See **solstice.**

sonar—**1.** (JCS) A sonic device used primarily for the detection and location of underwater objects. **2.** A system for determining distance of an underwater object by measuring the interval of time between transmission of an underwater sonic or ultrasonic signal and return of its echo. This term is derived form the words "sound navigation and ranging."

sonic depth finder—See **echo sounder.**

sonic navigation—See **acoustic navigation.**

sortie—(JCS) An operational flight by one aircraft; also, photography obtained on a flight.

sortie plot—(JCS) An overlay representing the area on a map covered by imagery taken during one sortie. Also called **photo index.**

sounding—**1.** The measured or charted depth of water. **2.** A measurement of the depth of water expressed in feet or fathoms and reduced to the tidal datum shown in the chart title. Also called **hydrographic sounding.** See also **depth; depth number. 3.** In geophysics, any penetration of the natural environment for scientific observation. See also **echo sounder; echo sounding; lead line; off soundings; on soundings; wire drag.**

sounding datum—The plane to which soundings are referred.

sounding pole—A round, wooden, 15-foot-long pole, used for shoal water soundings. It is graduated in feet and half-feet from the center toward both ends and numbered consecutively from the ends toward the center.

source map—The map used for the selection of map or chart detail.

source material—Data of any type required for the production of MC&G products including, but not limited to, ground control, aerial and terrestrial photographs, sketches, maps, and charts; topographic, hydrographic, hypsographic, magnetic, geodetic, oceanographic, and meteorological information; intelligence documents and written reports pertaining to natural and manmade features of the area to be mapped or charted.

south declination—See **declination,** definition 3.

south geographical pole—The geographical pole in the Southern Hemisphere, at latitude 90°S.

south geomagnetic pole—The geomagnetic pole in the Southern Hemisphere.

south magnetic pole—The magnetic pole in the Southern Hemisphere.

south point—See **celestial meridian.**

south polar circle—See **Antarctic Circle.**

southbound node—See **descending node.**

southing—See **latitude difference.**

space coordinates—(photogrammetry) Any general three-dimensional coordinate system used to define the position of a point in the object space, as distinguished from the image of the point on a photograph.

space motion—Motion of a celestial body through space. See also **proper motion.**

space-polar coordinates—A system of coordinates by which a point on the surface of a sphere is located in space by (1) its distance from a fixed point at the center, called the pole; (2) the colatitude or angle between the polar axis (a reference line through the pole) and the radius vector (a straight line connecting the pole and the point); and (3) the longitude or angle between a reference plane through the polar axis and a plane through the radius vector and polar axis.

spacecraft—Devices, manned and unmanned, which are designed to be placed into an orbit about the Earth or into a trajectory to another celestial body.

spading—Removing scribe coating from the base material by use of a wide flat blade.

spatial model—See **stereoscopic model.**

speaking rod—See **self-reading leveling rod.**

special area (SA) annotation—A structure or group of structures possessing unique physical characteristics, but whose area or linear dimensions do not qualify for application of one of the radar significance analysis codes.

special job-cover map—(JCS) A small-scale map used to record progress on photographic reconnaissance tasks covering very large areas. As each portion of the task is completed, the area covered is outlined on the map.

special meander corner—(USPLS) A corner established at: (1) the intersection of a surveyed subdivision-of-section line and a meander line; (2) the intersection of a computed center line of a section and a meander line. In the latter case, the centerline of the section is calculated and surveyed on a theoretical bearing to an intersection with the meander line of a lake (over 50 acres in area) which is located entirely within a section.

special-purpose map—Any map designed primarily to meet specific requirements.

special-subject map—See **topical map.**

specific force—The difference between the inertial acceleration and gravitation acting on a body. The physical gravity sensed by accelerometers. All spirit levels are normal to the specific force vector.

specifications—The rules, regulations, symbology, and a comprehensive set of standards which have been established for a particular map or chart series or scale group. Specifications vary with the scale and the purpose of the graphic.

spectral band—A set of adjacent wavelengths in the electromagnetic spectrum with a common characteristic, such as the visible band.

spectrophotometer—A device for the measurement of spectral transmittance, spectral reflectance, or relative spectral emittance.

spectroradiometer—A device for the measurement of spectral distribution of radiant energy.

specular reflection—(optics or microwave theory) The type of reflection characteristic of a highly polished plane surface from which all rays are reflected at an angle equal to the angle of incidence. See also **diffuse reflection.**

speed—(photography) The response or sensitivity of the material to light, often expressed numerically according to one of several systems (e.g., H and D, DIN, Scheiner, and ASA exposure index). See also **relative aperture.**

speed of lens—See **relative aperture.**

sphere—A body or the space bounded by a spherical surface. See also **celestial sphere; oblique sphere; parallel sphere; right sphere; terrestrial sphere.**

spherical aberration—An aberration caused by rays from various zones of a lens coming to focus at different places along the axis. This results in an object point being imaged as a blurred circle.

spherical angle—The angle between two intersecting great circles.

spherical coordinates—A system of polar coordinates in which the origin is the center of a sphere and the points all lie on the surface of the sphere. The polar axis of such a system cuts the sphere at its two poles. In photogrammetry, spherical coordinates are useful in defining the relative orientation of perspective rays or axes and make it possible to state and solve, in simple forms, many related problems.

spherical excess—The amount by which the sum of the three angles of a triangle on a sphere exceeds 180°. In geodetic work, in the computation of triangles, the difference between spherical angles and spheroidal angles is generally neglected; spherical angles are used, and Legendre's theorem is applied to the distribution of the spherical excess. That is, approximately one-third of the spherical excess of a given spherical triangle is subtracted from each angle of the triangle.

spherical harmonics—Trigonometric terms of an infinite series used to approximate a two- or three-dimensional function of locations on or above the Earth.

spherical lens—A lens in which all surfaces are segments of spheres.

spherical triangle—The closed figure formed when any three points on the surface of a sphere are joined by arcs of great circles.

spheroid—**1.** (general) Any figure differing slightly from a sphere. **2.** (geodesy) A mathematical figure closely approaching the geoid in form and size and used as a surface of reference for geodetic surveys. See also **Airy spheroid (ellipsoid); Australian National spheroid; Bessel spheroid (ellipsoid); Clarke spheroid (ellipsoid) of 1866; Clarke spheroid (ellipsoid) of 1880; ellipsoid of rotation; equilibrium spheroid; Everest spheroid (ellipsoid); Hayford spheroid (ellipsoid); International spheroid (ellipsoid); Krasovsky spheroid (ellipsoid); oblate spheroid; prolate spheroid; reference spheroid.**

spheroid junction—An accentuated line on a map or chart, separating two or more major grids which are based on different spheroids.

spheroid of reference—See **reference spheroid.**

spheroidal angle—An angle between two curves on a spheroid; measured by the angle between their tangents at the point of intersection.

spheroidal excess—The amount by which the sum of the three angles of a triangle on a spheroid exceeds 180°. See also **spherical excess.**

spheroidal triangle—A triangle on the surface of a spheroid.

spherop—An equipotential surface in the normal gravity field of the Earth. Also called **spheropotential surface.**

spheropotential surface—See **spherop.**

spider templet—A mechanical templet which is formed by attaching slotted steel arms, representing radials, to a central core. The spider templet can be disassembled and the parts used again. Also called **mechanical arm templet.**

spider-templet plot—See **spider-templet triangulation.**

spider-templet triangulation—A graphical radial triangulation made by the use of spider templets. Also called **spider-templet plot.**

spiral curve—(route surveying) A curve of uniformly varying radius connecting a circular curve and a tangent, or two circular curves whose radii are, respectively, longer and shorter than its own extreme radii. Also called **easement curve; transition curve.**

spiral to spiral (SS)—A common point between two spirals.

spirit level—A closed glass tube (vial) of circular cross section, its center line also forming a circular arc, its interior surface being ground to precise form; it is filled with ether or liquid of low viscosity enough free space being left for the formation of a bubble of air and gas. Also called **bubble level.** See also **chambered spirit level; circular level; hanging level; latitude level; level trier; plate level; reversible level; rod level; striding level; telescope level.**

spirit level axis—The line tangent to the surface of a spirit-level tube (vial) against which the bubble forms, at the center of the graduated scale of the level, and in the plane of the longitudinal axis of the tube (vial) and its center of curvature. Also called **axis of level; axis of the level bubble; bubble axis.**

spirit level wind—Lack of parallelism between the axis of a spirit level vial and the line joining the centers of its supports. When wind (pronounced to rhyme with find) is present, the bubble will respond with a longitudinal movement when the spirit level is rocked on its supports.

spirit leveling—The determination of elevations of points with respect to each other or with respect to a common datum, by use of a leveling rod and an instrument using a spirit level to establish a horizontal line of sight.

split cameras—(JCS) An assembly of two cameras disposed at a fixed overlapping angle relative to each other. [Mainly used for reconnaissance purposes.] Also called **split-vertical camera.**

split photography—See **split-vertical photography.**

split-vertical camera—See **split cameras.**

split-vertical photography—(JCS) Photographs taken simultaneously by two cameras mounted at an angle from the vertical, one tilted to the left and one to the right, to obtain a small side overlap. Also called **split photography.**

spoking—(JCS) (radar) Periodic flashes of the rotating time base on a radial display. Sometimes caused by mutual interference.

spot elevation—(JCS) A point on a map or chart whose elevation is noted. [Elevations are shown, wherever practicable, for road forks and intersections, grade crossings, summits of hills, mountains, and mountain passes, water surfaces of lakes and ponds, stream forks, bottom elevations in depressions, and large flat areas.] Also called **spot height.** See also **checked spot elevation; elevation; unchecked spot elevation.**

spot height—See **spot elevation.**

spot prediction—A single heading radar prediction intended to portray, as nearly as possible, a comprehensive analysis of the radarscope at a precise geographic location.

spot size—(JCS) The size of the electron spot on the face of the cathode-ray tube.

spring balance—An accessory of taping and base measuring apparatus which is used in applying proper tension to a tape.

spur line of levels—A line of levels run as a branch from the main line of levels, either for the purpose of determining the elevations of marks not conveniently reached by the main line of levels or to connect with tidal bench marks or other previously established bench marks in obtaining checks on old leveling either at the beginning or end of a line of levels or at intermediate junctions along the new line of levels.

spur traverse—Any short traverse that branches off the established traverse to reach some vantage point or position. Also called **stub traverse.**

stabilized mount—A mount controlled by a gyroscope vertical reference unit designed to maintain a mapping or positional camera or other devices such as TPR antenna in a near vertical orientation independent of aircraft pitch and roll.

stable base—A general term applied to mapping materials possessing a high degree of dimensional stability.

stable-base film—(JCS) A particular type of film having high stability in regard to shrinkage and stretching. [Suitable for aerial mapping photography and map production. Usually referred to by its commercial name.]

stable gravimeter—A gravimeter having a single weight or spring such that the sensitivity is proportional to the square of its period.

stable-type gravimeter—A gravimeter which uses a high order of optical and/or mechanical magnification so that a change in position of a weight or associated property is measured directly.

stadia—A graduated rod used in the determination of distance by observing the intercept on the rod subtending a small known angle at the point of observation. In practice, the angle is usually defined by two fixed lines in the reticle of a telescope (transit or telescopic alidade).

The term **stadia** is also used in connection with surveys where distances are determined with a stadia, as stadia survey, stadia method, stadia distance, etc.; also used to designate parts of the instrument used, as stadia wires. Also called **stadia rod.** See also **horizontal stadia.**

stadia circle—See **Beaman arc.**

stadia constant—(leveling) The ratio which is multiplied by the stadia interval to obtain the length of a sight in meters. Also, the ratio by which the sum of the stadia intervals of all sights of a run is converted to the length of the run in kilometers.

stadia diagram—A chart or drawing which provides a means for rapid field reduction of stadia readings. Usually it is prepared on cross-section paper and drawn to the scale of the survey being performed.

stadia intercept—See **stadia interval.**

stadia interval—(leveling) The length of rod subtended between the top and bottom crosswires in the leveling instrument as seen projected against the face of the leveling rod. Also called **stadia intercept.**

stadia rod—See **stadia.**

stadia slide rule—The most rapid method of reducing stadia readings is by the use of a slide rule which has, in addition to the ordinary scale of numbers (logarithms of the distances), two scales especially constructed for stadia work, one consisting of values of log $\cos^2 a$ and the other of log ½ sin 2 *a* for different values of *a*. On some rules, the values of *a* range from 0°34′ to 45°; on others, from 0°03′ to 45°. In some forms the horizontal distance is read directly; in others the horizontal correction $(1 - \cos^2)$ or $\sin^2$ is given. A 10-inch slide rule gives results sufficiently accurate for all ordinary purposes.

stadia traverse—A traverse in which distances are measured by the stadia method.

stadia trigonometric leveling—A technique of extending supplemental vertical control in areas of moderate or low relief. Distances are measured by stadia methods and can be done with planetable, transit, or theodolite. Field work is reduced to usable form by trigonometric computations.

stadimeter—An instrument for determining the distance to an object of known height by measuring the angle subtended at the observer by the object. The instrument is graduated directly in distance. See also **range finder.**

staff gage—The simplest form of tide or stream gage consisting of a graduated staff securely fastened to a pole or other suitable support. It is so designed that a segment of the staff will be below lowest low water when mounted and the remainder will be above water and positioned for direct observations from shore or some other vantage point.

Stampfer level—A type of leveling instrument having the telescope tube so mounted that it could be moved in a vertical plane about a horizontal axis, involving the use of a striding level and a micrometer screw.

standard—(JCS) An exact value, a physical entity, or an abstract concept, established and defined by authority, custom, or common consent to serve as a reference, model, or rule in measuring quantities or qualities, establishing practices or procedures, or evaluating results. A fixed quantity or quality.

standard-accuracy map—A map which complies with the U.S. National Map Accuracy Standards.

standard automatic tide gage—A chronograph used where extended time readings of tidal changes are required. The rise and fall of the tide is communicated by a wire (attached to a float) to a worm screw on the gage, which moves a pen transferring the data to a permanent paper record.

standard-content map—A map that represents natural and manmade features according to current standards and specifications.

standard corner—(USPLS) A senior corner on a standard parallel or base line.

standard deviation—See **standard error.**

standard elevation—An adjusted elevation based on the sea level datum of 1929 (now the National Geodetic Vertical Datum of 1929).

standard error (σ)—The square root of the quantity obtained by dividing the sum of the squared errors by the number of errors minus one. The square root of the variance of the set of observations. Also called **standard deviation.**

standard error of the mean—See **error of the mean.**

standard gravity—A starting value for the combination of the acceleration of gravity and the vertical component of centrifugal force acting on an object on the Earth's surface, as reduced to mean sea level. Potsdam gravity datum was previously used: previous values 981.274 gals; recently changed to 981.260 gals IGSN 1971 (International Gravity Standardization Net 1971).

standard indexing system (SIS)—A system developed for use within the Department of Defense for the indexing of all aerial photography held at national level. Aerial photographic missions are plotted on acetate sheets covering 1° squares of the world at a scale of 1:250,000.

standard meridian—**1.** The meridian used for determining standard time. **2.** A meridian of a map projection, along which the scale is as stated.

standard of length—A physical representation of a linear unit that is approved by competent authority.

standard parallel—**1.** (JCS) A parallel on a map or chart along which the scale is as stated for that map or chart. **2.** (USPLS) An auxiliary governing line established along the astronomic parallel, initiated at a selected township corner on a principal meridian, usually at intervals of 24 miles from the base line, on which standard township, section, and quarter-section corners are established. Also called **correction line.** Standard parallels, or correction lines, are established for the purpose of limiting the convergence of range lines from the south. **3.** A parallel of latitude which is used as a control line in the computation of a map projection.

standard port—See **reference station.**

standard quadrangle—A quadrangle of a specific series, conforming with the systematic pattern of the series.

standard station—See **reference station.**

standard survey—A survey which, in scale, accuracy, and content, satisfies criteria prescribed for such a survey by competent authority.

standard tension—(taping) That tension or pull at which a tape was standardized.

standard time—Mean solar time for a selected meridian adopted for use throughout a belt or zone.

standardization—The comparison of an instrument or device with a standard to determine the value of the instrument or device in terms of an adopted unit.

star chart—A chart or map of the celestial sphere showing principal stars which are useful for observations for navigation or field astronomy.

star finder—A device to facilitate the recognition of stars, particularly for purposes of navigation and geodetic astronomy.

star occultation method—A means of determining the distance between two observers at approximately the same latitude by observations of the times of the eclipsing (or occultation) of a star by the same point on the limb of the Moon.

star trail—A streak-like image of a star recorded on a stellar plate by a photographic time exposure caused by the rotation of the Earth.

starting control—Control available for the absolute orientation of the first plate pair along a line of flight for which control is to be extended.

state base map—A base map of the area of a state as the unit used as a base upon which data of a specialized nature are compiled or overprinted.

state coordinate systems—The plane-rectangular coordinate systems established by the National Geodetic Survey, one for each state in the United States, for use in defining positions of geodetic stations in terms of plane-rectangular (x and y) coordinates. Also called **state system of plane coordinates.**

state system of plane coordinates—See **state coordinate systems.**

static gravity meter—A type of gravity instrument in which a linear or angular displacement is observed or nulled by an opposing force.

static markings—(photogrammetry) Marks on photographic negatives or other sensor imagery caused by unwanted discharges of static electricity.

station—**1.** (surveying) A definite point on the Earth whose location has been determined by surveying methods. It may or may not be marked on the ground. A station usually is defined by the addition of a term which describes its origin or purpose. Usually marked on the ground by a monument of special construction, or by a natural or artificial structure. **2.** (route surveying) Any point whose position is given by its total distance from the starting hub; also, each stake set at 100-foot intervals along a route survey. See also **A-station; air station; astronomic station; auxiliary station; B-station; base station; control station; data-acquisition station; drift station; eccentric station; gravity reference stations; gravity station; ground station; horizontal control station; in-and-out station; intersection station; Laplace station; magnetic station; main-scheme station; master station; oceanographic station; plus station; primary tide station; projector station; radio range station; reference station; resection station; satellite triangulation stations; secondary station; secondary tide station; set-up; slave station; stream-gaging station; subordinate station; subsidiary station; supplementary station; taping station; tide station; tracking station; traverse station; triangulation station.**

station adjustment—The adjustment of angle measurements at a triangulation or traverse station to satisfy local requirements (such as horizon closure) without regard to observations or conditions at other points. Also called **local adjustment.**

station error—See **deflection of the vertical.**

station mark—A mark on the ground, either a monument of special construction, or a natural or artificial object, which pinpoints the location of a survey station. See also **mark,** definition 2.

station pointer—See **three-arm protractor.**

stationary field—Any natural field of force, as a gravimetric or magnetic field.

stationary orbit—An orbit in which the satellite revolves about the primary at the angular rate at which the primary rotates on its axis. From the primary, the satellite appears to be stationary over a point on the primary. See also **synchronous satellite.**

statoscope—A sensitive form of barometer used in aerial photography for measuring small differences in altitude between successive air stations. Usually recorded automatically on the film at the instant of exposure. See also **recording statoscope.**

stellar aberration—The displacement of the observed position from the position where the body was geometrically located at the instant of observation due to the motion of the observing platform.

stellar camera—A camera for photographing the stars.

stellar magnitude—See **magnitude,** definition 1.

stellar map matching—A process during the flight of a vehicle by which a chart of the stars set into the guidance system is automatically matched with the position of the stars observed through telescopes so as to give guidance to the vehicle. See also **map-matching guidance.**

stellar parallax—See **annual parallax.**

stellar plate—A precisely ground glass plate coated with a photographic emulsion used for recording satellite images against a stellar background.

step cast—The negative or positive reproduction of the stepped terrain base of a relief model.

step tablet—See **step wedge.**

step wedge—A strip of film or a glass plate whose transparency diminishes in graduated steps from one end to the other; often used to determine the density of a photograph. Also called **gray scale; step tablet.** See also **continuous tone gray scale.**

Stephenson leveling rod—A speaking rod having graduations forming a diagonal scale, with horizontal lines through the tenth-of-foot marks. This rod is read to hundredths of a foot.

stepped terrain base—In relief model making the acetate three-dimensional representation, in stepped form, of the contours appearing on the base map.

steradian—The unit of measure of a solid angle.

stereo—1. Contracted or short form of stereoscopic. **2.** The orientation of photographs when properly positioned for stereoscopic viewing. Photographs so oriented are said to be "in stereo."

stereo oblique plotter—A device which permits continuous plotting of planimetric detail from oblique photographs. Essentially, the device consists of two photoangulators linked under a stereoscope and is provided with plotting arms.

stereo pair—See **stereoscopic pair.**

stereo triplet—A stereogram composed of three photographs, the center photo having a common field of view with the two adjacent photos, arranged in such a manner as to permit complete stereoscopic viewing of the center photograph.

stereocomparagraph—A relatively simple and mobile stereoscopic instrument used for the preparation of topographic maps from photography. Differences in elevation are determined by measuring parallax difference on a stereoscopic pair.

stereocomparator—A stereoscopic instrument for measuring parallax; usually includes a means of measuring photograph coordinates of image points.

stereocompilation—See **compilation,** definition 2.

stereogram—(JCS) A stereoscopic set (pair) of photographs or drawings correctly oriented and mounted (or projected) for stereoscopic viewing. See also **stereo triplet.**

stereograph—A stereometer with a pencil attachment which is used to plot topographic detail from a properly oriented stereogram.

stereographic chart—A chart on the stereographic projection.

stereographic horizon map projection—A stereographic projection having the center of the projection on some selected parallel of latitude other than the Equator.

stereographic map projection—A perspective, conformal map projection on a tangent plane, with the point of projection at the opposite end of the diameter of the sphere from the point of tangency of the plane. Also called **azimuthal orthomorphic map projection.**

stereographic meridional map projection—A stereographic projection having the center of the projection on the Equator.

stereometer—A measuring device containing a micrometer movement by means of which the separation of two index marks can be changed to measure parallax difference on a stereoscopic pair of photographs. Also called **parallax bar.**

stereometric camera—A combination of two cameras mounted with parallel optical axes on a short rigid base, used in terrestrial photogrammetry for taking photographs in steroscopic pairs.

stereometric map—See **photogrammetric map.**

stereomodel—See **stereoscopic model.**

stereophotogrammetry—Use of stereo images, such as overlapping photographs, in the science of photogrammetry. See also **photogrammetry.**

stereoplanigraph—A precise stereoscopic plotting instrument, especially valuable for extension of control, and capable of handling most types of stereoscopic photography, including terrestrial.

stereoscope—(JCS) A binocular optical instrument for helping an observer to view photographs or diagrams, to obtain a three-dimensional mental impression (stereoscopic model). [The design of stereoscopic viewing instruments utilizes lenses, mirrors, and prisms, or a combination thereof.]

stereoscopic—Of or pertaining to stereoscopy.

stereoscopic base—The distance and direction between complimentary image points on a stereoscopic pair of photographs.

stereoscopic cover—(JCS) Photographs taken with sufficient overlap to permit complete stereoscopic examination.

stereoscopic exaggeration—See **hyperstereoscopy.**

stereoscopic fusion—The mental process which combines two perspective views to give an impression of a three-dimensional model.

stereoscopic image—See **stereoscopic model.**

stereoscopic model—(JCS) The mental impression of an area or object seen as being in three dimensions when viewed stereoscopically on photographs. Also called **spatial model; stereomodel; stereoscopic image.**

stereoscopic pair—(JCS) Two photographs with sufficient overlap of detail to make possible stereoscopic examination of an object or an area common to both. Also called **stereo pair.**

stereoscopic parallax—See **absolute stereoscopic parallax.**

stereoscopic plotting instrument—An instrument for compiling a map or obtaining spatial solutions by observation of stereoscopic models formed by stereoscopic pairs of photographs. See also **double-projection direct-viewing stereoplotter; radial plotter; single-model instrument; stereo oblique plotter; stereocomparagraph; stereoplanigraph.**

stereoscopic principle—The formation of a single, three-dimensional image by binocular vision of two photographic images of the same terrain taken from different exposure stations.

stereoscopic vision—The particular application of binocular vision which enables the observer to obtain the impression of depth, usually by means of two different perspectives of an object (as two photographs taken from different camera stations).

stereoscopy—(JCS) The science which deals with three-dimensional effects and the methods by which they are produced.

stereotemplet—A composite slotted templet adjustable in scale and representative of the horizontal plot of a stereoscopic model. An assembly of stereotemplets provides a means of aerotriangulation for horizontal positions with a stereoscopic plotting instrument not designed for bridging.

stereotemplet triangulation—Aerotriangulation by means of stereotemplets. The method permits scale solutions by area and is not restricted to solutions along flight strips.

stereotopographic map—See **photogrammetric map.**

stereotriangulation—A triangulation procedure that uses a stereoscopic plotting instrument to obtain the successive orientations of the stereoscopic pairs of photographs into a continuous strip. The spatial solution for the extension of horizontal and/or vertical control using these strips (or flight) coordinates may be made by either graphical or computational procedures. Also called **bridging; instrument phototriangulation; multiplex triangulation.** See also **vertical stereotriangulation.**

stick-up—Adhesive-backed or wax-backed film or paper on which map names, symbols, descriptive terms, etc., have been printed, for application in map and chart production.

stilling device—Any device or structure placed in the vicinity of a gage to reduce wave action and afford more accurate reading of the gage.

stipple—A random dot pattern used to depict certain topographic features such as sand.

Stokes' formula—A formula for computing geoid heights from gravity data.

stone bound—A substantial stone post set into the ground with its top approximately flush with the ground surface to mark accurately and permanently the important corners of a land survey.

stop—See **aperture stop.**

stop numbers—See **relative aperture.**

straight line graver—A variation of the rigid tripod graver so designed that the scribing point, the vertical vane, and one supporting leg are all directly in line; used with a straightedge for scribing long, straight lines.

strategic map—(JCS) A map of medium scale, or smaller, used for planning of operations, including the movement, concentration, and supply of troops.

strategic-planning model—Small scale terrain models depicting only the general character of the terrain and features of considerable prominence. They generally embrace continental areas, countries, extensive land mass areas, or principal island masses and are most frequently used in high-echelon planning activities.

stratosphere—(JCS) The layer of the atmosphere above the troposphere in which the change of temperature with height is relatively small. See also **atmosphere.**

stream-gaging station—A point along a stream at which periodic measurements of velocity or discharge are made, and at which daily or continuous records of the stage of height of the water surface above a given datum is obtained.

strength of figure—(triangulation) The comparative precision of computed lengths in a triangulation net as determined by the size of the angles, the number of conditions to be satisfied, and the distribution of base lines and points of fixed position. Strength of figure in triangulation is not based on an absolute scale but rather is an expression of relative strength. Also applicable to the individual geometric figures within a given net.

stretching apparatus—See **tape stretcher.**

striding level—A spirit level so mounted that it can be placed above and parallel with the horizontal axis of a surveying or astronomic instrument, and so supported that it can be used to measure the inclination of the horizontal axis to the plane of the horizon.

strip—See **flight strip.**

strip adjustment—Similar to a block adjustment, but limited to a single strip of photographs.

strip coordinates—The coordinates of any point in a strip, whether on the ground or actually on air station, referred to the origin and axes of the coordinate system of the first overlap.

strip film—A photographic film in which the emulsion membrane can be removed from its temporary base after exposure and processing;

the membrane is then transferred to a new base. Principally used in correction work. Also called **stripping film.**

strip mosaic—A mosaic consisting of one strip of aerial photographs taken on a single flight.

strip plot—(JCS) A portion of a map or overlay on which a number of photographs taken along a flight line is delineated without defining the outlines of individual prints.

strip prediction—A single heading prediction intended to convey the general nature and pattern of radar returns continuously along a specific flightpath.

strip radial plot—See **strip radial triangulation.**

strip radial triangulation—A direct radial triangulation in which the photographs are plotted in flight strips without reference to ground control and the strips are later adjusted together and to the ground control. Also called **strip radial plot.**

strip width—The average dimension, measured normal to the flight line, of a series of neat models in a flight strip. Strip width is generally considered as equal to width between flights.

stripping—The cutting, attachment, and other operations for assembling cut film sections to produce a flat.

stripping film—See **strip film.**

stub traverse—See **spur traverse.**

style sheet—A graphic guide for the format and portrayal of grid and marginal information. Also called **mock-up.**

subaqueous reconnaissance survey—A hydrographic survey which is a rapidly executed preliminary survey of a region to provide advance information to meet immediate military needs. Normally made at small scale, it is usually not controlled by triangulation, and may be little more than a sketch with only a few critical soundings shown.

subaqueous running survey—A hydrographic survey of an exploratory nature along an unknown or hostile coast made from shipboard to determine the general form of the coast and the nature of the area.

subastral point—See **substellar point.**

subdivision survey—A type of land survey in which the legal boundaries of an area are located, and the area is divided into parcels of lots, streets, right-of-way, and other accessories. All necessary corners or dividing lines are marked or monumented.

subgravity—A condition in which the resultant ambient acceleration is between zero and one g.

sublunar point—The geographical position of the Moon; that point on the Earth at which the Moon is in the zenith at a specified time. See also **subsatellite point.**

submarine relief—Variations in elevation of the ocean floor, or their representation by depth curves, tints, or soundings.

subordinate station—**1.** One of the places for which tide or tidal current predictions are determined by applying a correction to the predictions of a reference station. **2.** A tide or tidal current station at which a short series of observations has been made, which are reduced by comparison with simultaneous observations at a reference station.

subsatellite point—The point at which a line from the satellite perpendicular to the ellipsoid intersects the surface of the Earth. See also **sublunar point.**

subsidiary station—A station established to overcome some local obstacle to the progress of a survey, and not to determine position data for the station point. The term **subsidiary station** is usually applied to A-stations of a traverse survey. Subsidiary stations usually are temporary in character and not permanently marked. If serving the additional purpose of supplying control for a local survey, such station may be permanently marked and it is then a supplementary station.

subsolar point—The geographical position of the Sun; that point on the Earth at which the Sun is in the zenith at a specified time. See also **substellar point.**

substellar point—The geographical position of a star; that point on the Earth at which the star is in the zenith at a specified time. Also called **subastral point.** See also **subsolar point.**

substitute center—A point which, because of its ease of identification on overlapping photographs, is used instead of the principal point as a radial center.

subsurface float—A hollow cylinder, with its axis held vertical, at a constant depth by the buoyant effect of an indicating surface float; used to determine current velocities in streams or channels having a relatively uniform depth.

subtense bar—A horizontally held bar of precisely determined length, used to measure distances by observing the angle it subtends at the distance to be measured.

subtense-bar traverse—A traverse method in which course lengths are measured by use of a subtense bar.

subtense-base traverse—A traverse method in which distances are determined by precisely measuring, at one end of the course, the angle subtended by a precisely measured base at the other end of the course and approximately normal to it.

subtense method—A procedure by which distance measurements are obtained by use of a subtense bar.

subtracting tape—A calibrated surveyor's tape with the first foot (or meter) at each end graduated in tenths or hundredths. Also called **cut tape.** See also **adding tape.**

summer solstice—**1.** That point on the ecliptic occupied by the Sun at maximum northerly declination. Also called **first point of Cancer.**

2. That instant at which the Sun reaches the point of maximum northerly declination, about June 21.

Sun—The luminous celestial body at the center of the solar system, around which the planets, planetoids, and comets revolve. It is an average star. See also **apparent sun; dynamical mean sun; fictitious sun.**

Sun-zenith distance—The angle between the zenith and the Sun's disk.

super-wide-angle lens—A lens having an angle of coverage greater than 100°. A lens whose focal length is approximately less than one-half the diagonal of the format. Also called **ultra-wide-angle lens.**

superior conjunction—The conjunction of a planet and the Sun when the Sun is between the Earth and the other planet.

superior planets—The planets with orbits larger than that of the Earth: Mars, Jupiter, Saturn, Uranus, Neptune, and Pluto.

superior transit—See **upper transit.**

supplemental control—Points established by subordinate surveys, to relate aerial photographs used in mapping with the system of geodetic control. The points must be positively photoidentified, that is, the points on the ground must be positively correlated with their images on the photographs.

supplemental control point—A photoimage point for which an elevation or a horizontal position, or both, is to be, or has been determined. See also **control point.**

supplemental elevation—A point whose vertical position has been determined by photogrammetric methods and is intended for use in the orientation of other photographs. Also called **vertical pass point.**

supplemental photography—Noncartographic aerial and terrestrial photography that is used to enhance specific characteristics of mapping and charting photographic products. Primarily, supplemental photography is obtained with a reconnaissance camera using a relatively long focal length to provide greater image detail than is available in photographs obtained with mapping cameras.

supplemental plat—(USPLS) A plat prepared entirely from office records designed to show a revised subdivision of one or more sections without change in the section boundaries and without other modification of the subsisting record.

supplemental position—A point whose horizontal position has been determined by photogrammetric methods and is intended for use in the orientation of other photographs. Also called **horizontal pass point.**

supplemental posts for survey monuments—See **identification posts.**

supplemental station—Those stations established only for supplemental vertical control. They normally are not permanently marked (some are merely photoidentified) and accuracy does not have to be of the same order as the horizontal control to which it is tied. Also called **vertical-angle station.**

supplementary bench mark—See **temporary bench mark.**

supplementary contour—A contour line between intermediate contour lines to increase the topographic expression of an area, usually in areas of extremely low relief. Also called **auxiliary contour.**

supplementary instructions—New information, amendments, or changes to specifications or compilation instructions affecting the production of a specific map or chart, or a series of maps or charts.

supplementary station—An auxiliary survey station, established to increase the number of control stations in a given area, or to place a station in a desired location where it is impracticable or unnecessary to establish a principal station. Supplementary stations are permanently marked, and are established with an accuracy and precision somewhat lower than is required for a principal station, since they do not serve as bases from which extensive surveys are run. Also called **secondary station.**

surface anomalies—Irregularities at the Earth's surface, in the weathering zone, or in near surface beds which interfere with geophysical measurements.

surface chart—See **weather map.**

surface corrections—Corrections of geophysical measurements for surface anomalies and ground elevations.

surface float—A device, specially designed or improved, used in hydrographic surveys to determine surface movement of a stream.

surprint—See **overprint,** definition 1.

survey—1. The act or operation of making measurements for determining the relative positions of points on, above, or beneath the Earth's surface. **2.** The results of such operations. **3.** An organization for making surveys. See also **aerial survey; airborne control system; airborne electronic survey control; area survey; astronomic surveying; boundary survey; cadastral survey; city survey; compass survey; control survey; control survey classification; dependent resurvey; electronic survey; engineering survey; exploratory survey; field inspection; first-order work; geodetic survey; geoelectric survey; geographic survey; geologic survey; gravimetric survey; ground survey; hydrographic survey; independent resurvey; inventory survey; land survey; location survey; magnetic survey; magnetometer survey; metes-and-bounds survey; mine survey; mineral survey; oceanographic survey; photographic survey; photogrammetric survey; plane survey; preliminary survey; reconnaissance survey;**

rectangular surveys; resurvey; route survey; satellite surveying; second-order work; standard survey; subaqueous reconnaissance survey; subaqueous running survey; subdivision survey; third-order work; topographic survey; town-site survey; transit-and-stadia survey; trilinear surveying.

survey coordinates—See **rectangular space coordinates.**

survey net—1. (horizontal control) Arcs of triangulation, sometimes with lines of traverse, connected to form a system of loops or circuits extending over an area. Also called **horizontal control survey net; traverse net; triangulation net.** See also **triangulation system; trilateration net. 2.** (vertical control) Lines of spirit leveling connected to form a system of loops or circuits extending over an area. Also called **control net; framework of control; level net; net.** See also **area triangulation; U.S. control survey nets.**

survey photography—See **mapping photography.**

survey signal—A natural or artificial object or structure whose horizontal and sometimes vertical position is obtained by surveying methods. Signals are given special designations according to the kind of survey in which they are determined, or which they may later serve.

survey tower—A structure designed for rapid construction and removal to raise the survey instrument and observer above obstructions such as trees and buildings to permit a line-of-sight as required in higher order triangulation, trilateration, or traverse. See also **Bilby steel tower.**

surveying accessories—Those surveying devices which assist in making measurements with a surveying instrument.

surveying altimeter—An aneroid barometer with a dial graduated to read feet or meters of altitude, used to determine approximate differences in elevation between points.

surveying camera—See **mapping camera.**

surveying instruments—Those surveying devices with which measurements are made. See also **electronic distance-measuring equipment; leveling instrument; tachymeter; theodolite; transit.**

surveying sextant—A sextant intended primarily for use in hydrographic surveying. Also called **hydrographic sextant.** See also **marine sextant.**

surveyor's arrow—See **pin.**

surveyor's chain—See **Gunter's chain.**

sweep bar—A heavy section of steel rail suspended at a predetermined depth by two vertical cables and towed by a vessel for precise determination of navigation obstructions during a hydrographic survey.

swing—1. The rotation of a photograph in its own plane about its camera axis. **2.** On trimetrogon obliques, the angle between the principal line and the y-axis, or the angle between the isometric parallel and the x-axis. See also **relative swing. 3.** The angle at the principal point of a photo measured clockwise from the positive y-axis to the principal line at the nadir point. **4.** (triangulation) See **eccentric reduction.**

swing offset—The perpendicular distance from a point to a transit line found by holding the zero point of a tape at the given point and swinging the tape in an arc until the minimum (horizontal) distance is obtained.

swing-swing method—A technique for clearing y-parallax during relative orientation by applying identical swing (or y-motion) to both projectors of a pair at the same time. This method has the advantage of affecting y-parallax correction without the use of translational motions.

swivel graver—A scribing instrument with a swivel mechanism that permits changes in direction of scribing.

symbol—A diagram, design, letter, character, or abbreviation placed on maps, charts, and other graphics which by convention, usage, or reference to a legend is understood to stand for or represent a specific characteristic or feature.

symbolization—The method of portraying topographic features onto a manuscript. The symbols used on the manuscript are either a point (dot), a line, or an area (a delimiting line closing upon itself).

synchronous satellite—An Earth satellite moving eastward in an equatorial, circular orbit at an altitude (approximately 35,900 kilometers) such that its period of revolution is exactly equal to (synchronous with) the rotational period of the Earth. Such a satellite will remain fixed over a point on the Earth's Equator. Also called **fixed satellite; 24-hour satellite.** See also **stationary orbit.**

synodic period—The interval of time between any planetary configuration of a celestial body with respect to the Sun, and the next successive same configuration of that body, as from inferior conjunction to inferior conjunction.

synodical month—The average period of revolution of the Moon about the Earth with respect to the Sun, approximately 29½ days. Also called **lunar month; lunation.**

synoptic chart—See **weather map.**

synthetic aperture radar (SAR)—A system in which a synthetically long apparent or effective aperture is constructed by integrating multiple returns from the same ground cell, taking advantage of the Doppler effect to produce a phase history film or tape that may be optically or digitally processed to reproduce an image.

system of astronomic constants—An interrelated group of values constituting a model of the Earth and the motions which together with the theory of celestial mechanics serves for the calculation of ephemerides.

systematic error—An error that occurs with the same sign, and often with a similar magnitude, in a number of consecutive or otherwise related observations. For example, when a base is measured with a wrongly calibrated tape, there will be systematic errors. In addition, random errors will occur. Repetition does little or nothing to reduce the ill effect of systematic errors, which are a most undesirable feature of any set of observations. Much of the care in making observations is directed toward eliminating or correcting systematic errors. Also called **regular error.** See also **accumulative error.**

syzygy—A point of the orbit of a planet or satellite at which it is in conjunction or opposition. The term is used chiefly in connection with the Moon, when it refers to the points occupied by the Moon at new and full phase. See also **equinoctial colure; solstitial colure.**

T

table of meridional parts—A table listing lengths of the meridian from the Equator to the various parallels of latitude increased in the proportion required to show lengths along the parallels equal to the corresponding length along the Equator.

tachymeter (tacheometer or tachometer)—A surveying instrument designed for use in the rapid determination of distance, direction, and difference of elevation from a single observation. There are several forms of these instruments that may be classed as tachymeters: (1) An instrument in which the base line for distance determination is an integral part of the instrument. The term **tachymeter** is usually applied to this group. (2) An instrument equipped with stadia wires or gradienter, the base for distance determination being a graduated rod held at the distant point. See also **auto-reducing tachymeter.**

tachymetry (tachometry)—A surveying method used to quickly determine distance, direction, and relative elevation of a point with respect to the instrument station by a single observation. An example of **tachymetry** in the United States (where the term is less familiar) is the stadia method.

Tactical Commanders' Terrain Analysis (TacCTA)—A special map and graphic intelligence aid consisting of photographs, text, and maps overprinted with intelligence data; usually covering a small geographic area.

tactical map—(JCS) A large-scale map used for tactical and administrative purposes.

Tactical Pilotage Chart (TPC)—A 1:500,000 scale, coordinated series of multicolored charts which are produced in selected areas of interest. Designed to satisfy visual and radar navigation of high speed tactical aircraft operating at low altitude. Also used for detailed preflight planning and mission analysis. This series will replace the Pilotage Chart (PC) series.

tactical-planning model—Medium- or large-scale models providing considerable detailed terrain information; generally used for planning operations of a tactical nature.

tan alt—See **shadow factor.**

tangent—(surveying) **1.** That part of a traverse or alignment included between the point of tangency of one curve and the point of curvature of the next curve. **2.** (USPLS) A great circle line tangent to a parallel of latitude at a township corner. **3.** Sometimes applied to a long straight line of a traverse, especially on a route survey, whether or not the termini of the line are points of curve.

tangent conical map projection—See **conic map projection.**

tangent distance—The distance from the point of intersection (vertex) of a curve to its point of tangency or point of curvature.

tangent plane—A plane that touches a curved surface of double curvature at one and only one point or that touches a curved surface of single curvature along one or more parallel straight lines which are elements of the surface, without intersecting the surface. In geodetic work, a plane tangent to the spheroid at any point is perpendicular to the normal at that point.

tangent plane grid system—(engineer surveying) A grid system in a tangent plane with origin at the point of tangency. Usually the origin is designated 10,000 N and 10,000 E, or some similar amounts, to keep all coordinates positive. This system never extends for any great distance. See also **plane rectangular coordinates.**

tangent to spiral (TS)—The point at the end of a tangent and the beginning of a spiral.

tangential distortion—Linear displacement of image points in a direction normal to radial lines from the center of the field.

tape—(surveying) A ribbon of steel, Invar, specially made cloth, or other suitable material on which graduations are placed for the measurement of lengths or distances. See also **adding tape; base tape; instantaneous-reading tape; Invar tape; Lovar tape; piano-wire tape; subtracting tape.**

tape corrections—Quantities applied to a taped distance to eliminate or reduce errors due to the physical condition of the tape and to the way in which it is used. See also **alignment correction; grade correction; length correction; sag correction; temperature correction,** definition 3; **tension correction.**

tape gage—A device consisting of a tagged or indexed chain, tape, or other line attached to a weight which is lowered to touch the water surface, whereupon the gage height is read on a graduated staff or index. Also called **chain gage.**

tape rod—A rod consisting of a frame with rollers at both ends over which an endless, graduated metal tape moves. It is designed to permit direct readings by the instrument man, eliminating all addition and subtraction functions required by other types of rod readings. Also called **automatic rod.**

tape stretcher—A mechanical device which facilitates holding a tape at a prescribed tension and in a prescribed position. Also called **stretching apparatus.**

tape thermometer—A precision thermometer fitted in a specially designed case to clip on and against a metal tape in order to determine temperature corrections for precision base or traverse tape measurements.

taping—The operation of measuring distances on the ground with a tape or chain. Formerly the words **chaining** and **taping** were used synonymously, but the word **taping** is now preferred for all surveys except those of the public-land system. For the latter, because of historical and legal reasons, the term **chaining** is preferred.

taping arrow—See **pin.**

taping buck—See **taping stool.**

taping pin—See **pin.**

taping station—The stake marking each interval (one tape length) along a traverse from the initial point along road centerlines and similar survey operations. See also **plus station.**

taping stool—A metal stool used for precise taping operations. Stools are portable and provide a stable elevated table on which the positions of the survey tape ends can be accurately marked. Also called **taping buck.**

tare—An abrupt offset in the gravimeter normal reading level.

target—**1.** (JCS) A geographical area, complex, or installation planned for capture or destruction by military forces. **2.** An object which reflects a sufficient amount of a radiated signal to produce an echo signal on detection equipment. **3.** The distinctive marking or instrumentation of a ground point to aid in its identification on a photograph. In photogrammetry, **target** designates a material marking so arranged and placed on the ground as to form a distinctive pattern over a geodetic or other control-point marker, on a property corner or line, or at the position of an identifying point above an underground facility or feature. A **target** is also the image pattern on aerial photographs of the actual mark placed on the ground prior to photography. See also **area target; pinpoint target.**

target acquisition—**1.** (JCS) The detection, identification, and location of a target in sufficient detail to permit the effective employment of weapons. See also **target analysis.** **2.** The process of optically, manually, mechanically, or electronically orienting a tracking system in direction and range to lock on a target.

target analysis—(JCS) An examination of potential targets to determine military importance, priority of attack, and weapons required to obtain a desired level of damage or casualities.

target area survey base—(JCS) A base line used for the locating of targets or other points by the intersection of observations from two stations located at opposite ends of the line.

target complex—(JCS) A geographically integrated series of target concentrations.

target concentration—(JCS) A grouping of geographically proximate targets.

target dossiers—(JCS) Files of assembled target intelligence about a specific geographic area.

target folders—(JCS) The folders containing target intelligence and related materials prepared for planning and executing action against a specific target.

target intelligence—(JCS) Intelligence which portrays and locates the components of a target or target complex and indicates its vulnerability and relative importance.

target leveling rod—A type of leveling rod, carrying a target, which is moved into position according to signals given by the instrument man; when the target is bisected by the line of collimation of the instrument, it is read and recorded by the rodman. See also **double-target leveling rod; single-target leveling rod.**

target materials—(JCS) Graphic, textual, tabular, or other presentations of target intelligence, primarily designed to support operations against designated targets by one or more weapon systems. Target materials are suitable for training, planning, executing, and evaluating such operations. Also called **target material graphics.** See also **Air Target Chart; Air Target Materials Program; Air Target Mosaic.**

target material graphics—See **target materials.**

Target Materials Program (TMP)—The coordinated plan of collecting, evaluating, correlating, and reproducing intelligence data in the form of specified target materials.

target positioning data—The accurate horizontal and vertical values which define the location of a target or point. See also **precise installation position; precise radar significant location.**

target system—(JCS) All the targets situated in a particular geographic area and functionally related. See also **target complex.**

target system component—(JCS) A set of targets belonging to one or more groups of industries and basic utilities required to produce component parts of an end product such as periscopes, or one type of a series of interrelated commodities, such as aviation gasoline.

taut-wire apparatus—A 100-meter stranded sounding wire, graduated at 25-meter intervals, used to measure the distances between offshore control buoys during a hydrographic survey.

telemeter—(surveying) An instrument for determining the distance from one point to another. Some such instruments employ a telescope and measure the angle subtended by a short base of known length. See also **electronic telemeter; telemetry.**

telemetry—The science of measuring a quantity or quantities, transmitting the measured value to a distant station, and there interpreting, indicating, or recording the quantities measured.

telescope—An optical instrument used as an aid in viewing or photographing distant objects, particularly celestial objects. See also **achromatic telescope; erecting telescope; inverting telescope; meridian telescope; zenith telescope.**

telescope level—A spirit level attached to a telescope, with its axis parallel to the telescope axis.

telescopic alidade—A usual designation for an instrument composed of a telescope mounted on a straightedge ruler, and used with a planetable in topographic surveying.

telescoping—See **transit,** definition 3.

telluroid—A surface near the terrain being the locus of points in which the spheropotential is the same as the geopotential of corresponding points on the terrain. Its distance from the spheroid is the normal height.

Tellurometer—A trade name for a microwave distance-measuring system in which the velocity of a radio wave is used to determine the distance between two instruments operating alternately as master station (interrogator) and remote station (responder).

temperature correction—1. (leveling) That correction which is applied to an observed difference of elevation to correct for the error introduced when the temperature at which the leveling rods are used in the field is different from the temperature at which they were standardized. **2.** (pendulum) The quantity that is applied to the period of vibration of a pendulum to allow for the difference in the length of the pendulum at the temperature of observation and its length at some other temperature which has been adopted for purposes of standardization or for combining or comparing corresponding values. **3.** (taping) The quantity applied to the nominal length of a tape to allow for a change in its effective length due to its being used at a temperature other than that for which its standard length is given.

templet (template)—1. A pattern or guide, usually constructed of paper, plastic, or metal, used to shape, delimit, or locate an area. **2.** A device used in radial triangulation to represent the aerial photograph; the templet provides a record of the directions of radials taken from the photograph. See also **calibration templet; double-model stereotemplet; hand templet; Hayford deflection templets; Hayford gravity templets; mechanical templet; slotted templet; spider templet; stereotemplet.**

templet cutter—A mechanical device for punching center holes and slots in templets. The slots are centered on points transferred from aerial photographs and are radial to the center hole. Also called **secator; slot cutter; radial secator.**

templet laydown—The process of assembling individual slotted templets into a radial control net.

templet method—Any of the various methods utilized in graphical radial triangulation.

templet ratiograph—(photogrammetry) A device for determining the ratio in decimals between two distances. One distance is that between the principal point and another designated point on the aerial photograph. The other is the corresponding distance between the principal point on a templet and the marked center of the stud for the designated point upon completion of the templet laydown. The ratiograph is designed for a specific templet cutter. See also **ratiometer.**

temporary bench mark (TBM)—A bench mark at a junction of sections of a line of levels, at which no permanent bench mark is established. Also called **nonmonumented bench mark; supplementary bench mark.**

tension correction—(taping) The correction applied to the nominal length of a tape to allow for a change in effective length due to its being used at a tension other than that for which its standard length is known.

terrain—An area of ground considered as to its extent and topography.

terrain contour matching (TERCOM)— An electronic technique for comparing terrain elevations measured in real time by radar and barometric altimeters with stored digital terrain elevation data. It is based on the premise that a geographic area can be uniquely defined by its terrain elevation contours. The technique is used to provide in-flight positioning information to correct or check inertial air navigation or guidance systems. See also **terrain correlation.**

terrain correction—A positive correction used in conjunction with other corrections in making gravity reductions. It takes into account actual deviations from level terrain in the area surrounding a station by removing masses above the horizon and filling in mass-deficiencies below. Also called **topographic correction.**

terrain correlation—A process used by a vehicle's guidance system in evaluating the elevations of the terrain it is flying over and comparing it with prestored digital terrain elevation data. See also **terrain contour matching.**

terrain emboss—A model-making technique for portraying relief on a chart. A photographic process is used to produce the shaded relief effect from an embossed model.

terrain following—The flight mode by which a vehicle maintains a specified altitude above the Earth's surface.

terrain intelligence—(JCS) Processed information on the military significance of natural and manmade characteristics of an area.

terrain model—A three-dimensional graphic representation of an area, showing the conformation of the ground, modeled to scale and usually handpainted to depict realistically manmade and natural physical features. The vertical scale is usually exaggerated, without severe distortion, to accentuate the aspect of relief.

terrain profile photography—Cartographic photography obtained simultaneously with positional camera photography and recording of data relating to profile elevation information of the terrain along or near the ground track of the aircraft. The terrain profile recorder is normally used as the measuring device.

Terrain Profile Recorder (TPR)—An electronic instrument that emits a pulsed-type radar signal from an aircraft to the Earth's surface, measuring vertical distances in order to obtain a profile beneath the track of the aircraft. Also called **Airborne Profile Recorder.** See also **laser terrain profile recorder.**

terrain profiling—Obtaining an elevation profile of the Earth's terrain along or near the ground track of the aircraft by use of a Terrain Profile Recorder.

terrain study—(JCS) An analysis and interpretation of natural and manmade features of an area, their effects on military operations, and the effect of weather and climate on these features.

terrestrial camera—A camera designed for use on the ground. Also called **ground camera.**

terrestrial coordinates—See **geographic coordinates.**

terrestrial equator—See **astronomic equator; geodetic equator.**

terrestrial globe—A sphere, on the outer surface of which, by means of symbols and reference lines, the features of the surface of the Earth are shown in relative positions.

terrestrial latitude—Latitude on the Earth; angular distance from the Equator.

terrestrial longitude—Longitude on the Earth; the arc of a parallel, or the angle at the pole, between the prime meridian and the meridian of a point on the Earth.

terrestrial magnetism—See **geomagnetism.**

terrestrial meridian—See **astronomic meridian.**

terrestrial perturbations—The largest gravitational perturbations of artificial satellites which are caused by the fact that the gravity field of the Earth is not spherically symmetrical.

terrestrial photogrammetry—Photogrammetry utilizing terrestrial photographs. Also called **ground photogrammetry.**

terrestrial photograph—A photograph taken by a camera located on the ground. Also called **ground photograph.**

terrestrial planet—A planet that approximates the Earth in size (Mercury, Venus, Mars, and Pluto) and physical makeup.

terrestrial pole—See **geographical pole.**

terrestrial refraction—The refraction by the Earth's atmosphere of light from a terrestrial source. The path of light from a terrestrial source is usually not far from horizontal; it passes through only the lower strata of the atmosphere and suffers refraction throughout its entire length. See also **atmospheric refraction; horizontal refraction; lateral refraction.**

terrestrial sphere—The Earth.

terrestrial triangle—A triangle on the surface of the Earth, especially the navigational triangle.

tesla—(geomagnetism) The electromagnetic unit of magnetic induction. 1 tesla = 1 weber/m^2 = 10^{-9} nanotesla. See also **gauss.**

tesseral harmonics—The set of all spherical harmonics that are functions of both latitude and longitude. Sectorial harmonics are a special subset of tesseral harmonics.

test chart—See **resolving power target.**

texture—In a photo image, the frequency of change and arrangement of tone.

thematic map—See **topical map.**

theodolite—A precision surveying instrument consisting of an alidade with a telescope. It is mounted on an accurately graduated circle and is equipped with necessary levels and reading devices. Sometimes, the alidade carries a graduated vertical circle. See also **cine-theodolite; direction instrument theodolite; gyro theodolite; phototheodolite; repeating theodolite.**

theodolite-magnetometer—An instrument used in magnetic surveys consisting of a theodolite and a magnetometer modified to fit into a common base, which permits the determination of the true meridian and the magnetic meridian in a single observation.

theoretical corner—A term adopted by the U.S. Geological Survey to designate the corners on the map for which no marks are identified on the ground. The locations are determined by adjustment and are indicated on the map only by the intersection of the subdivision lines.

theoretical error—A systematic error arising from natural physical conditions, beyond the control of the observer. See also **external error.**

theoretical gravity—The value of gravity calculated for a particular latitude according to an accepted formula. See also **formula for theoretical gravity.**

theory of anharmonic ratio—A theory principally concerned with the processes of transformation and rectification whereby projectively related figures possess certain metric characteristics which are invariant under projection. Also called **theory of cross ratio.**

theory of cross ratio—See **theory of anharmonic ratio.**

thermal imagery (infrared)—(JCS) Imagery produced by measuring and recording electronically the thermal radiation of objects.

thermometric leveling—The determination of elevations above sea level from observed values of the boiling point of water. A type of indirect leveling.

thick lens—A term used in geometrical optics to indicate that the thickness of a lens is considered and that all distances are being measured from the nodal points instead of the lens center.

thin lens—A term used in geometrical optics to indicate that the thickness of a lens is ignored and that all distances are measured from the lens center; used for approximate computations.

third-order leveling—Spirit leveling which does not attain the quality of second-order leveling, but does conform to the current specifications for third-order leveling per "Classification Standards of Accuracy and General Specifications of Geodetic Control Surveys." Recommended for most general vertical control purposes within a limited area.

third-order traverse—A survey traverse which extends between adjusted positions of other control surveys which conform to the current specifications for third-order (class I or class II) triangulation per "Classification Standards of Accuracy and General Specifications of Geodetic Control Surveys." Recommended for most general horizontal control purposes within a limited area.

third-order triangulation—Formerly known as tertiary triangulation, these surveys conform to current specifications for third-order (class I or class II) triangulation per "Classification Standards of Accuracy and General Specifications of Geodetic Control Surveys." Recommended for most general horizontal control purposes within a limited area.

third-order work—This is the lowest order of control surveys for which monumentation is authorized.

three-arm protractor—A full-circle protractor, equipped with three arms, the fiducial edges (extended) of which pass through the center of the circle. The middle arm is fixed and reads 0° on the graduated circle. The other arms are movable, and their positions on the circle are read with the aid of verniers. The two movable arms are equipped with clamps and may be set at any angle with respect to the fixed arm, within the limits of the instrument. It is used for finding a (ship's) position graphically when the angles between three known fixed points are available. Also called **station pointer.**

three-body problem—That problem in classical celestial mechanics which treats the motion of a small body, usually of negligible mass, relative to and under the gravitational influence of two other finite point masses.

three-point method—See **resection.**

three-point problem—The determination of the horizontal position of a point of observation from data comprising two observed horizontal angles between three objects of known position. The problem is solved graphically by the use of a three-arm protractor, and analytically by trigonometrical computation. See also **resection; triangle-of-error method.**

three sigma—The 99.73 percent confidence level of a normal distribution. Three sigma is three times the value of 1 sigma. See also **standard error.**

three-wire leveling—A method of leveling applied when the reticle of the level has three lines. The rod is read at each of the three lines and the average is used for the final result with an accuracy as great as if three lines of levels had been run and the results averaged.

ticks—See **register marks.**

tidal bench mark—A bench mark set to reference a tide staff at a tidal station and the elevation of which is determined with relation to the local tidal datum.

tidal constituent—See **constituent.**

tidal correction—A correction applied to gravitational observations to remove the effect of Earth tides on gravimetric observations.

tidal current—The alternating horizontal movement of water associated with the rise and fall of the tide caused by the astronomic tide-producing forces.

tidal current chart—A chart showing, by arrows and numbers, the average direction and speed of tidal currents at a particular part of the current cycle. A number of such charts, one for each hour of the current cycle, usually are published together.

tidal datum—Specific tide levels which are used as surfaces of reference for depth measurements in the sea and as a base for the determination of elevation on land. Many different datums have been used, particularly for leveling operations. Also called **tidal datum plane.**

tidal datum plane—See **tidal datum.**

tidal day—See **lunar day.**

tidal variation of gravity—Periodic deviations from normal of the gravity on Earth and the direction of the plumb line caused by the attraction of the Moon and the Sun's mass.

tide—The periodic rise and fall of the surface of the ocean resulting from the gravitational attraction of the Moon and Sun acting upon the rotating Earth. See also **age of diurnal inequality; age of parallax inequality; age of phase inequality; amphidromic point; amphidromic region; annual inequality; anomalistic tide cycle; apogean tides; constituent; constituent day; corrected establishment; cotidal hour; degenerate amphidromic system; diurnal constituent; diurnal inequality; ebb tide; establishment of the port; flood tide; harmonic constants; height of the tide; high water; high water line; higher high water; higher high water interval; higher low water; higher low water interval; Indian spring low water; Indian tide plane; international low water; low water; low water full and change; low water line; lower high water; lower high water interval; lower low water; lower low water interval; lowest low water; lowest low water springs; lunar day; lunar tide; lunitidal interval; mean diurnal high water inequality; mean diurnal low water inequality; mean high water; mean high water springs; mean higher high water; mean higher high water springs; mean low water; mean low water springs; mean lower low water; mean lower low water springs; mean range; mean**

river level; mean sea level; mean tide level; nodal line; parallax inequality; phase inequality; set; semidiurnal constituent; tidal correction; tidal current; tidemark.

tide gage—A device for measuring the height of tide. It may be simply a graduated staff in a sheltered location where visual observations can be made at any desired time; or it may consist of an elaborate recording instrument making a continuous graphic record of tide height against time. Such an instrument is usually actuated by a float in a pipe communicating with the sea through a small hole which filters out shorter waves. See also **float gage; nonrecording gage; portable automatic tide gage; pressure gage; self-registering gage; staff gage; standard automatic tide gage.**

tide level—See **mean tide level.**

tide-over run—A reprint of a chart or map necessitated by unusual conditions before extensive revisions can be accomplished. Also called **emergency run.**

tide-producing force(s)—The slight local difference between the gravitational attraction of two astronomic bodies and the centrifugal force that holds them apart. These forces are exactly equal and opposite at the center of gravity of either of the bodies, but, since gravitational attraction is inversely proportional to the square of the distance, it varies from point to point on the surface of the bodies. Therefore, gravitational attraction predominates at the surface point nearest to the other body, while centrifugal repulsion predominates at the surface point farthest from the other body. Hence, there are two regions where tide-producing forces are at a maximum, and normally there are two tides each lunar day and solar day.

tide station—A place at which tide observations are made. See also **primary tide station; secondary tide station.**

tidemark—**1.** A high water mark left by tidal water. **2.** The highest point reached by a high tide. **3.** A mark placed to indicate the highest point reached by a high tide, or occasionally, any specified state of tide.

tie—A survey connection from a point of known position to a point whose position is desired. A tie is made to determine the position of a supplementary point whose position is desired for mapping or reference purposes, or to close a survey on a previously determined point. To "tie-in" is to make such a connection. See also **tie point,** definition 2.

tie flight—See **control strip.**

tie-in—See **tie.**

tie point—**1.** Image points identified on oblique photographs in the overlap area between two or more adjacent strips of photography. They serve to tie the individual sets of photographs into a single flight unit and to tie adjacent flights into a common network. **2.** Point of closure of a survey either on itself or on another survey.

tie strip—**1.** (cartography) An overlay containing all planimetric and relief features in the area along the edge of a map or chart. It is used to insure the matching of these features on adjoining sheets. Also called **match strip. 2.** (aerial photography) See **control strip.**

tier—(USPLS) Any series of contiguous townships situated east and west of each other; also sections similarly situated within a township.

tilt—The angle at the perspective center between the photograph perpendicular and the plumbline, or other exterior reference direction; also, the dihedral angle between the plane of the photograph and the horizontal plane. Also called **angle of tilt.** See also **cross tilt; direction of tilt; pitch; relative tilt; roll; x-tilt.**

tilt angle—See **tilt.**

tilt circle—In a tilted aerial photograph, a circle passing through the isocenter and having a diameter lying along the principal line. When this diameter is drawn to a convenient linear scale, then any chord through the isocenter gives the component of tilt for that particular direction.

tilt displacement—Displacement radial from the isocenter of the photograph caused by the tilt of the photograph.

tilt slide rule—A device which facilitates the determination for settings on a fixed-lens rectifier when certain tilt factors of an aerial photograph are known.

tilting-lens rectifier—A class of rectifiers in which the principal point is fixed on its axis of swing, and cannot be displaced.

tilting level—A leveling instrument in which the telescope with its attached bubble tube can be leveled by a fine screw at the eyepiece end of the telescope independently of the vertical axis, thus avoiding the need for careful leveling of the instrument as a whole. This type of level was first designed for precise work, but the principle has come into popular use for ordinary levels. In many cases the tilting screw is a micrometer.

time—The measurable aspect of duration. See also **A1 time; apparent sidereal time; apparent solar time; astrograph mean time; astronomic time; atomic time; civil time; day; ephemeris time; equation of time; Greenwich apparent time; Greenwich lunar time; Greenwich mean time; Greenwich sidereal time; Greenwich time; international atomic time; local apparent time; local astronomic time; local lunar time; local mean time; local sidereal time; local time; lunar time; mean sidereal time; mean solar time; month; rise time; sidereal time; solar time; standard time; universal time; universal time coordinate; UT0 time; UT1 time; UT2 time; WWV time; year; zone time.**

time diagram—A diagram in which the celestial equator appears as a circle, and celestial meridians and hour circles as radial lines; used to facilitate solution of time problems and others involving arcs of the celestial equator or angles at the pole, by indicating relations between various quantities involved. Conventionally, the relationships are given as viewed from a point over the South Pole, westward direction being counterclockwise. Also called **diagram on the plane of the equinoctial.** See also **diagram on the plane of the celestial meridian.**

time distance—Time required for any object to travel between two given points at a given rate of speed.

time-gamma curve—See **characteristic curve.**

time meridian—Any meridian used as a reference for reckoning time, particularly a zone or standard meridian.

time zone—An area in all parts of which the same time is kept. In general, each zone is 15° of longitude in width, centered on a meridian whose longitude is exactly divisible by 15°.

time zone chart—A small-scale chart of the world designed to show the legal time kept on land.

times (X) enlargement—The multiplication factor by which an original is to be enlarged in reproduction. A two-times (2X) enlargement of a 4- by 5-inch original would be 8 by 10 inches. See also **diameter enlargement; scale of reproduction.**

timing correction—A correction applied to the length of a trilateration measurement to compensate for the delay of the radar signal as it passes through the ground transponder unit of an electronic distance-measuring device.

tints—Color gradations used on maps to designate depth or height. See also **hypsometric tinting.**

tip—See **pitch.**

tipped panoramic distortion—In a panoramic camera system, the displacement of images of ground points from their expected vertical panoramic positions caused by the tipping of the scan axis within the vertical plane of the flightpath. This distortion is additive and modifies again the image positions of points already influenced by panoramic distortion, scan positional distortion, and image motion compensation distortion.

title block—A space on a nonstandard graphic, such as a mosaic, photograph, or plan devoted to identification, reference, and scale information.

titling (title information)—That information lettered on aerial photographic negatives for identification purposes. Also, the placing of such information on the negatives. Also called **film titling; negative titling.**

Tokyo datum—This datum has its origin in Tokyo. It is defined in terms of the Bessel ellipsoid and oriented by means of a single astronomic station. By means of triangulation ties through Korea, the Japanese datum is connected with the Manchurian datum. Unfortunately, Tokyo is situated on a steep geoid slope and the single-station orientation has resulted in large systematic geoid separations as the system is extended from its initial point. See also **preferred datum.**

tolerance—The maximum allowable variation from a standard or from specified conditions.

tone—(JCS) Each distinguishable shade variation from black to white on imagery.

tone copy—That material in which tones or shades of solid color appear.

topical map—A map designed to portray a special subject; e.g., administrative subdivisions, railroads, telecommunications, power lines, navigable waterways. Also called **special subject map; thematic map.**

topoangulator—An instrument used to measure vertical angles in the principal plane of an oblique photograph.

topocentric—Of measurements or coordinates, referred to the position of the observer on the Earth as the origin.

topocentric coordinates—Coordinates whose origin is on the Earth's surface as distinguished from geocentric coordinates whose origin is at the center of the Earth.

topocentric equatorial coordinates—A coordinate system centered at the observer's position on the surface of the Earth with one coordinate plane parallel to the Equator and one axis parallel to the north polar axis of the Earth.

topocentric horizon—See **apparent horizon.**

topographic base film—An aerial photographic film with a dimensionally stable base used primarily for mapping.

topographic correction—See **terrain correction.**

topographic deflection—That part of the deflection of the plumb line which is caused by the gravitational pull exerted by topographic masses. Topographic deflection is not the same as deflection of the plumb line or station error, but is the theoretical effect produced by the resultant gravitational pull of the unevenly distributed topographic masses around the station, no allowance being made for isostatic compensation. Also called **indirect effect on the deflections.**

topographic expression—The effect achieved by shaping and spacing contour lines so that topographic features can be interpreted with ease and fidelity. Good expression is achieved by delineating the contours in appropriate relationship to each other, with due consideration given to the scale and contour interval of the map. Also called **configuration of terrain.** See also **topography.**

topographic feature—See **topography,** definition 1.

topographic map—(JCS) A map which presents the vertical position of features in measurable form as well as their horizontal positions.

Topographic Map of the United States—The recommended designation for the topographic map of the United States being prepared of quadrangle areas in atlas sheet form, chiefly by the U.S. Geological Survey. This map portrays all basic information about location, elevation, and extent of physical and cultural features that are required for preliminary economic and engineering studies, and for incorporation in a base for maps prepared for special purposes.

topographic plot—Representation, by means of contour lines, of the ground relief of an area, shown in a stereoscopic model. See also **compilation,** definition 2.

topographic survey—A survey which has for its major purposes the determination of the relief of the surface of the Earth and the location of natural and manmade features thereon.

topographical latitude—See **geodetic latitude.**

topography—**1.** The configuration of the surface of the Earth, including its relief, the position of its streams, roads, cities, etc. The Earth's natural and physical features collectively. A single feature such as a mountain or valley is termed a topographic feature. Topography is subdivided into hypsography (the relief features), hydrography (the water and drainage features), culture (manmade features), and vegetation. **2.** The science of delineation of natural and manmade features of a place or region especially in a way to show their positions and elevations. The term includes the scientific and technical fields of surveying, geodesy, geophysics, military geography, photogrammetry, cartography, graphic arts, and related activities to the extent that they are essential to the accomplishment of the military mapping, geodesy, and military geographic intelligence mission. **3.** In oceanography, the term is applied to a surface such as the sea bottom or a surface of given characteristics within the water mass.

toponym—A name applied to a physical or cultural topographic feature. For U.S. Government usage, policies and decisions governing place names on Earth are established by the United States Board on Geographic Names. Also called **place name.** See also **descriptive name.**

toponymy—**1.** The study and treatment of toponyms. **2.** A body of toponyms.

topple—The vertical component of precession or wander, or the algebraic sum of the two.

topple axis—That horizontal axis, perpendicular to the (horizontal) spin axis of a gyroscope, around which topple occurs.

torsion balance—A device for measuring combinations of the second derivatives of the gravity potential, which are closely related to the horizontal components of the deflection of the vertical. It consists of a bar suspended horizontally by an elastic filament, one end of the bar being subjected to the influence of the attracting force to a greater degree than the other end. The attracting force is balanced and its comparative strength measured by the torsional reaction of the filament.

total departures—See **abscissa.**

total drift—The algebraic sum of drift due to precession and that due to wander. Also called **drift.**

total latitudes—See **ordinates.**

total magnetic intensity—The vector resultant of the intensity of the horizontal and vertical components of the Earth's magnetic field at a specified point.

touch plate—See **kiss plate.**

township—(USPLS) The unit of survey of the public lands; normally a quadrangle approximately 6 miles on a side with boundaries conforming to meridians and parallels within established limits, containing thirty-six sections, some of which are designed to correct for the convergence of meridians or range lines. See also **fractional township.**

township corner—(USPLS) A corner of a township. See also **closing township corner.**

township lines—(USPLS) The township boundaries that run north and south are termed **range lines;** with few exceptions the range lines are run on cardinal and have been intended to be on cardinal. The boundaries running east and west are termed **township lines.** By law, they were intended to be on true parallels of latitude.

town-plan inset—See **inset.**

town-site survey—The marking of lines and corners within one or more regular units of the township subdivision by which the land is divided into blocks, streets, and alleys as a basis for the disposal of title in parcels of land.

trace—See **selection overlay.**

track—(JCS) The actual path of an aircraft above, or a ship on the surface of the Earth. The **course** is the path which is planned; the **track** is the path which is actually taken.

track adjustment—Adjustment to a ship's track resulting from set and drift of the vessel.

track chart—A chart showing recommended, required, or established tracks, and usually indicating turning points, courses, and distances.

tracking camera—See **ballistic camera.**

tracking station—A ground-based complex set up to track an object moving through the atmosphere or space, by visual, photographic, photoelectric, or electronic methods.

traffic-circulation map—(JCS) A map showing traffic routes and the measures for traffic regulation. It indicates the roads for use of certain classes of traffic, the location of traffic control stations, and the directions in which traffic may move. Also called **circulation map.**

traffic separation schemes—Portrayal on nautical charts of schemes aimed at reducing the risk of collision in congested and/or converging areas by separating traffic moving in opposite, or nearly opposite, directions.

trajectory—In general, the curve that a body describes in space. An orbit is a trajectory which does not intersect the Earth.

transcriber—See **point-transfer device.**

transcription—**1.** The process of recording the sounds and/or grammatical elements of a language in terms of a specific writing system. **2.** An item of a language which has undergone this process.

transducer—Any device for converting energy from one form to another (electrical, mechanical, or acoustic). In sonar, it usually combines the functions of a hydrophone and a projector.

transformation—**1.** (photogrammetry) The process of projecting a photograph (mathematically, graphically, or photographically) from its plane onto another plane by translation, rotation, and/or scale change. The projection is made onto a plane determined by the angular relations of the camera axes and not necessarily onto a horizontal plane. See also **rectification.** **2.** (surveying) The computational process of converting a position from UTM or other grid coordinates to geodetic, and vice versa; from one datum and ellipsoid to another using datum shift constants and ellipsoid parameters. The survey position of a point is frequently given in several different grids or ellipsoids; local datum and Doppler-derived WGS 72 are common requirements.

transformed print—A photographic print made by projection in a transforming printer.

transforming printer—A specially designed projection printer of fixed geometry used for transforming the oblique components of a coupled camera installation, a multiple-lens camera, or a panoramic camera onto a plane perpendicular to the axis of the system. See also **rectifier; universal transforming printer.**

transit—**1.** The apparent passage of a star or other celestial body across a defined line of the celestial sphere, as a meridian, prime vertical, or almucantar. The apparent passage of a star or other celestial body across a line in the reticle of a telescope, or some line of sight. The apparent passage of a smaller celestial body across the disk of a larger celestial body. The transit of a star across the meridian occurs at the moment of its culmination, and the two terms are sometimes used as having identical meanings; such usage is not correct, even where the instrument is in perfect adjustment. At the poles, a star may have no culmination but it will transit the meridians. See also **culmination; lower transit; meridian transit; upper transit.** **2.** A surveying instrument composed of a horizontal circle graduated in circular measure and an alidade with a telescope which can be reversed in its supports without being lifted therefrom. See also **theodolite.** **3.** The act of reversing the direction of a telescope by rotation around its horizontal axis. Also called **plunge; inverting; telescoping.** **4.** An astronomic instrument having a telescope which can be so adjusted in position that the line of sight may be made to define a vertical circle. Also called **astronomic transit.** See also **broken-telescope transit; solar transit.**

transit-and-stadia survey—A survey in which horizontal and vertical directions or angles are observed with a transit and distances are measured by transit and stadia.

transit instrument—See **transit,** definition 4.

transit line—Any line of a traverse which is projected, either with or without measurement, by the use of a transit or other device. It is not necessarily an actual line of final survey but may be an accessory line. Also called **traverse line.**

transit micrometer—A form of registering micrometer with its movable wire placed in the focal plane of an astronomic transit and at right angles to the direction of motion of the image of a star which is observed at or near culmination. Also called **impersonal micrometer,** because it almost completely eliminates the effect of the personal equation on time observations made with it.

transit micrometer contact correction—A quantity applied to the chronograph record of a star transit observed with the aid of a transit micrometer to allow for the time required for the contact spring to cross one-half of the width of a contact strip in the head of the micrometer.

transit rule—A method of balancing a survey. Corrections corresponding to the closing errors in latitude and departure are distributed according to the proportion: latitude and departure of each line of the traverse to the arithmetical sums of the latitudes and departures of the entire traverse. The transit rule is used when it is assumed that the closing errors are due less to the errors in the observed angles than to errors in the measured distances.

transit traverse—A survey traverse in which the angles are measured with an engineer's transit or theodolite and the lengths with a metal tape. A transit traverse is usually executed for the control of local surveys and is of second-order or third-order quality.

transition curve—See **spiral curve.**

translation—**1.** The process of rendering oral or written text of one language in terms of text of corresponding meaning of another language. See also **transcription; transliteration; romanization.** **2.** Movement in a straight line without rotation.

translational movement—The systematic movement of projector assemblies in line-of-flight directions in a stereoplotting instrument.

transliteration—**1.** The process of recording the graphic symbols of one writing system in terms of corresponding graphic symbols of a second writing system. **2.** An item of a language which has undergone this process.

translocation—The determination of the relative position between two points from simultaneous Doppler satellite observations.

translunar—Outside the Moon's orbit about the Earth. See also **cislunar.**

translunar space—As seen from the Earth at any moment, space lying beyond the orbit of the Moon.

translunar trajectory—A trajectory extending outside the Moon's orbit about the Earth.

transmission—(optics) The ratio of transmitted light to the incident light. If 100 units of light fall upon a translucent material and 10 of them succeed in passing through, then it can be said that the material has 1/10 or 10 percent transmission.

transparency—(JCS) An image fixed on a clear base by means of a photographic, printing, chemical, or other process, especially adaptable for viewing by transmitted light.

transponder—(JCS) A transmitter-receiver capable of accepting the electronic challenge of an interrogator and automatically transmitting an appropriate reply.

transverse—In cartography, pertaining to or measured on a map projection in which a meridian is used as a fictitious equator. Also called **inverse.**

transverse axis—The distance between the apsides. It is identical to the semimajor axis for elliptical orbits.

transverse chart—A chart on a transverse projection. Also called **inverse chart.**

transverse cylindrical orthomorphic chart (TCOC)—See **transverse Mercator chart.**

transverse cylindrical orthomorphic map projection—See **transverse Mercator map projection.**

transverse equator—The plane which is perpendicular to the axis of a transverse projection. Also called **inverse equator.**

transverse graticule—A fictitious graticule based upon a transverse projection. See also **fictitious graticule.**

transverse latitude—Angular distance from a transverse equator. Also called **inverse latitude.** See also **fictitious latitude.**

transverse longitude—Angular distance between a prime transverse meridian and any given transverse meridian. Also called **inverse longitude.** See also **fictitious longitude.**

transverse map projection—A map projection in which the projection axis is rotated 90° in azimuth.

transverse Mercator chart—A chart on the transverse Mercator projection. Also called **inverse cylindrical orthomorphic chart; inverse Mercator chart; transverse cylindrical orthomorphic chart.**

transverse Mercator grid—An informal designation for a state coordinate system based on a transverse Mercator map projection. Also called **Gauss-Kruger grid.**

transverse Mercator map projection—A conformal cylindrical map projection, being in principle equivalent to the regular Mercator map projection turned (transversed) 90° in azimuth. In this projection, the central meridian is represented by a straight line, corresponding to the line which represents the Equator on the regular Mercator map projection. Neither the geographic meridians (except the central meridian) nor the geodetic parallels (except the Equator) are represented by straight lines. Also called **inverse cylindrical orthomorphic map projection; inverse Mercator map projection; transverse cylindrical orthomorphic map projection.**

transverse meridian—A great circle perpendicular to a transverse equator. The reference transverse meridian is called **prime transverse meridian.** Also called **inverse meridian.** See also **fictitious meridian.**

transverse model datum—See **model datum,** definition 1.

transverse parallel—A circle or line parallel to a transverse equator, connecting all points of equal transverse latitude. Also called **inverse parallel.** See also **fictitious parallel.**

transverse pole—One of the two points 90° from a transverse equator.

transverse polyconic map projection—A polyconic map projection which is turned (transversed) 90° in azimuth by substituting for the central meridian, a great circle perpendicular to the geographic meridian to provide a control axis for the projection, along which axis will lie the centers of the circular arcs representing lines of tangency of cones with the surface of the sphere.

transverse position—A split-camera installation so positioned that the plane containing the camera axis is perpendicular to the line of flight.

transverse rhumb line—A line making the same oblique angle with all fictitious meridians of a transverse Mercator projection. Transverse parallels and meridians may be considered special cases of the transverse rhumb line. Also called **inverse rhumb line.** See also **fictitious rhumb line.**

traverse—(JCS) A method of surveying in which lengths and directions of lines between points on the Earth are obtained by or from field measurements, and used in determining positions of the points. [A survey traverse may determine the relative positions of the points which it connects in series, and if tied to control stations on an adopted datum, the positions may be referred to that datum. Survey traverses are classified and identified in a variety of ways: According to methods used, as astronomic traverse; according to quality of results, as first-order traverse; according to purpose served, as geographical exploration traverse; and according to form, as closed traverse, etc.] See also **angle-to-right traverse; azimuth traverse; closed traverse; connecting traverse; deflection angle traverse; first-order traverse; fourth-order traverse; geographical exploration traverse; interior angle traverse; loop traverse; open traverse; phototrig traverse; planetable traverse; random traverse; second-order traverse; spur traverse; stadia traverse; subtense-bar traverse; subtense-base traverse; third-order traverse; transit traverse.**

traverse adjustment—See **balancing a survey.**

traverse angle—Measurement of the horizontal angle from a preceding adjacent station to the following adjacent station.

traverse error of closure—See **error of closure,** definition 8.

traverse line—See **transit line.**

traverse net—See **survey net,** definition 1.

traverse station—A point on a traverse over which an instrument is placed (a set-up). Also, on a traverse, a length of 100 feet measured on a given line, either straight, broken, or curved.

traverse tables—Mathematical tables listing the lengths of the sides opposite the oblique angles for each of a series of right-angle plane triangles as functions of the length and azimuth (or bearing) of the hypotenuse.

traverse the instrument—To rotate a survey instrument about its vertical axis; that is, turning the instrument in azimuth.

triangle—See **astronomic triangle; celestial triangle; navigational triangle; preliminary triangle; spherical triangle; spheroidal triangle; terrestrial triangle.**

triangle closure—See **error of closure,** definition 7.

triangle error of closure—See **error of closure,** definition 7.

triangle of doubt—In a simple two-point problem, the triangle resulting when the check ray fails to pass through the point of intersection of the two intersecting rays.

triangle of error—The triangle formed when three plotted rays fail to intersect perfectly. The center of the triangle may be considered to be the adjusted position. See also **resection.**

triangle-of-error method—In surveying, a technique for solving the three-point problem graphically by a triangle of error. These methods are generally referred to by name, such as **Bessel's method, Coast-Survey method,** and **Lehmann's method,** each of which is based upon its own factors. See also **triangle of error.**

triangulation—A method of surveying in which the stations are points on the ground which are located at the vertices of a chain or network of triangles. The angles of the triangles are measured instrumentally and the sides are derived by computation from selected sides which are termed base lines, the lengths of which are obtained from direct measurements on the ground. See also **analytical three-point resection radial triangulation; arc triangulation; area triangulation; base net; direct radial triangulation; first-order triangulation; flare triangulation; graphical radial triangulation; hand templet triangulation; isocenter triangulation; mechanical templet triangulation; nadir-point triangulation; phototriangulation; radial triangulation; satellite triangulation; second-order triangulation; semi-analytical triangulation; ship-to-shore triangulation; slotted templet triangulation; spider-templet triangulation; third-order triangulation; trilateration.**

triangulation base line—The side of one of a series of connected triangles, the length of which is measured with prescribed accuracy and precision, and from which the lengths of the other triangle sides are obtained by computation. Base lines in triangulation are classified according to the character of the work they are intended to control, and the instruments and methods used in their measurement are such that prescribed probable errors for each class are not exceeded. These probable errors, expressed in terms of the lengths, are as follows: First-order base line, 1/1,000,000; second-order base line, 1/500,000; third-order base line 1/250,000.

triangulation net—See **survey net,** definition 1.

triangulation reconnaissance—A preliminary survey to select the location of stations to give the most feasible triangulation scheme.

triangulation signal—A rigid structure erected over or close to a triangulation station and used for supporting an instrument and observer, or target, or instrument and observer and target, in a triangulation survey. Also, any object, natural or artificial, whose position is obtained in a triangulation survey. The term may be applied to a structure whose position is determined by triangulation, but whose primary purpose is to serve later in a hydrographic or topographic survey, when it may become known as a **hydrographic** or **topographic signal.**

triangulation station—(JCS) A point on the Earth whose position is determined by triangulation. Also called **trig point.**

triangulation system—The main scheme, or net, of primary stations and the auxiliary stations. The main scheme is the framework of the system and is tied at several points to previously established triangulation stations of equal or higher order. See also **survey net,** definition 1.

triangulation theodolite—See **direction instru-**

ment theodolite.

triangulation tower—A structure used to elevate the line of sight above intervening obstacles. Usually consists of two independent structures, one within the other; the center structure supports the theodolite and the outer structure supports the observer and the signal. See also **Bilby steel tower.**

triaxial ellipsoid—An ellipsoid having three unequal axes, the shortest being its polar axis, while the two longer ones lie in the plane of its equator.

tricamera photography—(JCS) Photography obtained by simultaneous exposure of three cameras systematically disposed in the air vehicle at fixed overlapping angles relative to each other in order to cover a wide field. See also **fan-camera photography.**

trig control—See **field control.**

trig dossier—A detailed record on the triangulation of an area, giving the coordinates of the triangulation stations.

trig list—(JCS) A list published by certain Army units which includes essential information of accurately located survey points. [A publication containing all available positional data and elevations with the respective descriptions of horizontal and/or vertical control points, usually arranged according to the location of the control points within the limits of map sheets of large-scale series.]

trig point—See **triangulation station.**

trigonometric leveling—The determination of differences of elevations from observed vertical angles combined with lengths of lines. A type of indirect leveling.

trilateration—A method of surveying wherein the lengths of the triangle sides are measured, usually by electronic methods, and the angles are computed from the measured lengths. See also **triangulation.**

trilateration net—A network of points whose positions relative to one another are determined by measurement and adjustment of the length of the sides of the triangles formed by these points.

trilinear surveying—The determination of the position of a point of observation by measuring the angles at that point between lines to three points of known position. See also **resection.**

trimetrogon camera—A triple camera assembly with one vertical and two fixed-angle obliques whose imagery overlapped the vertical and with all three axes lying in a plane perpendicular to the line of flight. Most camera assemblies of this design were referred to as "trimetrogon cameras" because of the widespread usage of the Metrogon lens in early tricamera photography.

trimetrogon photography—(JCS) Not to be used. See **fan-camera photography.**

trim marks—Lines placed on original copy to serve as guides in cutting or trimming the printed sheets to their prescribed size.

trim size—(JCS) The size of a map or chart sheet when the excess paper outside the margin has been trimmed off after printing.

trimming and mounting diagram—A sketch showing how the prints of a transformed multiple-lens photograph should be corrected to obtain, in effect, a photograph made by a single lens. The information is given in the form of distances referred to the fiducial marks on the photograph, and is the result of the calibration test for the particular camera used.

tropical month—The average period of the revolution of the Moon about the Earth with respect to the vernal equinox, approximately 27 1/3 days.

tropical year—The interval of time between two successive passages of the vernal equinox by the Sun. The tropical year is the year of the seasons, and the basis of the conventional calendar year. Also called **astronomic year; equinoctial year; natural year; solar year.**

Tropic of Cancer—The northern parallel of declination, approximately 23°27′ from the celestial equator, reached by the Sun at its maximum declination, or the corresponding parallel on the Earth.

Tropic of Capricorn—The southern parallel of declination, approximately 23°27′ from the celestial equator, reached by the Sun at its maximum declination, or the corresponding parallel on the Earth.

tropopause—(JCS) The transition zone between the stratosphere and the troposphere. The tropopause normally occurs at an altitude of about 25,000 to 45,000 feet in polar and temperate zones, and at 55,000 feet in the tropics. See also **atmosphere.**

troposphere—(JCS) The lower layers of atmosphere, in which the change of temperature with height is relatively large. It is the region where clouds form, convection is active, and mixing is continuous and more or less complete. See also **atmosphere.**

tropospheric scatter—(JCS) The propagation of radio waves by scattering as a result of irregularities or discontinuities in the physical properties of the troposphere.

trough compass—See **declinatoire.**

Troughton level—An English instrument having the spirit level permanently attached to the top of the telescope tube.

true—**1.** Related to true north as opposed to magnetic north. **2.** Actual, as contrasted with fictitious, as true Sun. **3.** Related to a fixed point, either on the Earth or in space, as true wind; in contrast with relative, which is related to a moving point. **4.** Corrected, as true altitude.

true altitude—**1.** (JCS) The height of an aircraft as measured from mean sea level. **2.** The actual altitude of a celestial body above the celestial horizon. Also called **observed altitude.**

true amplitude—Amplitude relative to true east or west. See also **amplitude.**

true anomaly—See **anomaly,** definition 3.

true azimuth—The horizontal direction of any line measured clockwise from true north.

true bearing—The horizontal angle between the meridian line and a line on the Earth. The term **true bearing** is used in many of the early descriptions of land boundaries in the United States. It is associated with true north, referring to the direction of the north point as determined by astronomic observations. If an astronomically determined bearing is used, however, the term **astronomic bearing** is preferred over **true bearing.**

true depression angle—The setting of the oblique cameras in the photographic aircraft with relation to the true horizon. It is defined by a ray from the exposure station through the principal point of the oblique photograph and a ray to the true horizon.

true direction—Horizontal direction expressed as angular distance from true north.

true error—See **resultant error.**

true horizon—(JCS) The boundary of a horizontal plane passing through a point of vision, or in photogrammetry, the perspective center of a lens system. See also **horizon trace.**

true line—(USPLS) A line of constant bearing (rhumb line) between two corners of a survey.

true meridian—A term used to distinguish the great circle through the geographical poles from **magnetic meridian, compass meridian,** or **grid meridian.**

true north—(JCS) The direction from an observer's position to the geographical North Pole. The north direction of any geographic meridian. [The term was originally applied to **astronomic north** to distinguish it from **magnetic north.**]

true place—See **true position.**

true position—The position of a celestial body after all known corrections including precession and nutation have been made. Also called **true place.**

true prime vertical—The vertical circle through the true east and west points of the horizon, as distinguished from magnetic or grid prime vertical through the magnetic or grid east and west points, respectively.

true sidereal time—See **apparent sidereal time.**

true solar time—See **apparent solar time.**

true sun—See **apparent sun.**

true-to-scale—A condition where map measurements are in exact agreement with the stated map scale. Since all map projections involve some scale change, the scale is not true at all places on a map.

true value—That value of quantity which is completely free from errors. Since the errors to which physical measurements are subject cannot be known exactly, it follows that the true value of a quantity cannot be known with exactness. In survey work, the most probable value is used as best representing the true value of the quantity.

turning point (TP)—A point on which both a minus sight (foresight) and a plus sight (backsight) are taken on a line of direct levels. Also, in topographic surveys, any point on which the rod is held while the instrument is moved to another station. These turning points are often marked for future use as tie or check points.

turning-point pin—A steel pin about one foot long. In leveling operations, the turning point is driven into the ground where it is necessary to establish a point that will be stable in elevation for a short period of time. When it has served its usefulness at one point, it is removed and carried by the rodman for subsequent reuse.

turret graver—A scribing instrument which permits the use of points of several weights interchangeably without the inconvenience of interrupting the scribing to replace points. The turret head is revolved to position the desired point, and scribing is resumed.

tusche—An ink for drafting printing areas on lithographic plates.

tusching—The operation of adding work to the image on a pressplate, correcting lines and lettering, and adding solids by means of tusche. Also called **lithographic drafting.**

24-hour satellite—See **synchronous satellite.**

two-base method—A technique of barometric leveling utilizing three barometers. Two barometers, one high established over a known elevation, and one low established over a known elevation, and a roving barometer operating between the two known positions. High and low barometers are read and recorded, with temperature, every five minutes. The roving barometer and temperature are read and recorded at each station occupied. Data are reduced to position and elevation by office computation. The only barometric leveling method able to consistently produce errors less than ±1 meter. See also **fly-by-method; leap-frog method.**

two-body orbit—The motion of a point mass in the presence of the gravitational attraction of another point mass, and in the absence of any other forces. This orbit is usually an ellipse, but may be a parabola or hyperbola.

two-body problem—That problem in celestial mechanics which treats of the relative motion of two-point masses under their mutual gravitational attraction.

two-dimensional pantograph—In relief model making, a machine permitting the cutting, at a predetermined scale, of the three-dimensional terrain base from the flat map contour drawing.

two-point problem—A problem in determining

the position of a point with the known factor being the length of one line that does not include the point to be located.

two-step enlargement/reduction—A technique of projecting and printing a small image; then copying and projecting it again to the required size. This is often necessary when copy size/copy camera limitations do not permit enlargement or reduction in a single operation.

two-transit method—A method of ship-to-shore triangulation whereby the position of the sounding boat or pole is determined by angle observations from two transits on shore set up over points previously positioned.

type—In printing (typography) a metal block having a raised letter or figure which, when inked, is used to make an impression on paper or other material. Type can also be in the form of negative or positive stripping film. Categories of type include hand-set cold type; hot type, such as Linotype; Monotype (punched tape to metal); phototype (film negative or positive); Photontype (tape to film). Type is identified by its style and size.

typography—The art of type composition and printing from raised type surfaces.

U

UT0 time—The mean universal time derived from observations of time of star transits. Since the directions of the meridians change with time because of the motion of the pole, the UT0 time will, thus, be affected and will, therefore, be irregular. See also **UT1 time; UT2 time.**

UT1 time—The true angular rotation of the Earth about its instantaneous spin axis in the mean equatorial system of dates. UT1 is obtained from UT0 by correcting UT0 for the difference between the instantaneous and mean longitude at the observing station. Since UT1 is keyed precisely to the instantaneous rotation of the Earth, which is not strictly uniform, UT1 does not progress uniformly. See also **UT0 time; UT2 time.**

UT2 time—The mean angular motion of the Earth, freed of predictable periodic variations but still affected by irregular variations and secular variations. UT2 is obtained from UT1 by correcting UT1 for seasonal variations in rotation rate. See also **UT0 time; UT1 time; universal time coordinated.**

ultra-wide-angle lens—See **super-wide-angle lens.**

unchecked spot elevation—Elevation determined by unchecked field survey methods, such as side shots on stadia lines, unchecked vertical angles, and barometric leveling. Also an elevation determined by repeated photogrammetric reading.

uncontrolled mosaic—(JCS) A mosaic composed of uncorrected prints, the details of which have been matched from print to print without ground control or other orientation. Accurate measurement and direction cannot be accomplished.

underground mark—A surveying mark set and plumbed below the center of a surface mark and separated therefrom so as to preserve the station in case of accident to the surface mark.

undevelopable—A surface, such as a sphere, that cannot be flattened to form a plane without compressing or stretching some part of it.

undulation of the geoid—See **geoid height.**

unidimensional magnification—Transformation of one rectangle into another of different proportions.

U.S. control survey nets—The two control survey nets being extended over the United States by the National Geodetic Survey for the control of nautical charts and topographic maps, and comprising: (1) The horizontal-control survey net consisting of arcs of first-order and second-order triangulation and lines of first-order and second-order traverse, a few of which have been executed by the United States Geological Survey, the Corps of Engineers, and other organizations. The data derived in this survey are being coordinated and correlated on the North American datum of 1927. The National Geodetic Survey is currently recomputing the horizontal control network to the North American datum of 1983. (2) The vertical-control survey net consisting of lines of first-order and second-order spirit leveling which determine the elevations of thousands of bench marks above a common datum, mean sea level. This net includes lines of levels run by the United States Geological Survey, the Corps of Engineers, and other organizations.

U.S. Engineer precise leveling rod—A speaking rod of T-shaped cross section, 12 feet long, graduated in centimeters.

U.S. Geological Survey level—A level of the dumpy type, constructed of stainless steel. It has an internal-focusing telescope; the level bubble is centered by the end-coincidence method, effected with the aid of a prism device and Stellite mirror which can be adjusted by the observor.

U.S. Geological Survey precise leveling rod—**1.** A speaking rod graduated in yards and fractions of a yard. It is read for each of three cross wires to the nearest thousandth of a yard. The sum of the three readings is then the mean reading in feet to the nearest thousandth. **2.** A target rod of plus-sign (+) cross section, a little over 12 feet in length. There are two forms of this rod: the single-target rod and the double-target rod.

U.S. National Map Accuracy Standards—**1.** Horizontal accuracy: For maps at publication scales larger than 1:20,000, 90 percent of all well-defined features, with the exception of those unavoidably displaced by exaggerated symbolization, will be located within 1/30 inch (0.85 mm) of their geographic positions as referred to the map projection; for maps at publication scales of 1:20,000 or smaller, 1/50 inch (0.50 mm). **2.** Vertical accuracy: 90 percent of all contours and elevations interpolated from contours will be accurate within one-half of the basic contour interval. Discrepancies in the accuracy of contours and elevations beyond this tolerance may be decreased by assuming a horizontal displacement within 1/50 inch (0.50 mm). Also called **map accuracy standards; national map accuracy standards.**

universal analog photographic rectification system—An electronic rectification system permitting the rapid transfer of detail from trimetrogon or any other type of aerial photography, to include panoramic coverage. The system consists of four basic components: input scanner, computer, console, and *x-y* plotter. Also called **electronic sketchmaster.**

universal instrument—See **alt-azimuth instrument.**

universal level—See **circular level.**

universal plotting sheet—A plotting sheet on which either the latitude or longitude lines are omitted, and are to be drawn in by the user, making it possible to quickly construct a plotting sheet for any part of the Earth's surface. See also **plotting chart; position plotting sheet.**

Universal Polar Stereographic (UPS) grid—A military grid system based on the polar stereographic projection, applied to maps of the Earth's polar regions north of 84°N and south of 80°S latitudes.

universal sketchmaster—A type of sketchmaster in which vertical or oblique photographs may be utilized.

Universal Space Rectangular (USR) Coordinate System—A right-handed orthogonal coordinate system with its origin at the center of the Earth, positive *x*-axis in the equatorial plane and passing through the 0° meridian, positive *y*-axis in the equatorial plane and passing through 90° east meridian, and positive *z*-axis along the rotational axis of the Earth and passing through the North Pole.

universal time (UT)—Standard universally accepted time based on the Greenwich meridian. See also **Greenwich mean time.**

universal time coordinated (UTC)—A time system based on an atomic second maintained within 1 second of UT1 or UT2 by the addition or deletion of leap seconds.

universal transforming printer—A specially designed printer for making glass diapositives in which a known camera distortion is eliminated or compensated for, or in some cases introduced. The glass diapositives may be reproduced at scale, enlarged, or reduced.

Universal Transverse Mercator (UTM) coordinates—Quantities which designate the location of a point on the Universal Transverse Mercator grid.

Universal Transverse Mercator (UTM) grid—(JCS) A military grid system based on the transverse Mercator projection, applied to maps of the Earth's surface extending to 84°N and 80°S latitudes.

unperturbed orbit—See **normal orbit.**

unstable-type gravimeter—A gravity meter which utilizes a moving system which approaches a point of instability such that small changes in gravity produce relatively large motions of the system. See also **astatized gravimeter.**

upper branch—That half of a meridian or celestial meridian from pole to pole which passes through a place or its zenith.

upper culmination—See **culmination.**

upper limb—That half of the outer edge of a celestial body having the greatest altitude, in contrast with the lower limb, that half having the least altitude.

upper motion—(surveying) Rotation of the upper plate of a repeating instrument.

upper transit—Transit of a celestial body over the upper branch of the celestial meridian. Also called **superior transit.** See also **culmination.**

V

vacuum box—The frame, containing its own vacuum unit, which encloses the mold for the forming of plastic relief maps.

Vaisala Comparator—A precise distance measuring instrument that uses optical interferometry for making measurements. It can measure distances as great as 864 meters with an accuracy of 1 part in 10^7.

value—See **absolute value; adjusted value; most probable value; observed value; true value.**

vanishing line—The straight line on a photograph upon which lie all the vanishing points of all systems of parallel lines parallel to one plane.

vanishing point—The image, in the plane of a photograph, of the point toward which a system of parallel lines in the object space converges.

variable contour interval—A non-uniform contour interval. It may result from the use of cartographic source materials which do not contain a constant contour interval or from adapting the contour interval to specific types of terrain for the optimum portrayal of relief features.

variable-perspective camera system—A system which, in its simplest form, consists of a standard-type view camera, a large aperture front-surface mirror of spherical configuration, and an easel used in the rectification of highly tilted long-focal-length photographs, and the transformation of maps and charts from one projection to another. When the camera component is replaced with a projector, it becomes possible to expedite the rectification of lunar photography taken by terrestrial observatories.

variable-ratio pantograph—See **pantograph.**

variance—The square of the standard error. Defined as the limit, as the number of observations becomes infinitely large, of the sum of the squares of the residuals divided by n: the mean of the mean of the squares of errors.

variation—See **magnetic variation.**

variation of coordinate method—A method of adjusting measurements in which the coordinates of geodetic points are varied so as to best fit the observations and retain mathematical homogeneity.

variation of latitude—A small change in the astronomic latitude of points on the Earth, due to variation of the poles.

variation of the poles—A small variation of the location of the instantaneous axis of rotation of the Earth with respect to the physical surface thereof. Also called **polar motion.** See also **conventional international origin.**

variational inequality—An inequality in the Moon's motion, due mainly to the tangential component of the Sun's attraction.

variometer—An instrument for comparing magnetic forces, especially of the Earth's magnetic field.

vectograph—A stereoscopic photograph composed of two superimposed images that polarize light in planes 90° apart. When these images are viewed through Polaroid spectacles with the polarization axes at right angles, an impression of depth is obtained.

vector—1. (JCS) A heading issued to an aircraft to provide navigational guidance by radar. **2.** Any quantity, such as force, velocity, or acceleration, which has both magnitude and direction, as opposed to a scalar which has magnitude only. Such a quantity may be represented geometrically by an arrow of length proportional to its magnitude, pointing in the assigned direction.

vectorial angle—The angle between the fixed line to which the direction is referred and the radius vector. See also **polar coordinates.**

velocity correction—A correction applied to the speed of light to obtain the true speed in consideration of humidity, temperature, and altitude for use in shoran operations.

Vening Meinesz formulas—Formulas for computing deflections of the vertical from gravity data.

verbal scale—See **equivalent scale.**

vernal equinox—That point of intersection of the ecliptic and the celestial equator, occupied by the Sun as it changes from south to north declination, on or about 21 March. Also called **first of Aries; first point of Aries; March equinox.** See also **mean equinox.**

vernier—A short, auxiliary scale situated alongside the graduated scale of an instrument, by means of which fractional parts of the smallest division of the primary scale can be measured accurately. See also **contact vernier; direct vernier; folding vernier; optical vernier; retrograde vernier.**

vernier closure—The difference between the initial and final vernier readings during the survey operation of closing the horizon.

vertex (vertices)—The highest point. The vertices of a great circle are the points nearest the poles. Also called **apex.**

vertex of curve—See **point of intersection.**

vertical—The line perpendicular to the geoid at any point. It is the direction in which the force of gravity acts. See also **local vertical; mass attraction vertical; normal,** definition 3.

vertical angle—1. An angle in a vertical plane. **2.** (surveying) One of the directions which form a vertical angle is usually either the direction of the vertical (zenith), and the angle is termed the zenith distance; or the line of intersection of the vertical plane in which the angle lies with the plane of the horizon, and the angle is termed the angle of elevation or angle of depression, or simply the altitude (plus or minus, as the case may be). The vertical angle between two directions, neither of which lies in the plane of the horizon or coincides with the vertical, is usually obtained from the combination of two vertical angles as defined above.

vertical-angle bench mark (VABM)—A bench mark with elevation established by vertical angle methods. See also **bench mark.**

vertical-angle station—See **supplemental station.**

vertical angulation—The process of obtaining differences of elevation by means of observed vertical angles, combined with lengths of lines. In geodetic work, trigonometric leveling is used with the same meaning.

vertical axis—(theodolite, transit) The line through the center of the instrument about which the alidade rotates. For an instrument in complete adjustment, this axis occupies a vertical position, passes through the center of the horizontal circle, and is perpendicular to its plane.

vertical bridging—See **bridging.**

vertical circle—1. A great circle of the celestial sphere, through the zenith and nadir. Vertical circles are perpendicular to the horizon. **2.** A graduated disk mounted on an instrument in such a manner that the plane of its graduated surface can be placed in a vertical plane. It is primarily used for measuring vertical angles in astronomic and geodetic work.

vertical collimator—A telescope so mounted that its collimation axis can be made to coincide with the vertical (or direction of the plumbline). The vertical collimator serves as an optical plumbline; it may be designed for use in placing a mark on the ground directly under an instrument on a high tower or in centering an instrument on a high tower directly over a mark on the ground. Also called **optical plummet.**

vertical comparator—(pendulum) A stand designed for the support of a pendulum, a bar of known length, and two micrometer microscopes, so placed with reference to one another that the length of the pendulum can be measured.

vertical control—The measurements taken by surveying methods for the determination of elevation only with respect to an imaginary level surface, usually mean sea level. See also **survey net,** definition 2.

vertical-control datum—Any level surface (as, for example, mean sea level) taken as a surface of reference from which to reckon elevations. Also called **vertical datum; vertical geodetic datum.** See also **datum level; reference level; reference plane.**

vertical control net—See **survey net,** definition 2.

vertical control point—See **control point; control station.**

vertical coordinates—The vertical distance of a point above or below a reference datum. Points may be plus or minus according to whether the point is above or below the datum.

vertical curve—A parabolic curve used to connect grades of different slope, and used at the vertex of a grade to avoid the sudden change in direction in passing from one grade to the other. This method of grade change is usually used when there is an algebraic difference of more than 0.2 percent in the two opposing grades.

vertical datum—See **vertical-control datum.**

vertical deformation—In relative orientation, the cumulative model warpage affecting the vertical datum from x-tilt error and y-tilt error.

vertical exaggeration—1. The change in a model surface created by proportionally raising the apparent height of all points above the base level while retaining the same base scale. **2.** The increase of the vertical scale over the horizontal scale of a terrain model or plastic relief map.

vertical extension—See **extension,** definition 2.

vertical geodetic datum—See **vertical-control datum.**

vertical intensity—The magnetic intensity of the vertical component of the Earth's magnetic field, reckoned positive if downward, negative if upward.

vertical obstruction data—Information on the height above the terrain of towers, buildings, and other natural or manmade obstacles to low-altitude flight.

vertical parallax—See **y-parallax.**

vertical pass point—See **supplemental elevation.**

vertical photograph—An aerial photograph taken with the axis of the camera being maintained as closely as possible to a truly vertical position with the resultant photograph laying approximately in a horizontal plane.

vertical plane—1. Any plane passing through a point on the Earth and containing the zenith and nadir of that point; also a plane containing a plumb line. **2.** (surveying) A plane at right angles to a horizontal plane and within which angles and distances are observed.

vertical sketchmaster—A type of sketchmaster in which vertical photographs are utilized.

vertical stereotriangulation—That portion of stereotriangulation concerned with the establishment of vertical data. Vertical stereotriangulation is often limited or precluded as an operation due to the more rigid accuracy standards established for vertical positions than for horizontal positions.

vibration—A single movement of a pendulum in either direction, to or fro. See also **oscillation.**

viewfinder—(aerial camera) An auxiliary device which shows the field of view of a camera. It is used in the taking of vertical aerial photography to correct crab angle and maintain forward lap (end lap).

vignetting—1. (photography) A gradual reduction in density of parts of a photographic image due to the stopping of some of the rays entering the lens. Thus, a lens mounting may interfere with the extreme oblique rays. An antivignetting filter is one that gradually decreases in density from the center toward the edges; it is used with many wide-angle lenses to produce a photograph of uniform density by cutting down the overexposure of the center of the photograph. **2.** (lithography) A photographic process which portrays a solid color in a screen which shades off gradually into the unprinted paper. Open water is often shown by this method.

virtual gravity—The force of gravity on an atmospheric parcel, reduced by centrifugal force due to the motion of the parcel relative to the Earth.

virtual image—An image that cannot be shown on a surface but is visible, as in a mirror.

virtual PPI reflectoscope (VPR) chart—A type of radar chart.

visible horizon—See **apparent horizon.**

visibility chart—A special-purpose map or other graphic showing which areas can be seen and those which cannot be seen from a given observation point.

visual acuity—A measure of the ability of the human eye to separate details in viewing an object. The reciprocal of the minimum angular separation, in minutes of arc, of two lines of detail which can be seen separately.

visual range—The limiting range of a light determined after taking into account both the geographic range and the luminous range. The geographic range is the maximum distance at which the curvature of the Earth permits a light to be seen from a particular height of eye without regard to the luminous intensity of the light. The luminous range is determined from the known nominal luminous range, called the nominal range, and the existing visibility conditions.

voting (TERCOM)—The technique of having a vehicle fly over the terrain covered by three unique but complementary maps within a finite distance, comparing the calculated positional accuracy within each of the maps and determining whether or not to update the vehicle navigation system based on how closely the three positional accuracies compare with each other.

vulgar establishment—See **establishment of the port.**

W

WWV time—Accurately controlled time signals transmitted from stations (radio) WWV in Colorado and WWVH in Hawaii. These stations broadcast UTC (universal time coordinated) which is offset from A1 by a variable amount determined annually before the start of the year. The epoch is shifted in increments of 100 milliseconds if it departs too far from UT2.

wading rod—A rod, graduated in feet and tenths of feet, used for stream-gaging in shallow water.

wall map—A special-purpose map of a large area designed to be displayed on a wall.

wander—See **apparent precession.**

want of correspondence—See **y-parallax.**

warped model—Any spatial model which, due to photographic distortions or orientation errors, has a model datum which is deformed or otherwise incapable of being leveled. See also **flat model.**

water leveling—A method of obtaining relative elevations by observing heights with respect to the surface of a body of still water.

water-stage recorder—An automatic recording instrument which records the rise and fall of the water surface at a stream-gaging station.

wavelength—Quantitative specification of kinds of radiant energy. See also **dominant wavelength.**

waving the rod—In leveling, a technique whereby the rodman slowly pivots the leveling rod toward and away from the instrument position. The least reading obtainable is the proper one to be recorded.

weather map—(JCS) A map showing the weather conditions prevailing, or predicted to prevail, over a considerable area. Usually, the map is based upon weather observations taken at the same time at a number of stations. Also called **synoptic chart; surface chart.**

weber—(geomagnetism) Unit of magnetic flux. In the mks system, 1 weber = 1 joule/amp = 1 $kgm^2/amp\ sec^2$.

wedge—(optics) A refracting prism of very small deviation, such as those used in the eyepieces of some stereoscopes. Also called **optical wedge.**

weight—The relative value of an observation, source, or quantity when compared with other observations, sources, or quantities of the same or related quantities. The value determined by the most reliable method is assigned the greatest weight.

weighted mean—A value obtained by multiplying each of a series of values by its assigned weight and dividing the sum of those products by the sum of the weights.

Werner map projection—A particular case of the Bonne map projection, in which the standard parallel is at the pole, and the tangent cone becomes a tangent plane. Any one geographic meridian is chosen as the central meridian and represented by a straight line, divided to exact scale. The geographic parallels are represented by circular arcs, also divided to exact scale, and the other meridians are curved lines.

west point—See **prime vertical plane.**

westing—See **departure,** definition 1.

wide-angle lens—A lens having an angle of coverage between 75° and 100°. A lens whose focal length is equal to approximately one-half the diagonal of the format.

wiggling-in—A survey procedure used when it is necessary to establish a point, exactly on line between two control points, neither of which can be occupied. It is essentially a trial-and-error technique where repeated fore and back readings are taken and the instrument shifted after each pair of readings until exactly in line with the stations. Also called **ranging-in.**

wiggling-in on line—See **double centering.**

wind rose—**1.** A diagram showing the relative frequency of winds blowing from different directions. It may also show average speed or frequency of occurrence of various speeds from different directions. **2.** A diagram showing the average relationship between winds from different directions and the occurrence of other meteorological phenomena.

wing photograph—A photograph taken by one of the side or wing lenses of a multiple-lens camera.

wing point—Three easily identified points along each side of an aerial photograph, one near each corner and one near the middle. Used in the extension of radial control in making controlled mosaics.

winter solstice—**1.** That point on the ecliptic occupied by the Sun at maximum southerly declination. Also called **first point of Capricornus; December solstice. 2.** That instant at which the Sun reaches the point of maximum southerly declination, about 22 December.

wire drag—A sounding device consisting of weighted wires which are maintained at a given depth by floats, and then dragged over any desired course.

witness corner—**1.** A marker set on a property line leading to a corner; used where it is impractical to maintain a monument at the corner itself. **2.** (USPLS) A monumented survey point usually on the line of survey near a corner established as a reference when the corner is so situated as to render its monumentation or ready use impracticable.

witness mark—A mark placed at a known distance and direction from a property corner or survey station to aid in its recovery and identification. Also called **witness post; witness stake.**

witness point—(USPLS) A monumented station

on a line of the survey, employed to perpetuate an important location without special relation to any regular corner, except that the bearing or distance may be known.

witness post—See **witness mark.**

witness stake—See **witness mark.**

Woodward base-line measuring apparatus—See **iced-bar apparatus.**

working-in on a line—See **double centering.**

working pendulum—A pendulum which is used (swung) in a determination of the intensity of gravity.

World Aeronautical Chart (WAC)—See **Operational Navigation Chart.**

World Geodetic System (WGS)—A consistent set of parameters describing the size and shape of the Earth, the positions of a network of points with respect to the center of mass of the Earth, transformations from major geodetic datums, and the potential of the Earth (usually in terms of harmonic coefficients).

World Geographic Reference System (GEOREF)—See **GEOREF.**

world polyconic grid—A grid system in which a grid network is mathematically derived from elements of a polyconic projection.

wrong-reading—A descriptive term for an image which is a reverted or mirror image of the original. Other terms, such as reverse reading, etc., are sometimes used to identify image direction, but are not recommended because of possible confusion in film negative-positive relationship.

wye level—A leveling instrument having the telescope, with attached spirit level, supported in Y-shaped bearings. The telescope can be rotated about its longitudinal axis (collimation axis) and it can be lifted and reversed, end for end, for testing and adjustment. Also called **Y level.**

x-axis—**1.** (JCS) A horizontal axis in a system of rectangular coordinates; that line on which distances to the right or left (east or west) of the reference line are marked, especially on a map, chart, or graph. **2.** The line joining the opposite fiducial marks in the direction which is most nearly parallel to the line of flight.

x-coordinate—See **abscissa.**

x-correction—The correction to an *x*-direction.

x-direction—An observed direction in a triangulation figure for which an approximate value is obtained and treated like an observed direction in the adjustment of the figure. The work of a least-squares adjustment of a triangulation figure sometimes requires the use of an *x*-direction, for which an approximate value is obtained by an inverse position computation, by the solution of the three-point problem, or by other means, and then using the *x*-direction in the adjustment and obtaining a correction (*x*-correction) for it, which makes it consistent with the adjusted values of the observed directions.

x-displacement—A component of image displacement. When a point image is to be located by coordinates with respect to rectangular axes, *x*-displacement represents the distance moved in the *x*-direction.

x-motion—In a stereoplotting instrument, that linear adjustment approximately parallel to a line connecting two projector stations; the path of this adjustment is, in effect, coincident with the flight line between the two relevant exposure stations.

x-parallax—See **absolute stereoscopic parallax.**

x-scale—(JCS) On an oblique photograph, the scale along a line parallel to the true horizon.

x-tilt—The component of tilt about the *x*-axis, which is most nearly in the direction of flight. Also called **list.**

x-y scaler—An instrument that provides *x*- and *y*-coordinates in digital form from analogue data.

Y

y-axis—**1.** (JCS) A vertical axis in a system of rectangular coordinates; that line on which distances above or below (north or south of) a reference line are marked, especially on a map; chart, or graph. **2.** The line which is perpendicular to the *x*-axis and passes through the origin.

y-coordinates—See **ordinates.**

y-displacement—A component of image displacement. When a point image is to be located by coordinates with respect to rectangular axes, *y*-displacement represents the distance moved in the *y*-direction.

Y level—See **wye level.**

y-parallax—The difference between the perpendicular distances of the two images of a point on a pair of photographs from the vertical plane containing the air base. The existence of *y*-parallax is an indication of tilt in either or both photographs, or a difference in flying heights, and interferes with stereoscopic examination of the pair. Also called **vertical parallax; want of correspondence.**

y-scale—(JCS) On an oblique photograph, the scale along the line of the principal vertical or any other line, inherent or plotted, which, on the ground, is parallel to the principal vertical.

y-swing method—See **one-swing method.**

y-tilt—See **pitch.**

yaw—**1.** (air navigation) The rotation of an aircraft about its vertical axis so as to cause the aircraft's longitudinal axis to deviate from the flight line. Also called **crab.** **2.** (photogrammetry) The rotation of a camera or a photograph coordinate system about either the photograph *z*-axis or the exterior *z*-axis. **3.** In some photogrammetric instruments and in analytical applications, the symbol κ may be used.

yaw angle—See **angle of yaw.**

year—The period of about 365¼ solar days required for one revolution of the Earth around the Sun. See also **anomalistic year; calendar year; eclipse year; fictitious year; great year; sidereal year; tropical year.**

Z

z-axis—In a three-dimensional rectangular coordinate system, the axis of reference that is perpendicular to both the *x*- and *y*-axes at their point of intersection.

z-motion—Movement of a stereoplotting projector in a vertical direction.

z-scale—(JCS) On an oblique photograph, the scale used in calculating the height of an object. Also, the name given to this method of height determination.

z-time—See **Greenwich mean time.**

zenith—That point of the celestial sphere vertically overhead. See also **geocentric zenith; geodetic zenith.**

zenith angle—See **zenith distance.**

zenith camera—A special camera so designed that its optical axis may be pointed accurately toward the zenith. It is used for the determination of astronomic positions by photographing the position of the stars. See also **photographic zenith tube.**

zenith distance—The vertical angle between the zenith and the object which is observed or defined. Zenith distance is the complement of the altitude. Also called **zenith angle.** See also **coaltitude.**

zenith telescope—A portable instrument adapted for the measurement of small differences of zenith distance, and used in the determination of astronomic latitude.

zenithal chart—See **azimuthal chart.**

zenithal map projection—See **azimuthal map projection.**

zonal harmonics—The set of spherical harmonics that are functions of latitude only and therefore do not affect the rotational symmetry of the surface about its polar axis.

zone time—The local mean time of a reference zone. See also **time zone.**

zoom system—See **pancratic system.**

zulu time—See **Greenwich mean time.**

☆ U. S. GOVERNMENT PRINTING OFFICE : 1981 730-392/22

Lightning Source UK Ltd.
Milton Keynes UK
UKHW040535291018
331380UK00001B/13/P